Sound Reception in Mammals

FRONTISPIECE: The right cochlea (×23) of a guinea-pig *Cavia porcellus* partly dissected to show the spiral organ, stained black by osmium tetroxide. The stapes is still present in the oval window. (Microdissection by Dr Ade Pye, photograph by Mr D. J. Connolly, Institute of Laryngology and Otology.)

SYMPOSIA OF THE ZOOLOGICAL SOCIETY OF LONDON

NUMBER 37

Sound Reception in Mammals

(The Proceedings of a Symposium organized jointly by The British Society of Audiology and The Zoological Society of London, held at The Zoological Society of London on 21 and 22 March, 1974)

Edited by

R. J. BENCH

Regional Audiology Research Unit, Royal Berkshire Hospital, Reading, England

ADE PYE

Institute of Laryngology and Otology, London, England

and

J. D. PYE

Department of Zoology and Comparative Physiology, Queen Mary College, London, England

Published for

THE ZOOLOGICAL SOCIETY OF LONDON

BY

ACADEMIC PRESS

1975

ACADEMIC PRESS INC. (LONDON) LTD
24/28 Oval Road
London NW1

U.S. Edition published by
ACADEMIC PRESS INC.
111 Fifth Avenue,
New York, New York 10003

Library of Congress Catalog Card Number: 74-5683

ISBN: 0-12-613337-9

PRINTED IN GREAT BRITAIN BY
J. W. ARROWSMITH LTD, BRISTOL

CONTRIBUTORS

ANDERSON, S. D., *Institute of Laryngology and Otology, 330 Gray's Inn Road, London W.C.1., England* (p. 69)

ARAN, J.-M., *Laboratoire d'Audiologie Expérimentale, Centre Régionale de Phono-Audiologie, Université de Bordeaux II, France* (p. 167)

BENCH, R. J., *Regional Paediatric Audiology Research Unit, Royal Berkshire Hospital, Reading, Berkshire, England* (p. 23)

BENNETT, M. J., *Department of Mechanical Engineering, Brunel University, Uxbridge, Middlesex, England* (p. 291)

BORG, E., *Department of Physiology II, Karolinska Institutet, Solnavägen 1, 104 01 Stockholm 60, Sweden* (p. 51)

BOSHER, S. K., *Ferens Institute of Otolaryngology, Middlesex Hospital Annexe, Cleveland Street, London W1P 5FD, England* (p. 11)

BROOKS, D., *Manchester Audiology Clinic, 45 Hardman Street, Manchester M3 3HD, England* (p. 41)

CHARLET DE SAUVAGE, R., *Laboratoire d'Audiologie Expérimentale, Centre Régionale de Phono-Audiologie, Université de Bordeaux II, France* (p. 167)

COLES, R. R. A., *Institute of Sound and Vibration Research, The University, Southampton SO9 5NH, England* (p. 107)

EVANS, E. F., *Department of Communication, University of Keele, Keele, Staffordshire ST5 5BG, England* (p. 133)

FRANCIS, R. L., *Department of Zoology and Comparative Physiology, Queen Mary College, Mile End Road, London E.1., England* (p. 237)

KNIGHT, J. J., *Institute of Laryngology and Otology, 330 Gray's Inn Road, London W.C.1., England* (p. ix)

MENTZ, D. L., *Regional Paediatric Audiology Research Unit, Royal Berkshire Hospital, Reading, Berkshire, England* (p. 23)

MICHAELS, L., *Institute of Laryngology and Otology, 330 Gray's Inn Road, London W.C.1., England* (p. 121)

PYE, A., *Institute of Laryngology and Otology, 330 Gray's Inn Road, London W.C.1., England* (p. 97)

PYE, J. D., *Department of Zoology and Comparative Physiology, Queen Mary College, Mile End Road, London E.1., England* (pp. 1 339)

ST JAMES ROBERTS, I., *Department of Psychology, University of Liverpool, PO Box 147, Liverpool L69 3BX, England* (p. 307)

SALES, G. D., *Department of Zoology and Comparative Physiology, Queen Mary College, Mile End Road, London E.1., England* (p. 187)

SALOMON, G., *Københavns Amts Sygehus, 1 Gentofte, Niels Andersens Vej 65, 2900 Hellerup, Denmark* (p. 205)

SMITH, J. C., *Department of Zoology and Comparative Physiology, Queen Mary College, Mile End Road, London E.1., England* (p. 317)
STEPHENS, S. D. G., *Institute of Sound and Vibration Research, The University, Southampton SO9 5NH, England* (p. 167)
VUORENKOSKI, V., *Department of Pediatrics, Keski-Pohjanmaa Central Hospital, SF-67200 Kokkola, Finland* (p. 331)
ZAKRISSON, J.-E., *Department of Oto-rhino-laryngology, University of Umeå, Umeå, Sweden* (p. 51)

ORGANIZERS AND CHAIRMEN

ORGANIZERS

R. J. Bench, Ade Pye and J. D. Pye
on behalf of The Zoological Society of London

CHAIRMEN OF SESSIONS

Beagley, H. A., *Institute of Laryngology and Otology, 330 Gray's Inn Road, London W.C.1., England*

Fisch, L., *Institute of Laryngology and Otology, 330 Gray's Inn Road, London W.C.1., England*

Hinchcliffe, R., *Institute of Laryngology and Otology, 330 Gray's Inn Road, London W.C.1., England*

Tumarkin, A., *Ty-Hwnt-I'r Afon, Rhos-y-Gwaliau, Bala, North Wales*

PREFACE

As Chairman of the British Society of Audiology it is my pleasure to welcome you, the speakers and participants, to this Symposium on Sound Reception in Mammals—our first meeting in association with the Zoological Society of London.

I wish to express our appreciation to the Zoological Society who enthusiastically took up our original suggestion of this meeting and have kindly, and generously, supported the recommendations of the small joint organizing committee consisting of Dr Bench, Dr Ade Pye and Professor Pye. Dr Vevers, the Assistant Director of the Zoological Society, and his staff have been particularly helpful in arranging the meeting. Of course the British Society of Audiology is by far the junior of the two societies in age and in membership, and I should like to express my great admiration of the expertise in detailed organization that accompanies your extra years of experience. It certainly gives a smaller society a model on which to work. You will see from the programme that the generosity of the Zoological Society has extended to giving the British Society of Audiology the nomination of the four chairmen of sessions, as well as to myself the privilege of giving this welcoming address.

For the benefit of our hosts today, may I briefly explain that the British Society of Audiology is just seven years old, but apparently is fulfilling a demand for an exchange of knowledge in the field of hearing in that it has the support of a very active membership of about 550, has branches in Scotland and the North of England, and arranges some 15 scientific meetings each year. It also has its own journal called the British Journal of Audiology. Because many disciplines have a contribution to make to audiology, and zoology is certainly no exception, two or three of these meetings are arranged jointly with other societies, as with those of the acousticians, the psychologists, the otologists, the teachers of the deaf and the speech therapists, to name a few. In these circumstances, our audience has exceeded 200 on many occasions and the papers have achieved a high standard of excellence.

Many such workers in audiology are participants today and, as well as zoologists, we are also to be addressed by a pathologist, a neurophysiologist and an engineer. Such is the breadth of audiology.

My welcome would be incomplete without particularly mentioning our three guest speakers from abroad. Dr E. Borg

(Stockholm, Sweden), Dr G. Salomon (Copenhagen, Denmark) and Dr V. Vuorenkoski (Kokkola, Finland). We especially acknowledge your interest which has caused you to take time away from your own busy activities to tell us of your work. We also have present a number of other distinguished workers from countries farther afield. Again, welcome, we are honoured by your presence.

Finally, I hope all participants will enjoy this symposium and find much of value to them academically, and perhaps socially as well, in these beautiful surroundings of the Zoological Gardens.

J. J. KNIGHT

London
March 1974

CONTENTS

Introductory Review

J. D. PYE

DEVELOPMENTAL ASPECTS

Chairman: L. Fisch

Morphological and Functional Changes in the Cochlea Associated with the Inception of Hearing

S. K. BOSHER

On the Measurement of Human Foetal Auditory Response

R. J. BENCH and D. L. MENTZ

Peripheral Aspects of the Physiology of Middle Ear Muscles

S. D. ANDERSON

COCHLEAR TRAUMA

Chairman: R. Hinchcliffe

Comparison of Histological Changes in the Cochlea of the Guinea-pig and the Bat *Rousettus* after Acoustic Trauma

ADE PYE

Relationships Between Noise-induced Threshold Shifts, Morphological Change and Social Handicap

R. R. A. COLES

The Pathology of Deafness Due to Trauma

L. MICHAELS

EIGHTH NERVE

Chairman: H. A. Beagley

Normal and Abnormal Functioning of the Cochlear Nerve

E. F. EVANS

Gross Responses From the Cochlear Nerve in Man and in the Guinea-pig

S. D. G. STEPHENS,
R. CHARLET de SAUVAGE and J.-M. ARAN

CENTRAL NERVOUS SYSTEM

Chairman: H. A. Beagley

Auditory Evoked Responses in Small Mammals

GILLIAN D. SALES

Automatic Evaluation of Human Auditory Evoked Potentials From the Vertex

G. SALOMON

BEHAVIOURAL METHODS OF THRESHOLD MEASUREMENT

Chairman: A. Tumarkin

Behavioural Audiometry in Mammals: Review and Evaluation of Techniques

R. L. FRANCIS

The Auditory Response Cradle: A Device for the Objective Assessment of Auditory State in the Neonate

M. J. BENNETT

Operant Audiometry with Children

I. ST JAMES-ROBERTS

BIOLOGICAL SIGNIFICANCE OF HEARING

Chairman: A. Tumarkin

Sound Communication in Rodents

JANE C. SMITH

Maternal Recognition of Infant Sounds

V. VUORENKOSKI

Symp. zool. Soc. Lond. (1975) No. 37, 1–9.

INTRODUCTORY REVIEW

J. D. PYE

Department of Zoology and Comparative Physiology, Queen Mary College, London, England

INTRODUCTION

The two-day symposia of the Zoological Society of London, held two or three times a year, have become something of a tradition, for this meeting forms number 37 of the series. A wide variety of topics has been discussed and many of the meetings have been organized jointly with other Societies representing diverse special interests. When Dr Fisch of the British Society of Audiology (the Chairman of this first session) suggested a joint meeting on hearing, the idea was warmly welcomed and an organizing "committee" was set up. The British Society of Audiology appointed two members of its Programmes Committee (one of them a zoologist) and the Zoological Society searched its membership for someone who might liaise with them. The task is not enviable because we three must accept responsibility for any shortcomings, whereas successes will be entirely due to the contributors; the actual organization, however, has been very straightforward because the Zoological Society, with its long experience in arranging such matters, has provided excellent advice and all the facilities for putting it into practice. Our sincere thanks are due to Dr Gwynne Vevers, Miss Unity McDonnell and Dr Marcia Edwards for smoothly organizing everything except the programme itself.

THE PROGRAMME

In planning this programme, we set ourselves a number of operating principles. First we decided to have a theme which would give some continuity to the list of papers presented. Considering the interests of members in both Societies, we chose to compare the techniques, the aims and the limitations of research interests in hearing which are either primarily clinical or primarily comparative (that is zoological-physiological) in approach. For this reason we have invited speakers who could present approaches to various aspects of the subject that may contrast but should also be complementary and stimulate discussion.

Secondly, we considered it important that the contributors should represent a wide range of personal experience. A programme composed only of well-known speakers describing completed investigations can easily tend to present simple reiterations of existing publications and to rather predictable confrontations in discussion. By emphasizing younger workers, some of whom have only recently started their research careers, we hoped to encourage a somewhat livelier atmosphere.

Thirdly, we designed the timetable to allow free discussion both on the floor and at a more personal level. The discussion periods have been reserved until the end of each session so that the Chairmen can participate more fully and encourage a comparative spirit. There are also generously long breaks for refreshment, when the catering facilities and the gardens themselves should provide a suitably stimulating and congenial environment for less formal discussions. This is not due to idleness or shortage of material—we all know the frustration of a meeting where the programme is so "busy" that individuals can only wave across the meeting hall. The word Symposium is derived from the Greek name for a drinking party and much important business is often achieved while honouring this etymology.

The scope of the meeting was largely determined by the theme. It would have been interesting to include a wider range of sound-sensitive animals such as the insects and the lower vertebrates. But the enormous variety of hearing mechanisms and the special problems of hearing in many of these cases would have led to a very superficial treatment with little interest on the clinical side. On the other hand the terrestrial mammals in this respect form a coherent group in which man is rather unexceptional. It has been said that man should have been mentioned in the title, but it never occurred to us that man is not a perfectly good mammal.

The mammals evolved from the now extinct synapsid group of reptiles (whereas modern reptiles belong either to the anapsid or diapsid groups) and the birds came from diapsid stock. The early divergence of these major groups is reflected in the different arrangements of their ears which suggest a large degree of somewhat parallel but largely independent evolution. All parts of the mammalian auditory system show peculiar features, and even within the mammals the special problems of hearing underwater are so different from those in air that thoroughly aquatic forms have been excluded from special attention in this programme.

THE MAMMALIAN AUDITORY SYSTEM

The basic plan of the auditory system in terrestrial mammals divides rather neatly into two main sections: a conductive section, before the transduction process, and a neural section after it. The former is shown diagrammatically in Fig. 1 and is concerned with

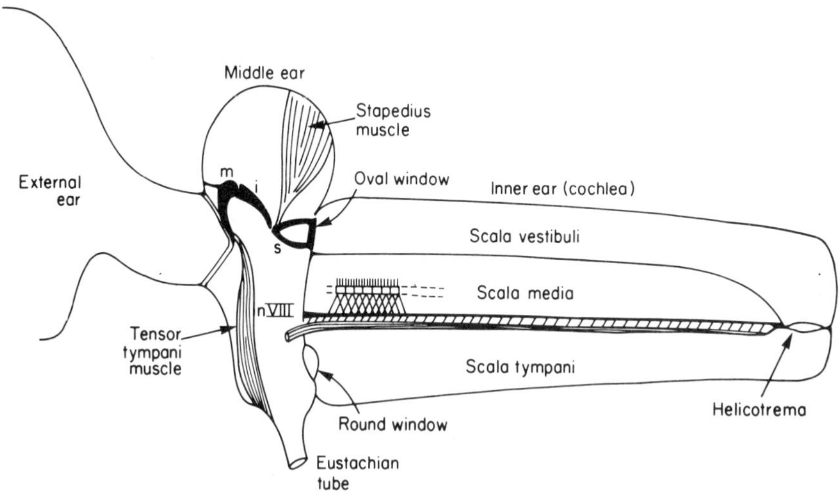

FIG. 1. A simplified diagram in two dimensions of the mammalian ear. It is unlikely that all the structures of the middle ear could be seen in one plane; the cochlea has been "uncoiled" and only a part of Corti's organ is indicated. Letters label the three middle ear ossicles—m: malleus attached to the conical tympanum; i: incus; s: stapes.

physical vibrations induced by pressure waves in the atmosphere. The outer ear is a mammalian "invention" for protecting the eardrum and sometimes for making it more directional. The middle ear is concerned with overcoming the change in acoustic impedance from the air to the liquids bathing the transducing hair cells; in mammals it is unique in having three ossicles and two muscles which allow variable attenuation and filtering in sound transmission. The inner ear contains the sensory hair cells connected to the auditory nerve and in mammals it is arranged so that different frequencies, or at least the higher ones, are spatially separated along the basilar membrane. Although the birds and some living reptiles, such as crocodiles, have well developed cochleae, these show some important differences from that of mammals. The lower vertebrates have nothing that is at all comparable with this organ.

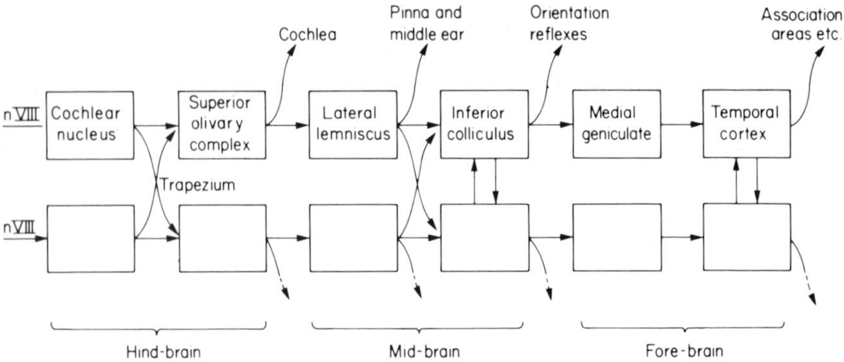

FIG. 2. A plan of the auditory nervous system of mammals showing the principal "centres". The arrows indicate the more important pathways and some probable output channels.

The neural levels, represented by the plan of Fig. 2, are concerned with the processing of auditory information after it has been coded in the form of action potentials. Here too, important differences occur, for in the fore-brain the mammals have developed the neocortex whereas birds have little cortex and a greatly enlarged corpus striatum. The auditory nervous system has a number of efferent pathways which at the lower levels control the sensitivity of hearing, at the middle levels operate various auditory reflexes and at the higher levels use the auditory information for initiating or modifying complex behaviour.

Responses at all neural levels can truly be said to represent sound reception by the entire animal. Responses before the neural level can indicate the probability of detection and even suggest the characteristics of the auditory system as a whole, but they do not establish conclusively that the animal can "hear" any given sound—unless, of course, they show changes due to efferent control from the neural levels. For example, the part played by cochlear microphonic potentials is still equivocal, although the relative levels recorded at different frequencies are usually in good agreement with auditory frequency response as measured by other methods; but changes in the microphonic response due to cochlear inhibition or to middle ear muscle action clearly indicate that the brain is receiving sensory information.

The argument about whether "hearing" must necessarily include cortical involvement is purely semantic and is rather barren in this context since activities of various kinds are initiated at subcortical levels. The word hearing was deliberately avoided in the title of this meeting although this should not preclude its free use in any desired sense. As Humpty Dumpty said, "When *I* use a word it means just what I choose it to mean—neither more or less".

VARIATION

There are about 5000 species of mammals of which a representative collection (about 230 species) may be seen alive in these gardens—and a few more at Whipsnade. Their habits are of course very diverse and so are their sensory requirements. It is one of the major aims of zoologists to compare the operation and performance of systems in different animals because, when related to their various ways of life, the differences between them are very revealing. Similarities that arise in animals that are not closely related but become adapted to similar ways of life (a process called convergent evolution) are equally or even more revealing. But with such diversity available it must be admitted that wide coverage has not yet been achieved in the study of auditory systems, and comparative work has been restricted to a fairly limited number of forms of mammals.

Nevertheless a few general principles emerge. Anatomically there are considerable variations in the details of the basic plan shown in Figs 1 and 2. In the conductive section these concern the relative development and arrangement of the various parts; fundamental differences and complete omissions (or additions) are rare. But the stapedius muscle, which is very well developed in most bats, appears to be absent in some rodents. The relative development of the various neural centres also varies, and in echolocating bats all centres up to and including the inferior colliculus are hypertrophied although the auditory cortex is not.

On the frequency scale the range of sensitivity of different mammals shows rather wide variations, even when similar methods are used to investigate them. The "best frequency", at which threshold is lowest, may differ markedly from that of man at 2–4 kHz and in many species there may be two such "best" frequencies, giving a bimodal hearing curve related to their auditory requirements. Also the upper frequency limit (and possibly the lower one too) may vary widely; bats are not the only animals to make extensive use of ultrasound. This is of course my own hobby-horse but the generalization is becoming increasingly stronger and it seems that man and the anthropoid apes have an unusually low upper cut-off frequency. The implications for all forms of comparative audiology in mammals can no longer be ignored.

It is striking that many studies which reveal the true threshold of hearing give much the same value for the lowest detectable intensity, regardless of the frequency at which this occurs in unrelated forms. As in man, it is often near to 2×10^{-4} μbar RMS or 0 dB sound pressure level (SPL). It may be that this represents some form of physical limit to the common transducing mechanism. It seems likely that thermal agitation or "Brownian movement" is important in determining this limit, although there have been misunderstandings about the difference between correlated movements of the basilar membrane as a whole and uncorrelated random agitation of its constituent molecules (thermal noise). The amplitudes of basilar membrane movement at the lowest threshold are very small indeed; the magnitudes have been disputed and the controversy has recently been reopened, but they appear to be of the order of thermal agitation amplitudes.

THE HUMAN SUBJECT

So man is just one of 5000 species of mammals, but the great attention paid to this species is not just narcissistic or even imbalanced, for human audiology fulfils an important social obligation. Clinicians are primarily concerned with alleviating the effects of hearing losses and with preventative measures based on a knowledge of their causes. All such work starts with the measurement of individual responses and this is also true of population studies. There has been built up a very detailed picture of the capabilities of this single system and its responses to a variety of accidental and self-inflicted circumstances.

It is clear that the comparative and the clinical approaches have much to offer and to interest each other. The zoologist has much greater experimental and surgical freedom as well as a variety of subjects, and can often provide basic information of value to clinical studies. The clinician generally has the advantage of subjects that are instructable and require little training. But of course this is not true of neonates, young children, geriatric cases, the mentally handicapped and deliberate malingerers, all of whom provide difficult but important branches of his work. Such problems naturally attract support for their solution on a scale that is seldom available for comparative work. Technical developments first applied to clinical problems may later find wider applications.

This is perhaps a recent trend due to the increasing complexity of apparatus, and the present pace of innovation makes meetings such as this very valuable.

TECHNICAL DEVELOPMENTS

Our contributors will describe and assess a number of current techniques, so it may be valuable to step back for a moment to take a broader view. There is no doubt that sounds are extremely difficult to handle by purely acoustic means, although early workers had very little alternative. Helmholtz was an outstanding example, for he used beautifully constructed but technically simple apparatus to achieve a remarkably detailed understanding of many aspects of acoustic physics and of hearing. A set of acoustic resonators, inserted one at a time into his own ears, and an array of tuning forks sufficed to analyse a wide variety of sounds and the response of our auditory systems to them. It is intriguing to wonder what he might have achieved with just one or two instruments that are now regarded as indispensable in even a modest modern laboratory; the qualities of intellect, ingenuity and perseverance are what really count in science.

All branches of acoustics benefited enormously from the development of electronics and of transducers for converting acoustic signals to electrical ones and vice versa, for electrical signals are very much easier to handle. It soon became a relatively simple matter to generate pure tones of accurately controlled frequency and amplitude, or complex sounds with a wide variety of predetermined spectral compositions. Intensities could be measured accurately to permit the calibration of stimuli, and sound analysis allowed other parameters to be monitored. Nevertheless, it is not yet possible to make transducers and associated instruments with properties that rival those of our own auditory systems, at least for the detection of very weak signals. It is the aim of the bionics engineer to learn from natural systems how to improve human technology and the ear may offer much in this respect if the normal progress from applied technology to biology can only be reversed.

At the same time electronics has provided a variety of instruments for the observation of physiological responses: electrodes, amplifiers and oscilloscopes for detailed measurements, or cameras, pen-writers and tape-recorders for more permanent records, are all rather recent developments. The last German edition of Helmholtz's *On the sensations of tone* was published in 1877 (it is still available in translation), yet electronics only emerged about 50 years ago. Even in the 1930s, neurophysiology was developed using string galvanometers because cathode ray tubes were then quite elaborate instruments with a vacuum pump attached.

All these techniques continue to improve in performance, reliability and simplicity of operation, but in addition there are now instruments that operate according to completely new concepts. Computers have greatly extended the ability to process data; they range from tiny pocket models, containing a "mere" 30 000 transistors on only five integrated circuit "chips", to large central machines that can handle prodigious quantities of information and perform an almost unlimited variety of statistical and other operations. Signal recovery techniques also allow tremendous improvements of signal-to-noise ratio (SNR) during actual recording, using averaging computers, box-car detectors or coherent "lock-in" amplifiers, according to what is known in advance about the signal to be observed. Lasers are now being used for the observation of vibration patterns by holography or for the measurement of vibrational velocities by Doppler shift. The atomic-nuclear Mössbauer effect is also exploited to measure velocities and hence the amplitudes involved in vibrations at physiological sound intensities. Changes in the transmission properties of the middle ear can be measured by observing the acoustic impedance of the tympanum. Stereotactic techniques, coupled with ever finer microelectrodes, can probe even tiny single cells in accurately predetermined regions of the nervous system.

INTERDEPENDENCE OF TECHNIQUES

To return to earth, however, it remains true that no single technique is capable on its own of providing all the information one needs in order to define the complete capabilities of the auditory system. There are always "ifs" and "buts". A single neuron can be investigated in great detail but although several hundred such studies may permit a general classification of responses, they still constitute an extremely small sample from the total population involved. Gross responses, recorded from a more general region of the nervous system, cannot distinguish between a large population of rather insensitive units and a smaller population of more sensitive units. The cochlear microphonic potentials may not be a reliable indication of neural sensitivity, although they generally do seem to be well correlated with it. Even behavioural tests only show which sounds the subject is prepared to respond to, although one can improve the motivation by carefully designed incentives. It may be possible to expose a malingerer by a single test but more often doubts remain and a number of different tests may be necessary before certainty can be approached.

In general, therefore, all possible tests of auditory sensitivity are complementary and do not form alternatives on an equal footing. This gives yet another justification for a meeting of this sort where different approaches can be compared and discussed. We are all involved in our various ways with the same type of system and we are all interdependent in a number of respects.

Symp. zool. Soc. Lond. (1975) No. 37, 11–22.

MORPHOLOGICAL AND FUNCTIONAL CHANGES IN THE COCHLEA ASSOCIATED WITH THE INCEPTION OF HEARING

S. K. BOSHER

Ferens Institute of Otolaryngology, Middlesex Hospital Annexe, London, England

SYNOPSIS

The inception of hearing in mammals, taken as the initial appearance of cortical responses to sound stimulation, depends on differentiation of the inner ear, as both the sound transmitting apparatus of the external and middle ears and the central auditory system are already capable of adequate function before cochlear responses arise. These first occur about the time of birth and, apart from minor differences in timing, the underlying developmental processes in most species are remarkably alike.

Cochlear maturation, as a whole, can be divided into four stages terminated chronologically by the appearance of the microphonic responses, followed two to three days later by the development of nerve action potentials. Then a marked increase in the sensitivity and frequency range of the reactions to sound occurs after a further two weeks, and, finally, the adult character is attained, although not for a year or more. The onset of any physiological function is always preceded by complete anatomical differentiation of the structures concerned and the normal specialized biochemical environment provided by the endolymph is an early and essential feature.

The general development of the labyrinth is characterized by short periods involving great changes and these almost certainly constitute critical phases in maturation. During them the evolving organ is likely to be especially susceptible to damaging influences and evidence of this is provided by certain features of genetically dependent deafness.

Noteworthily, the cochlea in man is fully differentiated morphologically about 13 weeks before birth and the available evidence suggests its functional development to be comparably early. More precise information about this aspect is needed, however, if future animal investigations are to be fruitfully related to the problems of human congenital deafness.

INTRODUCTION

The period of development, whether of an organ, a system or a physiological function, offers an unique opportunity to distinguish the nature and inter-relationship of its constituent features as these arise in the course of time. In the field of hearing, aspects as diverse as cell division in early otocysts and the acquisition of language have been investigated, but this paper is restricted to the changes in the cochlea associated with the onset of sound reception, for such a topic seemed better suited to the theme of the symposium. It is one, moreover, in which the more important observations have only become available during the last decade, from the wider use of the electron microscope and of modern electro-physiological techniques, despite over a century of scientific enquiry.

CENTRAL AUDITORY PATHWAYS

First, however, it is necessary to consider briefly the onset of function in the central auditory pathways, in order to provide the cochlear changes with an appropriate context. Here we are much indebted, amongst others, to Professor Marty and his colleagues at Montpellier for their extensive studies on both the guinea-pig and the cat (e.g. Pujol & Marty, 1968; König, Pujol & Marty, 1972; Pujol, 1972). These have revealed the maturation of the central system, as judged by the initial appearance of recordable activity to auditory nerve stimulation and later alteration in the characteristics of the responses, to progress centripetally in an orderly manner. This is in full agreement with the relevant histological findings, such as the degree of myelination, and means that development proceeds systematically from the cochlear nuclei in the brain stem to the cerebral cortex, although the hearing responses are still quite primitive at all levels even when auditory-evoked cortical activity first appears and the adult features are not attained until some months after birth.

What is particularly striking about these investigations and the earlier reports on the subject (e.g. Änggård, 1965), therefore, is that the final stage in the initial acquisition of hearing depends not on the cortex but on the appearance of action potentials in the VIIIth nerve, the lowest auditory neurone. This, nevertheless,

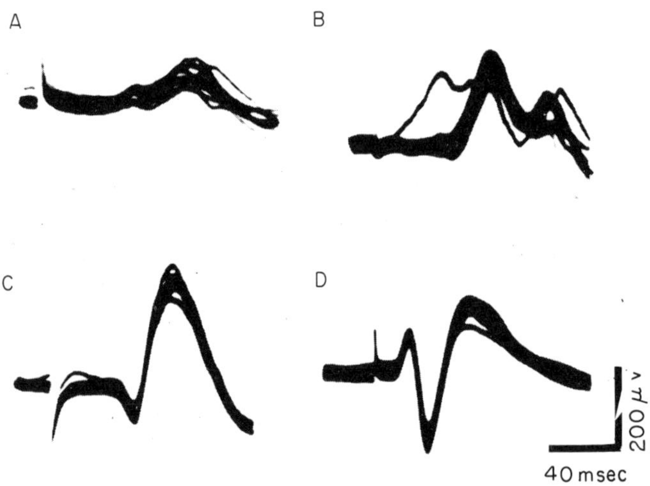

FIG. 1. Response of the primary auditory cortex to electrical stimulation of the contralateral auditory nerve in the cat; A aged 36 h, B 4 days, C 5 days and D 19 days. From Marty & Thomas (1963).

must not be taken as indicating atypical neuronal immaturity, for the nerve will respond to electrical stimulation before this period and, indeed, it is stimulation of this nature which is used to follow the early development of the higher centres (Marty & Thomas, 1963; Hassmannová & Mysliveček, 1965) as shown in Fig. 1.

Thus the onset of sound reception in all mammals so far investigated is due to the final maturation of the cochlea, the central auditory system already being capable of responding when this occurs.

ORGAN OF CORTI

With regard to the cochlea, a great deal of information has become available in recent years from the large number of physiological and anatomical studies undertaken upon a wide range of mammals (Änggård, 1965; Kikuchi & Hilding, 1965; Mikaelian & Ruben, 1965; Crowley & Hepp-Reymond, 1966; Hilding, 1969; Pujol & Marty, 1970; Sugiura & Hilding, 1970; Romand, Granier & Pujol, 1971; Sher, 1971; Ernstson, 1972; Pujol & Hilding, 1973; and others). The results obtained correspond to a quite remarkable degree, the only differences being minor variations in timing (Table I), and it is now possible to predict the functional capacity with reasonable certainty from the morphological appearance alone.

TABLE I

The ages, relative to birth, when the cochlear microphonic potentials first appear in seven common laboratory animals

Guinea-pig	−15	
Cat	− 2	Days before birth
Dog	+ 4	Days after birth
Rabbit	+ 4	
Rat	+ 8	
Mouse	+11	
Mink	+14	

The process of cochlear maturation may be conveniently divided into four functional phases. Of these, the initial one is constituted by the long period of early development before reactions to sound are present, and it is terminated about the time of

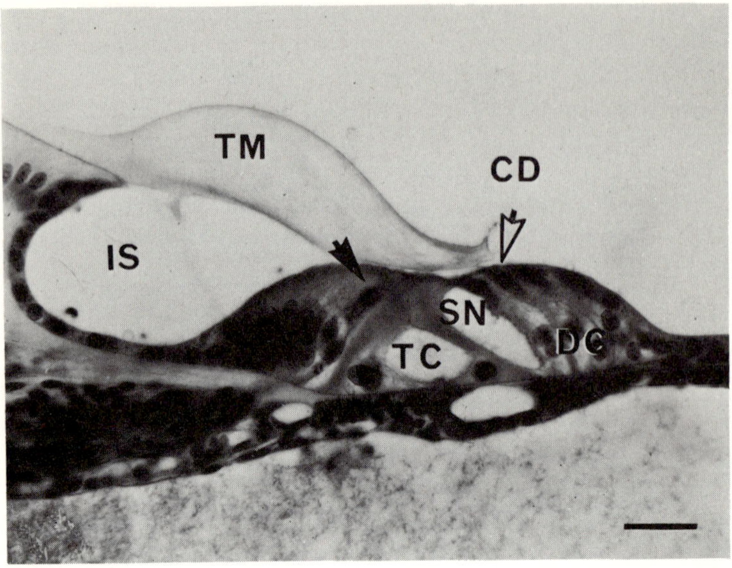

FIG. 2. Organ of Corti, basal turn, cat, at birth, end of first developmental phase. Evidence of immaturity in the absence of spaces around the outer hair cells, the relative squatness of Deiter's cells and the primitive cells still associated with the inner hair cell will be noted (cf. Fig. 3). Abbreviations: CD: cochlear duct; DC: Deiter's cells; IS: internal sulcus; SN: space of Nuel; TC: tunnel of Corti; TM: tectorial membrane. Solid arrow, inner hair cell; open arrow, outer hair cells. Marker, 40 μm.

birth when the first cochlear microphonic responses arise from sound stimulation.

Even though the organ of Corti, as a whole, is somewhat immature at this stage (Fig. 2), the tectorial membrane has become freed in its entirety by the resorption of the cells previously filling the internal sulcus and the tunnel of Corti has just become visible. The hair cells are not quite completely differentiated, but the apical or receptor poles, it must be stressed, have been fully developed, as judged by their electron microscopic structure, for as long as five or six days. In addition, both the efferent and afferent synapses at the bases of the hair cells have been most ably demonstrated by Professor Hilding (Pujol & Hilding, 1973) to be ultrastructurally completely mature at, or even slightly before, this time. No evidence, however, of the production of nerve action potentials in association with the microphonic responses has been obtained despite a most careful search by a number of workers.

The situation in respect of synaptic transmission, therefore, clearly resembles the state of affairs at the receptor pole and it

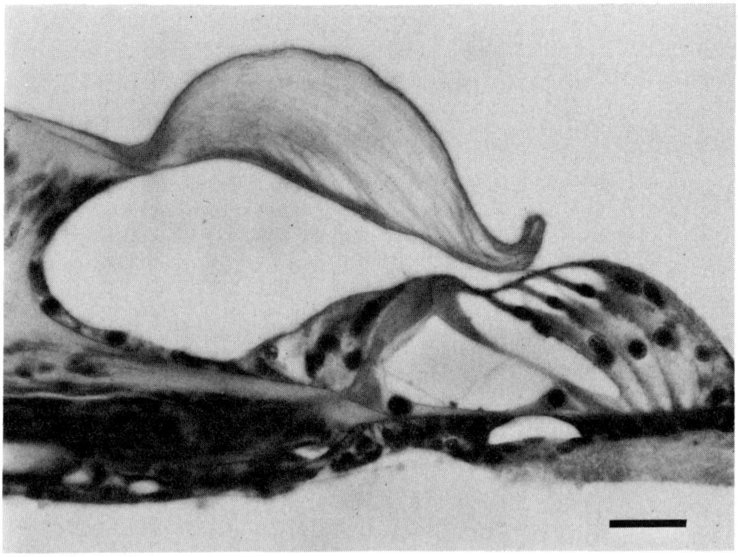

FIG. 3. Organ of Corti, middle turn, cat, 15 days after birth, latter part of third developmental phase. Complete anatomical differentiation has just been attained. For comparison with Fig. 2. Marker, 40 μm.

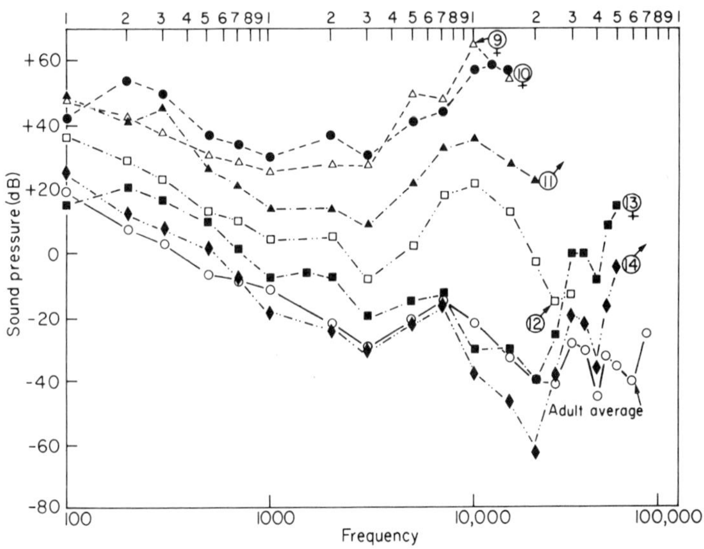

FIG. 4. Cochlear microphonic sensitivity curves from members of the same litter of infant rats. Each point represents the sound pressure in decibels, relative to 1 dyne/cm^2, required to produce a potential of 1 μV, the pinna and external meatus having been surgically removed before the recordings were made. The number inside the sex symbol gives the age in days after birth and the marked change between the eleventh and thirteenth days will be noted. From Crowley & Hepp-Reymond, 1966. Copyright (1966) by the American Psychological Association, reprinted by permission.

seems to be a general feature of cochlear differentiation that a definite interval is present between anatomical maturation and the onset of physiological function. The complex organization at the molecular level of the various membranes concerned would consequently appear to take place after they have attained their gross structural form.

Be this as it may, the two to three days when the microphonic but not the nerve action potentials are evident compose the relatively short second developmental period. During it, morphological differentiation continues steadily and the cochlea eventually become fully adult in form (Fig. 3) some two to three weeks after parturition, towards the end of the succeeding third phase, which commences with the first neural activity.

At the onset of this phase, Crowley, who has conducted an extensive series of highly informative investigations in the rat, has shown the physiological responses to be severely restricted. Progress, moreover, is slow until the last maturation period is heralded by a marked and relatively abrupt increase in both the sensitivity and frequency-response range of the microphonic potentials (Fig. 4), associated with a change in the peak sensitivity from 3 kHz to the usual value of over 40 kHz (Crowley & Hepp-Reymond, 1966). The period of sudden change again has no immediate anatomical correlate and the responses following it are still not quite mature in character. The final stage in the development process is unexpectedly long and occupies a year or more with respect to some criteria (Crowley, Schramm, Swain & Swanson, 1973).

ENDOLYMPH SYSTEM

The physiological events connected with the inception of hearing are obviously of the utmost importance and so the endolymph (the fluid filling the cochlear duct) cannot be neglected, since it provides the external environment for the hair cell receptor pole and is intimately connected with hair cell function. It is unique amongst extracellular fluids, for not only is its major cation potassium but it also has a positive potential of 90 to 100 mV, both of these features being considered to have a vital role in the normal transduction process.

In 1971, Bosher & Warren reported that the endocochlear potential in neonatal animals was very much lower than the adult level (Fig. 5). On the eighth day after birth in the rat, for example, the average potential was merely 14 mV and the rate of increase

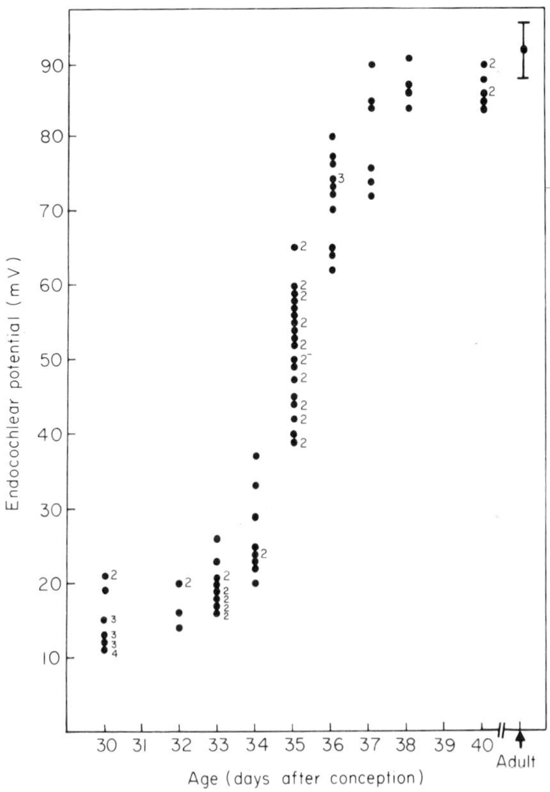

FIG. 5. The change in the endocochlear potential in the rat with age. The range and mean of the potential in adult animals is also shown. Numerals indicate the number of experiments giving the same result. The average length of gestation was 22 days. From Bosher & Warren (1971).

correspondingly slow. Thus 3 days later the potential was still only about 19 mV; but then it increased surprisingly rapidly, especially between the twelfth and fourteenth days when the rise was of the order of 25 mV a day, and this alteration, moreover, occurred simultaneously throughout the whole cochlea.

The exact cause of this dramatic change remains obscure but the endolymph system is clearly undergoing significant modification, probably the result of a marked increase in the electrical resistance of some part of the cochlear duct membranes. What seems especially relevant in the present context is that the period coincides with the phase of sudden development of the microphonic responses and there is good reason to believe the substantial improvement in hearing ultimately depends on the elevation of the endocochlear potential.

The analytical findings in the same animals, however, revealed a completely different situation, as shown in Fig. 6. The tendency

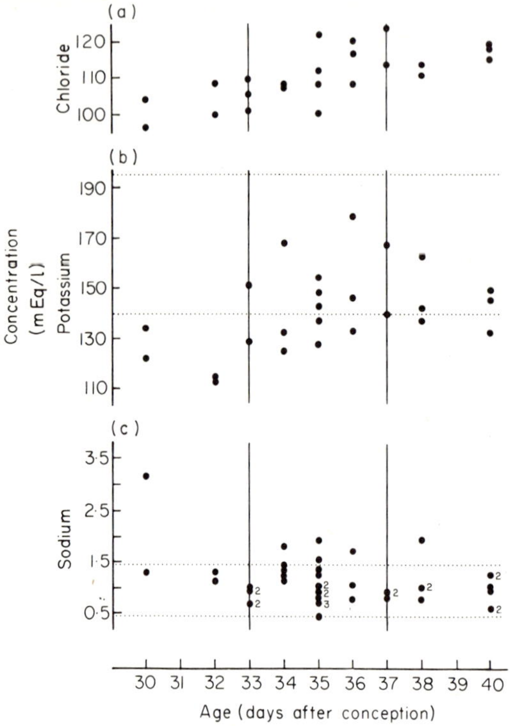

FIG. 6. The endolymphatic ionic composition in the neonatal rat. The horizontal lines define the normal adult range for sodium and potassium; the difference in scale will be noted. The vertical lines indicate the total period of rapid development of the endocochlear potential. Numerals indicate the number of experiments giving the same result. The average length of gestation was 22 days. From Bosher & Warren (1971).

towards somewhat lower values evident in the younger animals was proved to be due not to immaturity, but to diminution in the general osmotic pressure with consequent dilution of the inner ear fluids, and no deviation in any major ionic concentration was, in fact, demonstrated. As might be expected in such circumstances, the stria vascularis was already morphologically fully mature in the youngest animals studied, although its enzyme activity remains well below the adult level until the final developmental stage is reached (Kuijpers, 1974).

In this situation, it seems pertinent that the hair cells were also completely differentiated in the youngest animals studied, and that changes in the endolymphatic ionic composition are known to have a deleterious action on hair cells. The early attainment of the adult constitution, therefore, appears to be concerned with maintaining the integrity of the coincidently maturing hair cells and, recently, striking evidence in support of this concept has come to hand. Professor Rose and his colleagues at the University of Wisconsin (Sobkowićz, Bereman & Rose, 1974) have been able to follow the development of the organ of Corti in explants of the cochlear duct removed from the mouse at birth, when the cochlea is quite immature, and thereafter maintained by tissue culture; a really tremendous achievement. In particular, it has been noted that if the cochlear duct is torn during the manipulative procedures, then differentiation ceases and degeneration ensues, confirming a correct biochemical environment to be absolutely essential for normal cochlear maturation.

SURVEY OF NORMAL DEVELOPMENT

If, as seems reasonable, the appearance of cortical responses to sound is utilized to define the inception of hearing, then this depends on the final maturation of the hair cell synaptic mechanisms. Limitations of space prevent mention of the wealth of information concerning the external and middle ear, but the sound transmitting apparatus, like the central auditory system, appears capable of adequate function before the cochlea. Here the developmental process proceeds in an orderly and rather stereotyped fashion with morphological differentiation preceding the onset of the associated physiological responses.

A suitable biochemical environment is clearly a vital requirement and the whole process is punctuated by short periods of marked change. These are undoubtedly of great significance for they almost certainly constitute, in the inner ear, the critical phases of development during which an evolving organ or system is now generally believed to be peculiarly susceptible to damaging influences.

ABNORMAL DEVELOPMENT

While an extensive review of the vast number of studies on congenital and neonatal deafness currently available is outside the scope of the present paper, an example of this type of cochlear vulnerability seems of relevance.

Genetically determined deafness might well be thought commonly to arise from some derangement of the normal complex maturation process leading eventually to abnormal anatomical forms. Yet this is not the case, for the commonest cause of such deafness in animals is a degenerative condition, whose onset is delayed until the time after birth when the inception of hearing normally occurs, a finding which alone would make this period of great interest. The morphologically completely, or nearly completely, differentiated cochlea, therefore, is almost certainly unduly susceptible to some noxious influence during one of the critical phases in its terminal physiological development.

Furthermore, Bosher & Hallpike (1965) in their examination of the deaf white cat were able to prove conclusively that the toxic agent was environmental in origin. Its nature could not be defined with any precision, but some induced biochemical abnormality of the endolymph seemed the most likely possibility. The abnormal gene was thought to be responsible for a substantial increase in the vulnerability of the cochlea to the potentially damaging environmental circumstance. This accordingly appeared capable, if severe enough, of itself initiating degeneration in a genetically normal animal, a situation which reveals yet another relatively unexplored aspect of cochlear development.

HUMAN COCHLEAR DEVELOPMENT

Finally, the situation in man must be outlined in order to complete the picture presented, although this field correctly belongs to the following paper. Maturation of the human cochlea is essentially the same as that in the other mammals but for one important exception. Apart from the marsupials with their unique extrauterine method of protecting the young, the stage characterized by the initial appearance of the microphonic potentials occurs in animals about the time of birth and the differences in exact timing (Table I), while of general biological significance, are small. In man, however, the inner ear attains this stage as long as 13 weeks before the end of gestation.

The cochlea, in consequence, might well be better protected in man than in other mammals against adverse environmental circumstances during the critical phases in its differentiation. This, however, is not certain and, in addition, the early completion of its development is not without attendant disadvantages (Bosher, 1972).

CONCLUSION

Much further research is still needed if the processes underlying cochlear maturation and the inception of hearing are to be fully understood, but fortunately the requisite methods, such as the new organ culture techniques, are available. Nevertheless, if the animal investigations are to be fruitfully related to the problem of human deafness, more knowledge about the fundamental processes in man must be ascertained. In particular, answers are urgently required to two basic questions; namely, when exactly does the inception of hearing occur in the human foetal ear, and when do the various forms of deafness presenting in early life actually have their onset?

The matter is of more than theoretical interest, for in Waardenburg's syndrome, where the presence of the causative, potentially pathogenic gene can be accurately determined, the occurrence of deafness is surprisingly low—only about 20% (Waardenburg, 1951). Such a feature surely offers the tantalizing prospect of reducing the incidence of so-called congenital deafness in general if only more was known about the underlying processes concerned.

ACKNOWLEDGEMENTS

Grateful acknowledgement is made to Masson & Cie, Paris (Fig. 1), the American Psychological Association (Fig. 4) and the Editorial Board of the Journal of Physiology (Figs 5 and 6) for permission to reproduce the figures specified.

REFERENCES

Änggård, L. (1965). An electrophysiological study of the development of cochlear functions in the rabbit. *Acta otolar.* suppl. **203**.

Bosher, S. K. (1972). Neonatal changes in the endolymph and their possible relationship to peri-natal deafness. *Acta otolar.* **73**: 203–211.

Bosher, S. K. & Hallpike, C. S. (1965). Observations on the histological features, development and pathogenesis of the inner ear degeneration of the deaf white cat. *Proc. R. Soc.* B. **162**: 147–170.

Bosher, S. K. & Warren, R. L. (1971). A study of the electrochemistry and osmotic relationships of the cochlear fluids in the neonatal rat at the time of the development of the endocochlear potential. *J. Physiol., Lond.* **212**: 739–761.

Crowley, D. E. & Hepp-Reymond, M. C. (1966). Development of cochlear function in the ear of the infant rat. *J. comp. physiol. Psychol.* **62**: 427–432.

Crowley, D. E., Schramm, V. L., Swain, R. E. & Swanson, S. N. (1973). Age-related wave-form changes of VIIIth nerve action potentials in rats. *Laryngoscope* **83**: 264–275.

Ernstson, S. (1972). Cochlear physiology and hair cell population in a strain of the waltzing guinea-pig. *Acta otolar.* suppl. **297.**

Hassmannová, J. & Myslivečck, J. (1965). Maturation of the primary cortical response to stimulation of medial geniculate body. *Electroen. Neurophysiol.* **22**: 547–555.

Hilding, D. A. (1969). Electron microscopy of the developing hearing organ. *Laryngoscope* **79**: 1691–1704.

Kikuchi, K. & Hilding, D. A. (1965). The development of the organ of Corti in the mouse. *Acta otolar.* **60**: 207–222.

König, N., Pujol, R. & Marty, R. (1972). A laminar study of evoked potentials and unit responses in the auditory cortex of the postnatal cat. *Brain Res.* **36**: 469–473.

Kuijpers, W. (1974). Na-K-ATPase activity in the cochlea of the rat during development. *Acta otolar.* **78**: 341–344.

Marty, R. & Thomas, J. (1963). Réponse électrocorticale à la stimulation du nerf cochléaire chez le chat nouveau-né. *J. Physiol., Paris* **55**: 165–166.

Mikaelian, D. & Ruben, R. J. (1965). Development of hearing in the normal CBA-J mouse. *Acta otolar.* **59**: 451–461.

Pujol, R. (1972). Development of tone-burst responses along the auditory pathway in the cat. *Acta otolar.* **74**: 383–391.

Pujol, R. & Hilding, D. A. (1973). Anatomy and physiology of the onset of auditory function. *Acta otolar.* **76**: 1–10.

Pujol, R. & Marty, R. (1968). Structural and physiological relationships in the maturing auditory system. In *Ontogenesis of the brain*: 377–385. Jilek, L. & Trojan, S. (eds). Prague: Charles University.

Pujol, R. & Marty, R. (1970). Postnatal maturation of the cochlea of the cat. *J. comp. Neurol.* **139**: 115–126.

Romand, R., Granier, M. R. & Pujol, R. (1971). Potentiels cochléaires chez le foetus de cobaye dans la période prénatale. *J. Physiol., Paris* **63A**: 280–281.

Sher, A. E. (1971). The embryonic and postnatal development of the inner ear of the mouse. *Acta otolar.* suppl. **285**.

Sobkowićz, H. M., Bereman, B. & Rose, J. E. (1974). Development of the organ of Corti of the newborn mouse in culture. *J. acoust. Soc. Am.* **55**: 459.

Sugiura, A. & Hilding, D. A. (1970). Cochleo-saccular degeneration in Hedlund white mink. *Acta otolar.* **69**: 126–137.

Waardenburg, P. J. (1951). A new syndrome combining developmental anomalies of the eyelids, eyebrows and nose root with pigmentary defects of the iris and head hair and with congenital deafness. *Am. J. hum. Genet.* **3**: 195–253.

Symp. zool. Soc. Lond. (1975) No. 37, 23–40.

ON THE MEASUREMENT OF HUMAN FOETAL AUDITORY RESPONSE

R. J. BENCH and D. L. MENTZ

Regional Audiology Research Unit, Royal Berkshire Hospital, Reading, England

SYNOPSIS

Previous reports of human and animal foetal response to audio-frequency stimuli (foetal audiometry) are briefly reviewed. Attention is drawn to the methodological difficulties of work in this field and the need of controls for: reaction by the mother to the foetal stimulus; the psychological state of the mother; geometric and postural factors; uterine contractions; a suitable stimulus at the foetal ears; spontaneous "response" changes; and Law of Initial Value effects. Early work on the human foetus was conducted before term by recordings made via the maternal body wall. Advances in obstetrics now permit more direct recordings of foetal heart rate and electroencephalogram (EEG) from foetal scalp electrodes during birth, and the more recent work in foetal audiometry has been directed to this aspect. An experiment is described in which foetal heart rate responses to 38 two-second duration 100 dB sound pressure level (SPL) noise-band stimuli were recorded from a group of 14 human foetuses during labour, together with "responses" to an additional 38 no-sound control trials. Foetal heart rate was derived for the interbeat interval immediately preceding each stimulus or simulated stimulus onset, and for the 60 interbeat intervals thereafter. Using the difference between prestimulus and poststimulus (or precontrol and postcontrol) foetal heart rate averaged across foetuses and trials as an indicator of response, significant differences were found between stimulus and control applications. The response curves obtained from these data are compared with the outcome of similar work on babies studied after birth and on adults.

INTRODUCTION

The foetus is difficult to study because of its inaccessibility and its physiological dependence on its mother. This implies that a number of controls and other factors need to be considered when evaluating the results of tests of foetal sensory function. These factors are discussed at some length before an experiment on the human foetus during labour is described.

Early work

Although quite early claims were made (Peiper, 1925; Forbes & Forbes, 1927) for the responsivity of the human foetus *in utero* to auditory/vibratory stimuli, they were published as brief notes of occasional, sometimes accidental, observations on one, or a very few subjects, or as notes on patients' subjective reports. They provided some indication that the foetus may be responsive to audio-frequency stimuli, but the generality of their conclusions is limited because of the small number of cases reported. Moreover, little reliance can be placed on the validity of the findings since the reports lacked some necessary controls. Experimental work in this field began in the 1930s and 1940s with studies by Sontag & Wallace (1935a), Sontag (1936, 1944) and Spelt (1938, 1948), at about the same time as other workers were considering related aspects of foetal activity and physiology, both in human subjects, and in animals (Ray, 1932; Sontag & Wallace, 1935b; Hooker, 1936; Barcroft & Barron, 1937; Newbery, 1941; Windle, 1941; Lindsley, 1942; Norman, 1942; Rose, 1947).

Techniques

Perhaps because of the technical difficulties, the investigation of foetal responses to sounds (foetal audiometry) seems to have attracted relatively few investigators, even though the human foetal ear appears to be anatomically complete well before birth (Bast & Anson, 1949; Vasiliu, 1968;—see also the preceding paper, in this symposium, by Dr Bosher). However, Sontag's pioneering work on foetal auditory/vibratory responses *in utero* was continued by Murphy & Smyth (1962), who were concerned to develop techniques as aids to the assessment of the clinical status of the foetus in addition to the more specific aim of assessing foetal auditory function. The techniques were further refined by Johansson, Wedenberg & Westin (1964) who noted *inter alia* that where responses to sounds as opposed to vibratory responses were to be studied, the frequency of the stimuli should be above 1·5 kHz to reduce the possibility of tactile artefact (Knudsen, 1928; von Békésy, 1939). The method used in these studies was to stimulate the foetus with audio-frequency stimuli generated by an audio-oscillator, amplified, and fed to a transducer on the mother's abdomen immediately above the foetal head, as found by palpation. Murphy & Smyth (1962) used a small loudspeaker embedded in a foam rubber annulus as their transducer. However, with this method reflections at the air/maternal body interface are considerable and do not permit efficient transduction. It is thus noteworthy that Bench & Vass (1970) failed to replicate the results of Murphy & Smyth (1962). A more efficient technique (Johansson *et al.*, 1964) is to use a vibrator to energize a small plate in contact with the maternal abdomen.

The foetal response has often been assessed by monitoring gross changes in foetal movement by a displacement transducer on the mother's body, or by measuring changes in foetal heart rate computed from the foetal electro- or phonocardiogram, or, more recently, by means of an ultrasonic device (Grimwade, Walker, Bartlett, Gordon & Wood, 1971). Very recently Boddy & Robinson (1971) have described an ultrasonic technique for measuring foetal respiratory effort *in utero* both in the sheep and in pregnant women. Such methods have permitted study of sound-evoked foetal responses in the last trimester of pregnancy, but reliable work on the foetus before this is very difficult, if only because of problems associated with the detection of foetal signals of low amplitude at the maternal body wall.

Depression of the foetal nervous system

The studies cited above, all of which claimed that the human foetus did respond to sound or vibration, have been criticized on methodological grounds by Bench & Vass (1970), and by Bench (1972, 1973), who argued a case for regarding the foetal nervous system as being in a relatively depressed state. This view was supported by evidence from Barcroft & Barron (1937) working with sheep, from Riesen (1947) working with chimpanzees, and from others. It has also been supported by C. Kennedy (personal communication), who noted that in dogs the foetal nervous system is depressed metabolically at birth, but that this changes rapidly in the early weeks of life. But Kennedy also writes: "It is puzzling, however, that the central hearing pathways i.e. the cochlear nucleus, lateral lemniscus, superior olive, and inferior colliculus have selectively high blood flow even in the first few days of life". If the foetal nervous system is in a relatively depressed state, and if, as has been indicated, it is not too easy to monitor aspects of foetal response to sound, then it follows that experimental work must be carefully controlled to reduce the possibility of interpreting spontaneous or artefactual events as valid responses.

Controls

It is convenient to consider this matter of controls under two headings: procedural and statistical (Bench, 1973).

Procedural measures and controls

Exclusion of the mother from the foetal stimulus. It appears to be important to exclude the mother from the sound stimuli offered to the foetus, because a humoral or movement response by the mother could be transmitted to the foetus as an indirect stimulus. Although the possible effects of such an indirect stimulus would be delayed relative to the sound stimulus, particularly in the case of humoral factors, they could produce changes in foetal "state" which might confuse the interpretation of foetal response over a sequence of sound stimuli presented to the foetus. This problem was studied by Bench & Vass (1970), who presented loud tones to the ears of pregnant women. They found evidence of interaction between mothers/foetuses and effects: some foetuses appeared to show heart rate changes when loud tones were presented to their mothers, while others did not. As a precautionary measure, it would therefore seem best to exclude all mothers from the foetal stimulus. This can be done conveniently by presenting the mother with continuous masking noise via close-fitting earphones.

The psychological state of the mother. Although little reliable evidence is available, misplaced apprehension by the mother about experimental techniques could also affect foetal state via humoral mediation. It therefore seems appropriate to accustom her to the transducers to be applied to her body, and to the experimental treatments, before beginning the foetal stimulus tests proper. The considerate clinician/experimenter will normally do this as a matter of course in the interest of good relations with the patient.

Geometric and postural factors. As pregnancy advances, and especially during labour, the geometry of the mother's abdomen and of the foetus will change. This will have some effect on the sound pressure level of the stimulus at the foetal ears, especially where external transducers are employed. Also, the mother's posture will affect the tension of her abdomen, and hence the mechanical impedance of body tissues to which an external vibrator is coupled. This implies that sounds transduced across the body wall will change somewhat in sound pressure level and frequency spectrum with different maternal postures and types of support. However, although note should be taken of factors contributing to maternal geometry and posture, their effects are likely to be fairly small.

Uterine contractions. Contractions of the uterus may affect the geometric and postural factors and will also affect the foetus directly and indirectly in so far as the pressures on the foetus and umbilical cord may be considerable (Goodlin, 1971). It remains to be shown that such contractions have an appreciable effect on foetal responses to sound stimuli. In the meantime, a well conducted study will take their possible influence into account.

Estimation of the stimulus at the foetal ears. Early workers made no attempt to measure the properties of the sound stimulus reaching the foetal ear. However, techniques have recently been developed which permit a fair estimate to be made of stimulus properties and ambient noise levels in regions about the foetal head. Johansson *et al.* (1964) calibrated their stimuli on four patients who had just given birth by inserting a probe microphone into the uterus. This method provides useful data, but these may be in error because of geometrical and other factors resulting from an empty uterus. Bench (1968) calibrated his transducer with a probe microphone placed against the cervix in one patient at 37 weeks gestation, but, because the microphone was sited at some distance from the foetal ears, the results may have underestimated the sound levels within the uterus. In subsequent work, calibrations were attempted on goats (Bench, Anderson & Hoare, 1970) via a hydrophone implant. Walker, Grimwade & Wood (1971) measured intrauterine sound levels by inserting a small microphone into the uterus of pregnant women at term, but before the rupture of the membranes, although their data have been questioned by Henshall (1972). Taken together, these studies have shown that the ambient intrauterine noise is probably of the order of 85 dB SPL (Henshall, 1972), is mainly low frequency in character, and probably arises from turbulent blood flow and muscle movement, punctuated by maternal borborygmi at irregular intervals. In order to prevent masking of the stimulus sounds presented to the foetus it would seem advisable to ensure a stimulus SPL of 95–105 dB (Hackman & Bench, 1972), which does not seem likely to be so high as to damage the foetal ears (Bench, Hoare & Pye, 1973).

Statistical controls

Spontaneous changes. Biological organisms are remarkable for presenting the unwary experimenter with "responses" which may turn out to be spontaneous changes reflecting deep-seated and underlying variations in physiological state. Although the relative isolation of the foetus protects it from external factors which may contribute to such spontaneous changes, and although the latter are to a degree "buffered" by the physiological relation of the foetus with its mother, adequate controls for spontaneous responses are still required.

The Law of Initial Value. Foetal responses to sound stimuli have most often been assessed as a change in some physiological parameter (e.g. heart rate) with stimulation. Responses assessed in this way should be interpreted with care since it has been shown (Wilder, 1958) that there is a negative regression of magnitude of response change on the prestimulus level of activity. Stimulation at a low level of prestimulus activity results in a large positive response change, whereas stimulation at a high prestimulus level produces no change or a negative change. If foetal responses are assessed as changes of response magnitude, then cognisance should be taken of prestimulus activity levels.

Recent work

Some of the most recent work has been performed on the foetus during labour. Here, during the birth process, the accessibility of the foetal scalp permits more reliable measures to be made of heart rate and other variables, although it must be noted that the foetus is changing, or is about to change, its physiological state at this time. Thus it is possible to measure the foetal electroencephalogram (FEEG) via scalp electrodes, and this has opened up a new field for assessing foetal responses. Although Lindsley (1942) was the first to attempt to record the foetal electroencephalogram, he used electrodes placed on the maternal abdomen. Signals obtained in this way are of doubtful validity. Scibetta & Rosen (1969) developed a method for foetal electroencephalographic recording via an incision made in the body of pregnant guinea-pigs. Later (Rosen, Scibetta & Hochberg, 1970; Scibetta, Rosen, Hochberg & Chik, 1971) they and their colleagues used scalp electrodes to monitor human foetal electroencephalographic responses during labour to maternal medications and to sound stimuli.

The present study

The study which is reported here also capitalized on the accessibility of the foetal scalp during labour. Recordings were made of both foetal electroencephalogram and heart rate, of which only the latter will be considered here. A particular aim was to measure the pattern of the sound-evoked heart rate response over time. Previous reports had suggested that foetal heart rate changes occurred with sound stimulation, but the pattern of this response was usually not described in detail, although Dwornicka, Jasienska, Smolarz & Wawryk (1964) and Johansson *et al.* (1964), studying the human foetus in the last weeks of gestation, have given some description.

TABLE I

Sample of 14 mothers and babies

Subject number	Min from membrane rupture to test start	Min to delivery from test start	Post-menstrual age by date	Post-menstrual age by examination	Maternal age	Gravidity	Parity	Infant: 5 min Apgar score	Infant: birth weight (g)	Infant: sex
1	75	197	40	39	22	1	0	10	3840	F
2	c.120	322	40	37	18	1	0	9	2910	F
3	385	540	38	38	27	9	8	8	3060	M
4	960	367	42	40	20	1	0	10	3290	M
5	218	214	41	40	17	3	2	9	3320	F
6	66	310	41	40	18	1	0	8	3400	F
7	673	427	41	38	37	1	0	9	3480	F
8	59	522	44	41	18	1	0	9	3430	M
9	1425	39	39	41	22	3	1	9	2370	F
10	675	533	39	38	14	1	0	7	3080	M
11	25	148	42	42	29	7	5	10	4400	M
12	68	160	40	41	24	1	0	10	3100	F
13	30	283	41	40	16	1	0	10	2780	F
14	30	245	39	40	28	4	1	10	3880	M
Median	198	297	41	40	22	1	0	9	3255	—
Lowest	25	39	38	37	14	1	0	7	2370	—
Highest	1425	540	44	42	37	9	8	10	4400	—

METHOD

Subjects

The foetuses of 17 women in labour were taken for study at Cincinnati General Hospital, Ohio. For three of these, obstetric treatments curtailed the recording. Of the remaining 14, whose results are reported here, eight were Negro and six Caucasian. The amniotic membranes had ruptured spontaneously in seven cases and were ruptured surgically (amniotomy) in the other seven. In three cases, analgesics (meperidine and promethazine) were given to the mothers (105, 225, 375 min) before the study. These drugs did not appear from inspection of the polygraph recordings (see below) to affect foetal auditory responses. At postnatal examination no baby showed congenital abnormalities, but one showed signs of placental insufficiency. Other details are given in Table I. All the babies responded to a 95 dB SPL noise stimulus after birth.

Stimuli

Thirty-eight $\frac{1}{3}$ octave noiseband stimuli of 2 sec duration with 10 msec rise/decay times were recorded on magnetic tape, amplified, and fed to a modified Ling Altec V4M Vibrator secured to the maternal abdomen with an elastic belt. All stimuli were centred at 2 kHz apart from numbers 22, 32 and 37 which were centred at 400 Hz. The interval between stimuli was 52 sec, and the block of stimulus trials was preceded and followed by a block of 19 simulated stimulus presentations (controls). The study thus lasted about 1 h for each patient. Stimulus levels were estimated by the method of Hackman & Bench (1972) to be approximately 100 dB SPL at the foetal ears. An analysis of harmonics at the vibrator/maternal body interface showed some distortion of the original signal, increasing the stimulus bandwidth (e.g. harmonics only 12 dB down at one octave above the source frequency band). This effect tended to make the 400 Hz and 2 kHz proximate signals more alike than their originals, and in analysis of the results the stimuli were treated as equivalent.

Procedure

The women, in or about to undergo labour, were brought close to the labour theatres to a room where the ambient noise was 67 dB SPL (generated mainly by recording equipment). Amniotomy was performed as necessary and transducers were attached for continuous clinical recording of the foetal electroencephalogram (FEEG) and heart rate (FHR), intrauterine pressure (IUP), and maternal heart rate (MHR) via a Corometrics Fetal Monitoring System 101A, a Hewlett Packard 7700 polygraph, and a Hewlett Packard 3955 magnetic recording system. These recording techniques are in regular use or have been described elsewhere (Rosen *et al.*, 1970). They will not be discussed further. Some 15–20 min before the present experiment began, its aims were briefly explained to the mother, who was asked to lie as still as possible throughout. Then she was asked to wear a pair of close-fitting earphones through which a continuous $\frac{1}{3}$ octave noiseband centred at 2 kHz was presented at a level sufficient to prevent her from hearing the foetal stimulus. The vibrator was secured on the maternal abdomen above the foetal ear with an elastic strap, and 5 min later the study began with the first block of 19 simulated sound presentations.

RESULTS

Portions of the polygraph charts are shown in Fig. 1 (beginning of the sound stimulus trials) and Fig. 2 (end of the sound stimulus trials) for one subject who gave more obvious responses than most.

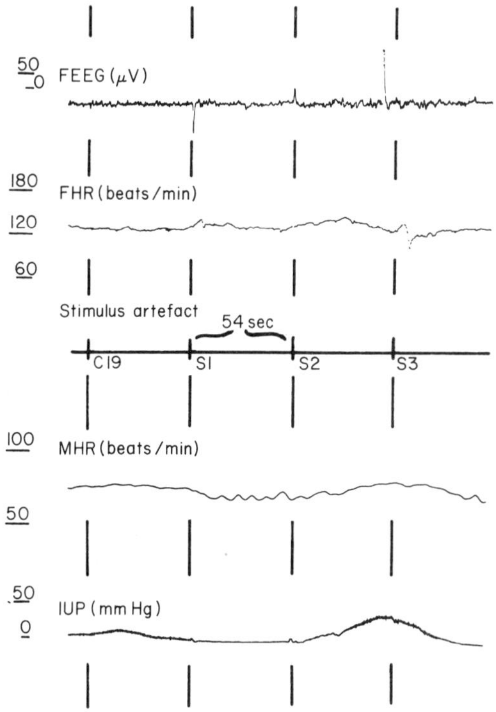

FIG. 1. Example of polygraph chart: beginning of sound trials.

The data to be considered will be restricted to foetal heart rate. Although it is likely that other variables (see introduction) may exert subtle effects on foetal heart rate, the present concern is with a survey of the heart rate responses only.

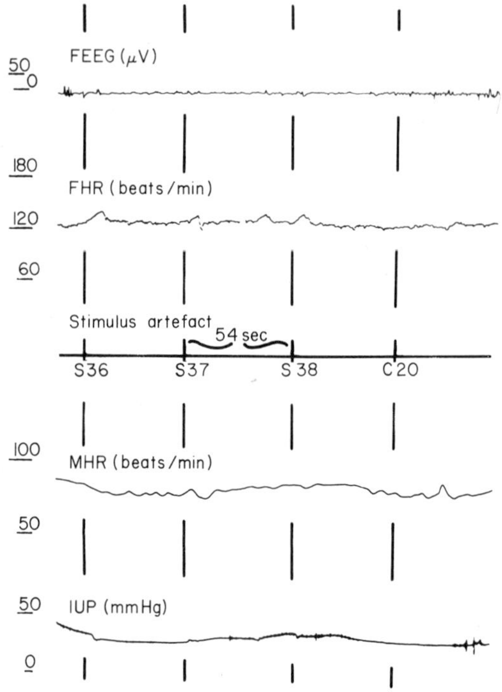

FIG. 2. Example of polygraph chart: end of sound trials.

Inspection of the polygraph recordings showed no obvious consistent heart rate change in most cases. It was apparent, however, that the type of response (i.e. whether an increase or decrease in heart rate or an increase followed by a decrease, etc.) might have changed from subject to subject and from trial to trial, and also that this change, if any, could have occurred at different times after the application of the stimulus. A further difficulty was the fact that large "systematic" changes in heart rate seemed to have occurred for some subjects, this change itself being a gradual increase in some cases, and a gradual decrease in others. Quantities which seemed unlikely to be greatly affected by these changes in foetal heart rate are the differences d_i between pre- and poststimulus heart rate levels,

i.e. $$d_i = S_i - \text{pre}S\ \text{FHR}$$

where S is a function of some subset (i) of the poststimulus heart rate.†

Foetal heart rate (in beats/min) was computed from the polygraph charts for interbeat intervals of interest. Sixty contiguous readings were taken in the period immediately following each sound stimulus or simulated stimulus onset, and one immediately before each such onset. In order to assess the magnitude of response at different times after the application of the stimulus, the poststimulus period of 60 interbeat intervals was divided into 12 equal periods of five interbeat intervals each, and the prestimulus measurement was subtracted from the mean heart rate of each period to give 12 difference values. These values, taken together for all 14 subjects, were found to be approximately normally distributed in both the control and the stimulus applications, although the variance of the stimulus values was consistently greater than that of the controls. The distributions of the differences are shown in Fig. 3, and their means and variances in Table II. It is clear from Fig. 3 that zero changes in d_i were more common

† This assumes that the Law of Initial Value effects are not marked.

TABLE II

Means and variances of the d_i *values*

d_i	Control		Stimulus	
	Mean	Variance	Mean	Variance
1–5	−0·39	0·018	−0·44	0·015
6–10	−0·58	0·040	0·92*	0·072
11–15	−0·47	0·072	1·99*	0·126
16–20	−0·52	0·097	2·45*	0·134
21–25	−1·13*	0·123	1·24*	0·144
26–30	−0·15	0·112	−0·48	0·173
31–35	−0·23	0·116	−1·42*	0·150
36–40	−0·15	0·109	−1·87*	0·148
41–45	−0·09	0·123	−2·41*	0·216
46–50	0·79*	0·138	−1·88*	0·163
51–55	−0·22	0·198	−1·47*	0·151
56–60	−0·32	0·160	−1·23*	0·150
Mean	−0·289	—	−0·383	—

* Indicates that the value of d_i is different from −0·289 (mean control value) at the 5% level.

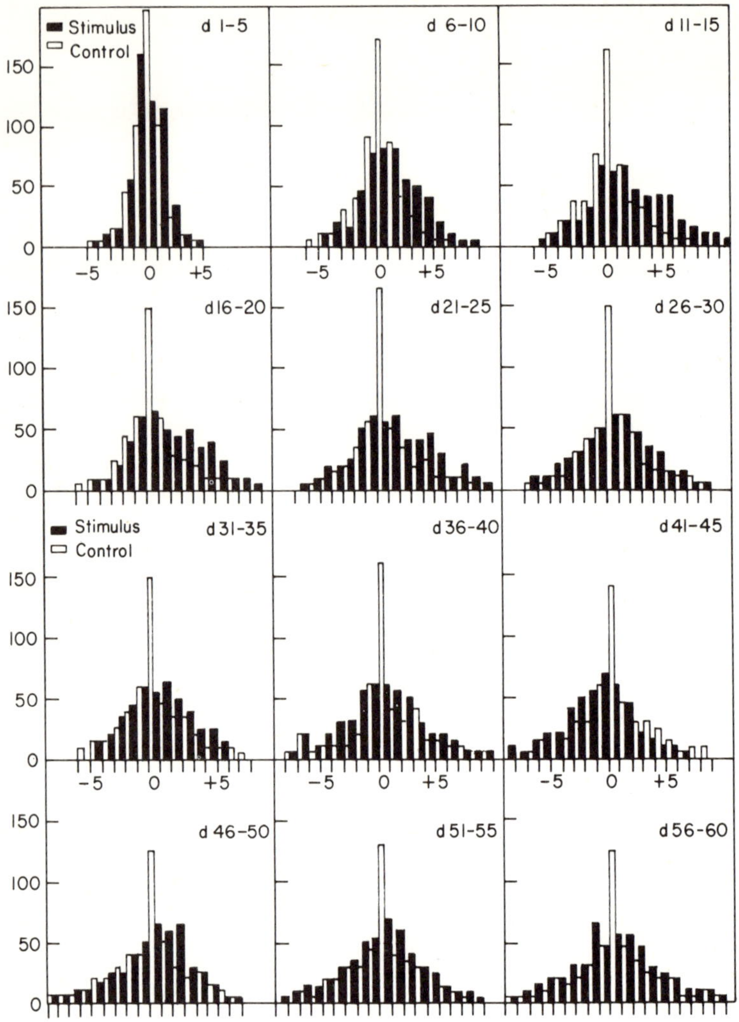

FIG. 3. Distributions of d_i for sound and control trials.

for control than for stimulus trials, i.e. stimulation more often resulted in displacement from the prestimulus level. There were about 532 observations (14 subjects × 38 stimulus or control applications) of each difference (the numbers vary slightly owing to missing observations where recording artefacts made measurements impossible).

The changes in heart rate during the 60 interbeat intervals plotted separately for stimulus and control trials are shown in Fig.

4, where the time scale is related to the overall mean heart rate for all subjects over the entire duration of the trials. This mean heart rate was 142 beats/min with a range of 35 beats/min amongst the subjects. Thus each period of five mean beats represented about 2·1 sec. It would appear from Fig. 4 that for stimulus trials the heart rate response change averaged over all 14 subjects and all trials was an increase rising to a maximum of approximately 2·5 beats/min approximately 8 sec after the stimulus onset, followed by a decrease to a minimum of approximately −2·4 beats/min about 11 sec later. This was reflected in a positive displacement of d_i distributions for stimulus trials (top half of Fig. 3) followed by a negative displacement (bottom half of Fig. 3). The heart rate "response" change for control trials in Fig. 4 was essentially "flat" with a small but unreliable negative bias.

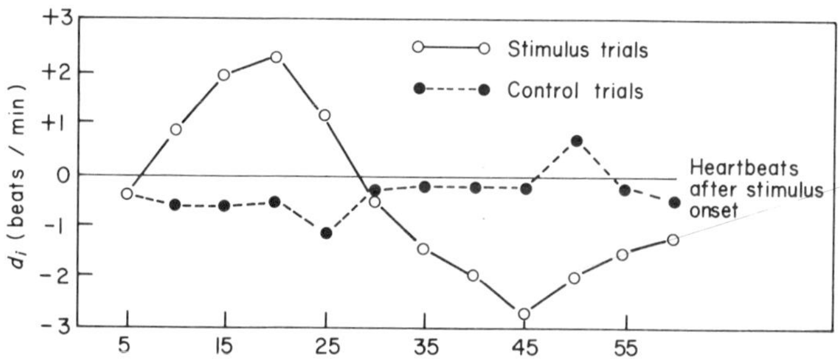

FIG. 4. The temporal pattern of averaged foetal heart rate changes for sound and control trials.

The overall mean value of the differences for the control application was −0·8 and it was against this value that the differences for each of the 12 five-beat intervals were tested. Only two of the 12 control values were found to be significantly different at the 5% level, and the conclusion is clearly that the fluctuations in the control d_i were due to chance. This was not the case for the stimulus applications, where all but two of the differences were significantly different from −0·289.

Comparison of blocks of stimulus trials, not reported in detail here, showed no tendency for the heart rate response change (for both $d_{i=16-20\,(\text{maximum})}$ and $d_{i=41-45\,(\text{minimum})}$) to vary significantly with time e.g. because of habituation or end-organ fatigue, although there was a tendency for the decelerative component to decrease with repeated stimulation. Also there was no marked difference in heart rate "response" changes for control trials between the first and second blocks of controls.

DISCUSSION

It will be apparent that the results given above present a preliminary level of analysis. They do not take into account certain aspects raised on pp. 26–28 and the consequences of averaging, which may require some revision of the outcome. It seems likely, however, that the effects of such aspects would have been randomly distributed amongst subjects and trials, so that, although the results as presented may eventually need some modification, this is likely to be necessary more for the magnitude of quoted values, rather than for the pattern of response.

It has already been noted that the response pattern of foetal heart rate change (Fig. 4) consisted of two parts: an accelerative phase rising to approximately 2·5 beats/min about 8 sec after stimulus onset, followed by a decelerative phase falling to approximately −2·4* beats/min about 19 sec after stimulus onset. It remains to explain the reasons for this pattern, and to locate it in the development of the organism.

Sokolov (1963) has discussed two general reflex systems: a defensive and an orienting reflex. The defensive reflex occurs in response to intense stimuli and serves to contain the effects of stimulation. The orienting reflex is evoked by "interesting" or "novel" stimuli and serves to enhance the effects of stimulation. Routtenberg (1968) has described analogous systems of arousal. He suggested that there are two mutually inhibiting arousal systems, of which one is evoked by intense stimuli and is associated with the reticular activating system (Moruzzi & Magoun, 1949) to limit the effects of stimulation, while the other is associated with the limbic system, and functions to enhance the effects of stimulation. Lacey (1959) has argued that heart rate acceleration occurs when (further) stimulation is to be rejected, and heart rate deceleration is found when stimulation is to be accepted. From here, it is easy to argue that the heart rate pattern of Fig. 4 reflects the operation of both reflexes or arousal systems outlined above. Figure 4 could be said to show that the foetus produced an initial heart rate acceleration, reflecting a relatively brief defensive response to the sudden (10 msec) onset of the stimulus (Graham & Clifton, 1966), followed by a deceleration of rather longer duration, reflecting an orienting reaction. It is interesting to note that the sound-evoked foetal heart rate curve of Fig. 4 is remarkably similar to an idealized curve produced by Davis, Crowell & Chun (1965) to illustrate the adult

* Averaging over trials may have led to an underestimate in view of the tendency for the decelerative component to decrease with repeated stimulus presentations.

heart rate response to brief sensory stimuli. The only notable difference is that the latencies of aspects of the foetal heart rate curve are much longer than for the adult, but this could be explained on the basis of the immaturity of the foetal nervous system or its relative depression.

Attractive though this observation is, it is not easy to accommodate to the results of work on infants studied a few days after birth. This raises problems when an attempt is made to locate the foetal heart rate response pattern into a developmental sequence. In a comprehensive review of infant heart rate reactions and arousal mechanisms, Graham & Jackson (1970) have summarized considerable evidence to show that the characteristic heart rate response in newborn infants is a relatively prolonged, short-latency, accelerative curve with no second decelerative phase.* Newborn heart rate decelerations following upon accelerations have proved to be small and very much the exception. Moreover, in an experiment especially designed to optimize the occurrence of a second decelerative phase, Jackson, Kantowitz & Graham (1971) failed to find convincing evidence for neonatal heart rate deceleration, which first begins to appear reliably in infants a few weeks or months old (Lipton, Steinschneider & Richmond, 1966). Comparison with such data as are available from foetuses studied a few weeks before birth does not cast much light on this problem. Dwornicka *et al.* (1964) produced sound-evoked heart rate response curves for foetuses studied in the last month before birth which displayed only an accelerative phase. However, they did not record the foetal heart rate long enough to detect a second decelerative component if such had been present. Results given for individual foetuses by Johansson *et al.* (1964) also do not help to identify the "general" foetal heart rate response, although they contain a suggestion that certain foetuses responded with a heart rate acceleration followed by a deceleration (see their Fig. 5).

If the present findings are reliable—and, as has already been mentioned, they represent the outcome of a preliminary investigation, which needs further replication and amplification—then it might be concluded that the heart rate response change curve of Fig. 4 is more likely to be governed by mechanisms subserving heart rate responses for older infants and adults than for

* However, it must be pointed out that most sound-evoked heart rate response studies on newborns have been conducted when the babies were drowsy or asleep (Graham & Jackson, 1970: 87–91). This makes the comparison of developmental heart rate response patterns difficult on the basis of data presently available.

newborns. A corollary of this position is that the newborn period may represent a temporary inhibition of aspects of autonomic nervous and/or limbic system function. It would be interesting to study the heart rate responses of babies very shortly after birth, since it is at this time that the transition between the foetal heart rate response pattern and that of the older neonate should occur. In this context, it is noteworthy that Vallbona, Desmond, Rudolph, Pap, Hill, Franklin & Rush (1963), investigating heart rate responses to a "thumping on the foot" stimulus in babies during the first hours of life, found acceleration and "rebound" deceleration. They concluded that there was considerable cardiovascular instability for at least the first six hours after birth.

Acknowledgements

Data reported in this paper were collected while J. Bench was a Visiting Scientist at the Fels Division of Pediatric Research, University of Cincinnati College of Medicine. Thanks are due to Dr T. P. Barden, Department of Obstetrics and Gynecology, U.C.C.M., who collaborated in the data collection and organized the obstetric treatments, to Dr R. W. Stander, also of the Department of Obstetrics and Gynecology, U.C.C.M., who kindly consented to accommodate the project within his Department, and to Dr R. W. Keith, Department of Otolaryngology and Maxillo-facial Surgery, U.C.C.M., for the loan of apparatus. Thanks are also due to Dr F. Falkner and Dr R. B. McCall, Fels Research Institute, Yellow Springs, Ohio, for generous assistance with funds and provision of working conditions. The Department of Medical Photography, Royal Berkshire Hospital, was of great assistance in preparing the illustrations.

References

Barcroft, T. & Barron, D. H. (1937). Movements in mid-foetal life in the sheep embryo. *J. Physiol., Lond.* **91**: 329–351.

Bast, T. H. & Anson, B. J. (1949). *The temporal bone and the ear.* Springfield, Illinois: Thomas.

von Békésy, G. (1939). Über die Vibrationsempfindung. *Acust. Z.* **4**: 316–334.

Bench, J. (1968). Sound transmission to the human foetus through the maternal abdominal wall. *J. genet. Psychol.* **113**: 85–87.

Bench, J. (1973). Foetal audiometry in relation to the developing foetal nervous system. In *Disorders of auditory function*: 115–119. Taylor, W. (ed.). London and New York: Academic Press.

Bench, R. J. (1972). Fetal audiometry: tests in sensory function in the fetus. In *Mental retardation: prenatal diagnosis and infant assessment*: 23–26. Douglas, C. P. & Holt, K. S. (eds). London: Butterworths.

Bench, R. J., Anderson, J. H. & Hoare, M. (1970). Measurement system for fetal audiometry. *J. acoust. Soc. Am.* **47**: 1602–1606.

Bench, R. J., Hoare, M. & Pye, A. (1973). Potential ear damage in foetal audiometry. *Br. J. Audiol.* **7**: 65–68.

Bench, R. J. & Vass, A. (1970). Fetal audiometry. *Lancet* **1970 1**: 91–92.

Boddy, K. & Robinson, J. S. (1971). External method for detection of fetal breathing in utero. *Lancet* **1971 1**: 1231–1233.

Davis, C. M., Crowell, D. H. & Chun, B. J. (1965). Monophasic heart rate acceleration in human infants to peripheral stimulation. *Am. Psychol.* **20**: 478 (abstract).

Dwornicka, B., Jasienska, A., Smolarz, W. & Wawryk, R. (1964). Attempt of determining the fetal reaction to acoustic stimulation. *Acta otolar.* **57**: 571–574.

Forbes, H. S. & Forbes, H. B. (1927). Fetal sense reaction: hearing. *J. comp. Psychol.* **7**: 353–355.

Goodlin, R. C. (1971). Intrapartum fetal heart rate responses and plethysmographic pulse. *Am. J. Obstet. Gynec.* **110**: 210–226.

Graham, F. K. & Clifton, R. K. (1966). Heart-rate change as a component of the orienting response. *Psychol. Bull.* **65**: 305–320.

Graham, F. K. & Jackson, J. C. (1970). Arousal systems and infant heart rate responses. In *Advances in child development and behavior* **5**: 60–117. Reese, H. W. & Lipsitt, L. P. (eds). New York & London: Academic Press.

Grimwade, J. C., Walker, D. W., Bartlett, M., Gordon, S. & Wood, C. (1971). Human fetal heart rate change and movement in response to sound and vibration. *Am. J. Obstet. Gynec.* **109**: 86–90.

Hackman, B. & Bench, J. (1972). Tonal stimulation of the human foetus. *Sound (Br. J. Audiol.).* **6**: 1–2.

Henshall, W. R. (1972). Intrauterine sound levels. *Am. J. Obstet. Gynec.* **112**: 576–578.

Hooker, D. (1936). Early fetal activities in mammals. *Yale J. Biol. Med.* **8**: 579–602.

Jackson, J. C., Kantowitz, S. R. & Graham, F. K. (1971). Can newborns show cardiac orienting? *Child Develpm.* **42**: 107–121.

Johansson, B., Wedenberg, E. & Westin, B. (1964). Measurement of tone response by the human foetus. *Acta otolaryng.* **57**: 188–192.

Knudsen, V. O. (1928). Hearing with the sense of touch. *J. gen. Psychol.* **1**: 320–352.

Lacey, J. I. (1959). Psychophysiological approaches to the evaluation of psychotherapeutic process and outcome. In *Research in psychotherapy*: 160–208. Rubinstein, E. A. & Parloff, M. B. (eds). Washington D.C.: Am. Psychol. Assoc.

Lindsley, D. B. (1942). Heart and brain potentials of human fetuses *in utero*. *Am. J. Psychol.* **55**: 412–416.

Lipton, E. L., Steinschneider, A. & Richmond, J. B. (1966). Autonomic function in the neonate: VII. Maturational change in cardiac control. *Child Develpm.* **37**: 1–16.

Moruzzi, G. & Magoun, H. W. (1949). Brain stem reticular formation and activation of the EEG. *Electroenceph. Clin. Neurophysiol.* **1**: 972–973.

Murphy, K. P. & Smyth, C. N. (1962). Response of foetus to auditory stimulation. *Lancet* **1962** **1**: 972–973.

Newbery, H. (1941). Studies in fetal activity. *J. comp. Psychol.* **32**: 521–530.

Norman, H. N. (1942). Fetal hiccups. *J. comp. Psychol.* **34**: 65–73.

Peiper, A. (1925). Sinnesempfindungen des Kindes vor seiner Geburt. *Mschr. Kinderheilk.* **29**: 236–241.

Ray, W. S. (1932). A preliminary report on a study of fetal conditioning. *Child Develpm.* **3**: 175–177.

Riesen, A. H. (1947). The development of visual perception in man and chimpanzee. *Science, N.Y.* **106**: 107–108.

Rose, D. (1947). Heart rate as a behavioral indicator in the fetal rat. *J. comp. Physiol. Psychol.* **40**: 157–164.

Rosen, M. G., Scibetta, J. J. & Hochberg, C. J. (1970). Human fetal electroencephalogram III. Pattern changes in presence of fetal heart rate alterations and after use of maternal medications. *Obstet. Gynec.* **36**: 132–140.

Routtenberg, S. (1968). The two-arousal hypothesis: reticular formation and limbic system. *Psychol. Rev.* **75**: 51–80.

Scibetta, J. J. & Rosen, M. G. (1969). Response evoked by sound in the fetal guinea pig. *Obstet. Gynec.* **33**: 830–836.

Scibetta, J. J., Rosen, M. G., Hochberg, C. J. & Chik, L. (1971). Human fetal brain response to sound during labor. *Am. J. Obstet, Gynec.* **109**: 82–85.

Sokolov, E. N. (1963). *Perception and the conditioned reflex*. New York: Macmillan.

Sontag, L. W. (1936). Changes in the rate of the human fetal heart in response to vibratory stimuli. *Am. J. Dis. Child.* **51**: 583–589.

Sontag, L. W. (1944). Differences in modifiability of fetal behavior and physiology. *Psychosom. Med.* **6**: 151–154.

Sontag, L. W. & Wallace, R. F. (1935a). The movement response of the human fetus to sound stimuli. *Child Develpm.* **6**: 253–258.

Sontag, L. W. & Wallace, R. F. (1935b). The effect of cigarette smoking during pregnancy upon fetal heart rate. *Am. J. Obstet. Gynec.* **29**: 77–82.

Spelt, D. K. (1938). Conditioned responses in the human fetus *in utero*. *Psychol. Bull.* **35**: 712–713.

Spelt, D. K. (1948). The conditioning of the human fetus *in utero*. *J. exp. Psychol.* **38**: 375–376.

Vallbona, C., Desmond, M. M., Rudolph, A. J., Pap, L. F., Hill, R. M., Franklin, R. R. & Rush, J. B. (1963). Cardiodynamic studies in the newborn II. Regulation of the heart rate. *Biol. Neonat.* **5**: 159–199.

Vasiliu, D. I. (1968). L'enfant est prêt à entendre même dans la vie intra-uterine. Contributions sur l'embryologenèse de l'oreille et la pneumatisation de la mastoide et du rocher. *Int. Audiol.* **7**: 181–185.

Walker, D. W., Grimwade, J. C. & Wood, C. (1971). Intra uterine noise: a component of the fetal environment. *Am. J. Obstet. Gynec.* **109**: 91–95.

Wilder, J. (1958). Modern psychophysiology and the Law of Initial Value. *Am. J. Psychotherapy.* **12**: 199–221.

Windle, W. F. (1941). Physiology and anatomy of the respiratory system in the foetus and newborn infant. *J. Pediat.* **19**: 437–444.

Symp. zool. Soc. Lond. (1975) No. 37, 41–50.

EXAMINATION OF NORMAL AND ABNORMAL MIDDLE EAR FUNCTION

D. BROOKS

Manchester Audiology Clinic, Manchester, England

SYNOPSIS

The middle ear functions as an acoustic transformer converting the air pressure changes in the auditory range to fluid movements in the cochlea. The mechanism is acoustically efficient and compact and has valuable additional features such as reduction of the low frequency sounds of the voice when listening and talking simultaneously and partial protection from loud sounds. Examination of the middle ear by visual means is limited. Audiometric evaluation of middle ear function is indirect, inferential and inexact. A third approach, acoustic impedance measurement, is now possible which is objective, informative and clinically simple. This is based on an investigation of the way in which the tympanic membrane reacts to the sound waves that impinge on it. This reaction will be dependent on conditions in the middle ear. In the normal ear the major part of the incident energy is transmitted to the cochlea, but disorders of the conduction mechanism will most probably change the acoustic efficiency of the system, and these changes can be observed and measured. Mechanical or electrical analogues of the middle ear system have been devised which reproduce the characteristics of the normal middle ear, and of certain well-defined pathological conditions. Further studies may assist in detecting rarer disorders of the conduction mechanism.

MIDDLE EAR FUNCTION

Before considering the examination of middle ear function it seems advisable to determine first what that function is. What purpose does the complex anatomical structure of bones, cavities, tissues, ligaments and muscles that we call the middle ear perform, and how does it operate?

Let us, for present purposes, consider sound as variations in the ambient air pressure which lie within the range of frequencies to which the ear is sensitive. These pressure variations are transmitted from the source through the tenuous elastic medium of air to the ear of the listener. The sensory receptors, the hair cells of the cochlea, function in a fluid environment, and fluid has a much greater density than air. If the sound waves impinged directly upon the cochlear fluids there would be a very inefficient transfer of acoustical energy from the transmission medium to the sense organ. At an air/water boundary only about $\frac{1}{16}$ of 1% of the incident energy enters the water, over 99·9% being reflected back from the surface of the denser medium. Air is very compliant and presents little opposition to the passage of vibrations in the auditory spectrum; water by contrast has a much greater stiffness and considerably greater force is required to produce the same degree of vibration in water as in air. In physical terms we say that air has a low and water a high specific acoustic impedance. At the air/water boundary we then have a mis-match between a low impedance source—the air, and a high impedance receptor—the fluids of the cochlea. The purpose of the middle ear is to act as a matching device; an acoustic transformer, to maximize the transmission of energy from the propagation medium of air to the fluid environment of the hair cells of the cochlea.

The principal way in which this is achieved is by collecting the energy over a relatively large area, the tympanic membrane, and transmitting it into the cochlea through a relatively small area, the footplate.

Small pressure variations acting over the relatively large surface of the membrane are thereby transformed into much greater pressure variations acting over the smaller area of the footplate suspended in the oval window of the inner ear by its elastic ligaments. At its very simplest a single ossicle could be used connected from the centre of the tympanic membrane to the footplate. Von Békésy (1960) indicated, however, that this system has disadvantages for the transmission of airborne sounds in human beings. With inward movement of the tympanic membrane the single ossicle will tend not only to move inwards, but also upwards. In the following rarefaction wave it will move into its mid-line position under tension, and on the next positive wave will tend to move downwards because its kinetic energy will carry it through the middle position. It is therefore evident that during a full period of movement of the eardrum, the ossicle makes an excursion to one side or, in other words, its frequency is one-half of that of the tympanic membrane. By this means subharmonics or undertones are produced in the system and this results in distortion. To suppress these lateral movements of the ossicle it is necessary to have an attachment to the bony wall as near as possible to the free end, the tympanic end of the ossicle. This principle is observed in the ear of the frog where there are two ossicles: the columella connecting the membrane to the footplate and a bony rod acting as a stabilizer for the system.

In the human middle ear we find a somewhat more complex method of achieving the connection between the tympanic membrane and the footplate, three bones being employed instead of only two. This additional complexity effects some highly desirable improvements in the performance of the system.

The effective length of the malleus from its point of rotation to its point of maximum motion at the centre of the tympanic membrane is approximately one and a quarter times greater than the effective length of the incus from the same axis of rotation to its articulatory joint with the stapes. The product of force and distance being constant, a force acting at the end of the longer ossicle, the malleus, will be converted to a larger force acting at the end of the short ossicle, the incus. Hence a smaller ear drum area is necessary with this transformer system than would be required if a simple columella were employed, and the overall size of the system is thereby reduced.

Secondly, though for normal sound pressures the malleus and incus act in combination as if they were a single unit, with higher sound pressures, or indeed for large static pressure changes, the joint between the malleus and incus tends to be somewhat elastic. All movements of the malleus are not transmitted to the incus but a slip takes place in the joint and this gives a degree of protection against high intensity sounds and pressure changes.

In passing it may be noted that what appears at first to be the superfluous mass of the head of the malleus and incus is in fact a necessary part of the system in order to make the centre of gravity of the ossicles coincide with the axis of rotation. This not only increases the efficiency of the system for air-conducted sounds, but also assists in protecting the ear against vocal sounds transmitted by bone conduction.

In answer to our original question then, the complex anatomical structure of bones, cavities, tissues, ligaments and muscles that we call the middle ear, acts as a highly efficient transformer of acoustical energy from the air through which sound is transmitted to the fluid of the cochlea. It is efficient over a wide range of frequencies and intensities, but at higher intensities its efficiency diminishes, thereby providing a protective role. In the region of high sensitivity the movement of the tympanic membrane is of the order of a thousand-millionth of a centimetre (10^{-11}m). Inevitably in a system as complex and yet as delicate as this, disorders can arise through such things as trauma, disease or neoplasm. In the clinical context we wish to know how best to determine the nature of the disorder.

METHODS OF EXAMINATION OF MIDDLE EAR FUNCTION

Until quite recently there have been two basic methods for studying disorders of the middle ear. The first of these is direct visual examination. In a living human being with a normal auditory system, however, it is not possible to see the conditions that exist behind the tympanic membrane. Where a membrane is perforated one may see some of the changes that have taken place in the transmission system, but perforations are the exception rather than the rule. In the presence of an intact membrane, direct visualization of the middle ear is not possible, though some information may be obtained about the crude mobility of the system by applying varying air pressures to the drum and observing its movement. The pressures employed in this technique are, however, many orders of magnitude greater than the highest pressures involved in hearing and consequently the method has limited practical applicability in assessing the proper function of the middle ear system. The second approach is indirect and depends on inference rather than observation. Hearing may be assessed by the conventional airborne sound path and by the bone conduction route which by-passes the middle ear. Information about the degree of mis-match in the system may be obtained in this way, but little information can be obtained about the actual cause of the mis-match. Audiometry rarely tells us anything of value about the underlying pathology of hearing loss originating in the middle ear system.

A third method of examining the function of the middle ear has been developed over approximately the last 30 years. Its historical origins go back more than a century, but the beginning of its clinical application was in 1946 when Otto Metz published the now classic monograph: *The acoustic impedance measured on normal and pathological ears.* In the human subject with a normal auditory system a sound wave entering the external auditory meatus progresses inwards until it reaches the tympanic membrane. Thence the majority of the energy is transmitted to the cochlea through the matching system of the middle ear. A small part of the incident energy is inevitably lost in overcoming friction and a small fraction is reflected. If the middle ear is in any way abnormal, the normal relationship between the three components of transmitted, reflected and frictional parts will probably be disturbed. For example, in exudative otitis media the middle ear space is filled with fluid and consequently the mobility of the tympanic membrane is grossly inhibited. A much greater proportion of the incident energy is reflected than in the normal system. Conversely, in a situation where the ossicular chain is broken between the incus and the stapes, tympanic membrane mobility tends to be higher than normal due to the restraint of the footplate being removed. In such a condition the component of the sound reflected from the membrane may well be diminished.

It is evident that by studying the reaction of the tympanic membrane to incident sound energy, it ought to be possible to make deductions about conditions behind the translucent membrane. A better understanding of the system can be achieved by resorting to the use of models or analogues. Physical models of the middle ear are exceedingly difficult to make and use, because of the minute dimensions of the system. Consequently resort is normally made to a mathematical or electrical analogue.

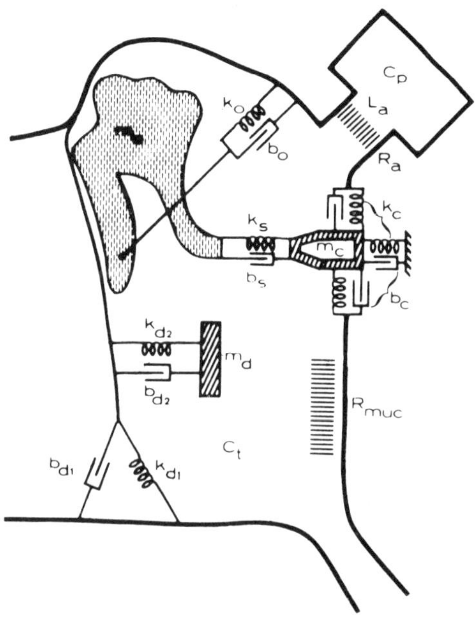

FIG. 1. Mechanical-acoustical model of the middle ear. (From Marquet *et al.*, 1973.) Elastic components are designated by the letters k or C (mechanical or acoustical); inertial components by the letters m or L; and frictional elements, b or R. Subscripts refer to the tympanic membrane (d), ossicles (o), incudo-stapedial joint (s), footplate and stapes (c), mastoid process (a and p), middle ear cavity (t) and membranous lining of the middle ear (muc).

Figure 1 illustrates a rather hybrid analogue in which we can see three basically different kinds of elements. There are elements of mass of inertance such as the mass of the ossicles or the tympanic membrane itself or the inertance of the air in the mastoid cavities (L_a). There are elastic or spring-like elements which reflect the elastic nature of the tympanic membrane, the suspensory ligaments or the air within the middle ear cavity which, when the tympanic membrane moves inwards or outwards, tends to restore it to its rest position because of the compression or rarefaction of the volume of air. Thirdly, we have resistive elements. These represent the frictional losses that are inevitable in any system.

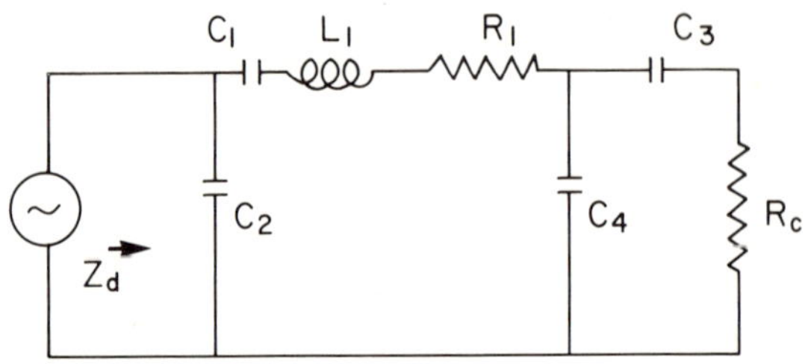

FIG. 2. Simple electrical analogue of the normal middle ear. C_1 L_1 R_1 correspond to the compliance, inertance and resistance of the major ossicles and their suspensory ligaments; C_2 corresponds to the compliance of the tympanic membrane; C_3 R_c correspond to the footplate and cochlear elements; C_4 corresponds to the incudo-stapedial joint compliance.

Zwislocki (1957) rationalized and simplified this analogue and produced the simple electrical analogue of Fig. 2. For such an analogue to be of practical value it is necessary to be able to quantify the components so that they bear a one-to-one relationship with the corresponding elements of the original. Zwislocki determined the numerical values of the elements in his analogue partly by anatomical evidence, for example by measuring the volume of the middle ear cavity in a series of cadavers, and partly by a trial and error fitting method. Those values were accepted which best fitted measurements of the acoustic impedance of normal ears and of ears with specific abnormalities, the effects of which on the analogue could be readily explained. An interesting fact that emerged from this exercise was that the inductor which represents the mass of the ossicles, when reconverted to mass terms, indicated a mass of only 12 mg. According to von Békésy & Rosenblith, the total mass of the malleus and incus is actually of the order of 50 mg, that is about four times as great as the effective mass determined by means of the analogue. Zwislocki (1962) pointed out that this agreed well with the observation that the ossicles tend to rotate about their centre of gravity so that the effective mass is reduced, thereby improving the characteristics of the system.

PRACTICAL APPLICATIONS

An unusual application of the analogue has recently been made by McCrae (1973) in Australia. The analogue indicates the input impedance of the cochlea as essentially resistive. McCrae suggested that in patients with acoustic neuromata this resistive component would be increased due to the increase in the stiffness of the cochlear duct. Measurements made on six cases having acoustic neuromata indicated that the acoustic impedance at low frequencies agreed well with the effect predicted by making appropriate changes to the analogue. Acoustic neuromata are, however, relatively rare conditions. The majority of conductive disorders affect the stiffness of the mechanism. For example, otosclerosis fixation of the footplate not only inhibits movement of the stapes but, because the joint between the stapes and incus is only moderately elastic, also affects the movement of the two large ossicles and consequently stiffens the tympanic membrane to some degree. By contrast, disarticulation or fracture of the incudo-stapedial joint results in an enhanced degree of freedom of the two large ossicles and an abnormally high mobility. From a diagnostic standpoint, therefore, it is probably of major interest to know how that part of the impedance, which is contributed by the elastic forces, has altered from the normal situation.

The analogue shows that the acoustic impedance of the middle ear is a complex quantity: complex not only because it has a large number of components but also because these components react differently to the initial stimulus depending on its frequency. Impedance theory shows that when the stimulating sound has a high frequency the elements which are associated with mass are dominant, and when the stimulating frequency is low the elements of stiffness dominate. It is possible to devise instrumentation that will measure the impedance over a wide range of frequencies, but such equipment tends to be expensive, complicated, difficult to use and time consuming. In practice, therefore, a compromise has been made and for clinical purposes most instruments operate at a fixed frequency which is low because of the major interest in disorders of stiffness.

In its simplest form such an instrument would give us a single measurement of the stiffness of the auditory system, but this alone would be of limited diagnostic value. If we found the system to be stiffer than normal this might be due, for example, to otosclerosis, retraction of the tympanic membrane, or adhesive processes within the middle ear. A method of increasing the information content was therefore required and a major step forward in measurement was made when facilities were incorporated for varying the pressure differential across the tympanic membrane.

By means of a small pump and its associated pressure gauge, it is possible to introduce into the external auditory meatus small controlled pressures either negative or positive relative to atmospheric pressure. Normally the pressure in the middle ear is the same as the atmospheric pressure, this being the function of the Eustachian tube. Under normal circumstances then, the radial fibres of the tympanic membrane are unstretched and the stiffness of the membrane is at a minimum, or we might say conversely its mobility or its compliance is at a maximum. If the pressure in the canal is increased or decreased, then the radial fibres of the tympanic membrane start to stretch, the stiffness of the drum increases and its compliance diminishes. When the pressure differential across the drum has reached a value such that the pressure on one side is 2% different from that on the other side, the fibres are stretched to their elastic limit and the membrane becomes effectively rigid. It is practicable and in fact quite simple to plot the mobility, or the compliance, of the tympanic membrane as a function of the pressure differential across it. This technique is called tympanometry and the graphical representation is called a tympanogram.

The tympanogram A in Fig. 3 shows that with positive or negative pressures in excess of 2% across the membrane it is

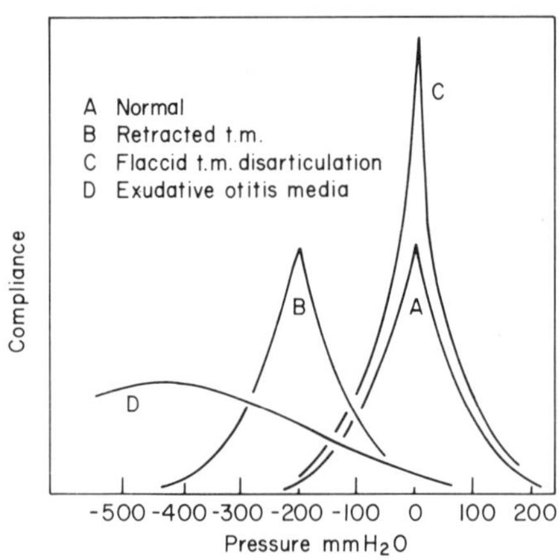

FIG. 3. Idealized tympanograms for the normal middle ear and specified pathological conditions. t.m.: tympanic membrane. A pressure change of 2% represents about 200 mmH_2O.

effectively rigid, but as the pressure differential is reduced so the mobility increases reaching a maximum value when the pressure differential is zero; that is when the middle ear pressure is equal to the atmospheric pressure in the external canal.

Perhaps the simplest of all abnormalities of the auditory system is a simple blockage of the Eustachian tube. When this happens the gases within the middle ear tend to be absorbed, and a partial vacuum is formed in the cavity. When a tympanogram is performed on such an ear a curve similar to that for a normal middle ear is obtained, except that the point of maximum compliance or mobility is reached now when the external canal pressure is the same as the pressure within the middle ear, at whatever negative value this may be (B in Fig. 3). With this technique we have a practical means of objectively measuring the degree of retraction of the tympanic membrane. The tendency with otosclerotics is for the tympanogram to be smaller than normal due to the increased stiffness or decreased mobility of the system.

Where there is a disarticulation of the ossicular joint the tendency is for the opposite effect to take place and for a higher than normal compliance to be shown (C in Fig. 3). It must be stressed, however, that these are tendencies and the technique is not of great diagnostic value for these conditions.

Blockage of the Eustachian tube is frequently followed by transudation of fluid into the middle ear cavity, particularly in children. The whole of the middle ear spaces may fill with fluid. The tympanogram produced in such a condition (D in Fig. 3) shows the negative pressure which has been the forerunner of the transudation and also shows a much reduced mobility and high

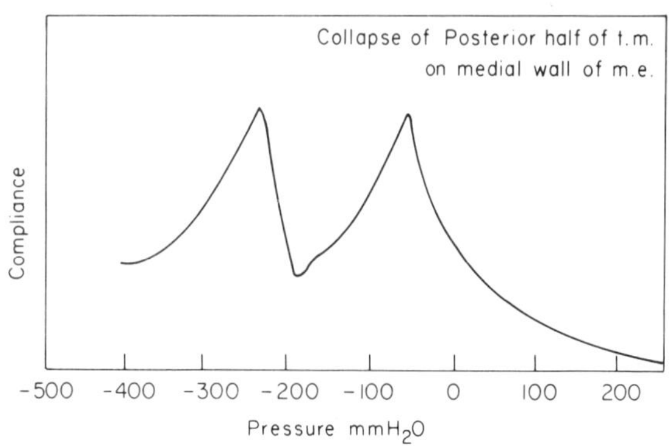

FIG. 4. Bimodal tympanogram obtained on a child with a history of middle ear disorder. t.m.: tympanic membrane; m.e.: middle ear.

damping characterized by flattening of the tympanogram. This has been shown to be one of the most useful aspects of tympanometry: identifying middle ear disorders in young children where subjective tests are difficult to perform because of their age.

In rare instances a bimodal tympanogram is obtained (Fig. 4). When this curve was first seen, the theoretical explanation was that it represented a two cavity system and this was borne out by microscopic examination of the middle ear shortly afterwards. The tympanic membrane had collapsed inwards on to the promontory and the middle ear had divided into two completely separate compartments (epitympanum and mesotympanum) which were at slightly different pressures.

CONCLUSION

To sum up, acoustic impedance measurements have helped in understanding the workings of the normal middle ear and in diagnosing disorders of middle ear function. It seems likely that in the near future with sophistication of technique, better diagnosis of middle ear disorders will be attainable.

REFERENCES

von Békésy, G. (1960). Bone conduction. In *Experiments in hearing*: 127–203. Wever, E. G. (ed.). New York: McGraw Hill Book Co. Inc.

von Békésy, G. & Rosenblith, W. A. Quoted by Zwislocki, J. (1962).

Marquet, J., Van Camp, K. J., Creten, W. L., Decraemer, W. F., Wolff, H. B. & Schepens, P. (1973). Topics in physics and middle ear surgery. *Acta Oto-Rhino-Laryngol. Belg.* **27**: 137–219.

McCrae, J. H. (1973). The acoustic impedance of the ear in a case of acoustic neuroma. *J. oto-laryngol. Soc. Austr.* **3**: 513–515.

Metz, O. (1946). The acoustic impedance measured on normal and pathological ears. *Acta otolaryng.* Suppl. **63**.

Zwislocki, J. (1957). Some impedance measurements on normal and pathological ears. *J. acoust. Soc. Am.* **29**: 1312–1317.

Zwislocki, J. (1962). Analysis of middle ear function. Part I—input impedance. *J. acoust. Soc. Am.* **34**: 1514–1523.

Symp. zool. Soc. Lond. (1975) No. 37, 51–68.

THE STAPEDIUS MUSCLE AND SPEECH PERCEPTION

E. BORG

Division of Physiological Acoustics, Department of Physiology II, Karolinska Institutet, Stockholm, Sweden

and

J.-E. ZAKRISSON

Department of Oto-rhino-laryngology, University of Umeå, Umeå, Sweden

SYNOPSIS

The present series of experiments concerned the stapedius muscle activity when subjects themselves vocalized and when they listened to the speech of others. The role of the stapedius muscle in the perception of speech sounds was also explored. In subjects with perforation of the eardrum, electromyographic activity was recorded from the stapedius muscle during vocalization and acoustic stimulation. The stapedius muscle was active over almost the whole range of vocal intensities and the degree of activity was approximately proportional to the vocal intensity. It was concluded, mainly on the basis of latency measurements, that the stapedius activation is part of the vocalization process and not merely an acoustic reflex. When a subject listened to speech the stapedius muscle was activated by phrases above 92 dB SPL, which was close to the reflex threshold for a 0·5 kHz pure tone. Discrimination score for nonsense monosyllables was determined in subjects with unilateral stapedius paralysis (Bell's palsy). For intensities of the test word at and above 100 dB SPL there was a significant decrease in discrimination score when the stapedius muscle was paralysed. The masking produced by the low frequencies of the speech on the high frequency components was suggested as the explanation for this lowered score during paralysis. This was supported by the finding that masking produced by 0·5 kHz narrow-band noise on high frequency pure tones rose up to 50 dB in the absence of stapedius reflex.

The stapedius muscle is thus active during vocalization and when listening to speech. It decreases the masking effect produced by the low frequency components of the speech sounds on the high frequency ones and is of importance for intelligibility of speech. The dynamic range of the ear is extended by 15–20 dB. Most likely the significance of the stapedius activity for perception of external sounds is greatest when one is talking.

INTRODUCTION

The middle ear muscles, the stapedius muscle and the tensor tympani muscle, have recently been widely studied both in animals and man. The main interest has been directed to the acoustic stapedius reflex which has become of great importance in clinical diagnosis. Several theories on the functional role of the middle ear muscles have been presented during the last century (see Jepsen, 1963 and Møller, 1972 for reviews). Most of them are based on measurements with pure tones and noises. Little is known about how the stapedius muscle activity affects the hearing of natural sounds. The work of Henson (1965, 1970) is an exception and is of great importance in this respect. According to his results, the stapedius muscle is activated in some bat species during vocalization, attenuating the energy of the vocalized sound that reaches the inner ear. Thereby the sensitivity of the ear is maintained and the cochlea is ready to receive the echo. It seems very likely that this function of the stapedius muscle is significant for the bat. Direct proof from behavioural experiments that the bat's performance is improved by middle ear muscle activity is still lacking, however.

It is doubtful whether the results obtained in the bat, which has a specialized auditory system suited to echolocation, are applicable to other species, especially human beings. To cast some light on the role of the stapedius muscle for the perception of speech, the most important acoustic human signal, the present authors undertook a series of experiments in man which have been described in full in three recent articles (Borg & Zakrisson, 1973, 1974, in press). These studies, which are to be discussed in what follows, explored: (a) the activation of the stapedius muscle during vocalization and when listening to speech; (b) the role of the stapedius muscle for discrimination of speech sounds with no background interference and speech sounds masked by low frequency noise; (c) the possible mechanisms behind the stapedius muscle influence on speech discrimination, especially with regard to the masking produced by low frequency sound upon high frequency sound.

MATERIAL

The experiments were performed on 35 subjects divided into five experimental groups: group I, eight subjects with chronic perforation of the eardrum, five used for recording of electromyograms (EMG) from the stapedius muscle and seven for visual observation of movements of the stapes and the stapedius tendon; group II, six subjects with normal hearing used for measurements of stapedius reflex response to speech sounds; group III, seven subjects with total unilateral paralysis of the stapedius muscle (Bell's palsy), used to determine discrimination scores; group IV, three subjects with total unilateral stapedius muscle paralysis and normal hearing up to 8·0 kHz, used for measuring of masking; group V, 19 subjects with total unilateral stapedius muscle paralysis in whom measurements were performed during paralysis and after the stapedius muscle function had completely recovered. The latter were used to determine attenuation provided by the stapedius reflex on sound transmission at 0·5 kHz as described by Borg (1968).

METHODS

Impedance change measurement

The impedance change of the ear at 800 Hz was measured in all subjects by the method of Møller (1960, 1961). This technique allows simultaneous recordings of ipsilateral and contralateral middle ear muscle activity. The rectified, low-pass filtered impedance change signal is largely proportional to the integrated electromyogram of the stapedius muscle within a wide range and is most likely a linear measure of the tension developed in the muscle (Zakrisson, Borg & Blom, 1974). The stimuli were pure tones of 0·5 or 2·0 kHz, or speech phrases used for discrimination measurements (see below).

Electromyography

Electromyograms were recorded from five subjects of group I with tungsten microelectrodes of 3 mm length and an impedance at 1·0 kHz of about 40 kOhms (Zakrisson, Borg & Blom, 1974). The recording electrode was introduced through the perforation of the eardrum and inserted into the tendon of the stapedius muscle. It was oriented as near to the muscle as possible. The reference electrode was a similar but longer tungsten needle inserted in the ipsilateral ear lobe. No local anaesthesia or sedation of the patient was used. The electromyographic signal recorded by the electrode was amplified by a Tektronix 2A61 amplifier. The frequency response was between 60 Hz and 6 kHz which gave the best signal-to-noise ratio (SNR). The amplified signals were recorded on one channel of a two-channel tape recorder (Revox, type A 77), the voice signal being simultaneously recorded on the other channel. The integrated electromyographic responses were obtained by low-pass filtering the full wave rectified electromyogram. Active low-pass filters with an attenuation rate of 18 dB/octave and a cut-off frequency (−3 dB) of 3·9 Hz were chosen.

Vocalization

The subjects of group I were trained to vocalize an [*a*:] of 1–3 seconds duration with distinct onset and constant amplitude. The mouth, opened some seconds prior to vocalization, was kept relaxed during the vocalization, as were the tongue and face muscles. This was essential since it had been found in control measurements that jaw and face movements were followed by electrical activity conducted passively to the recording electrode.

The vocal sounds were recorded with a Brüel & Kjaer condenser microphone (type 4133) placed 20 cm in front of the mouth. The intensity was measured by a noise level meter (Brüel & Kjaer, type 2606, fast setting) and recorded on a Brüel and Kjaer level recorder. The intensity of each vocalized sound was defined as the smoothed (by eye) maximum value of the recording from the noise level meter. The speech signal was also recorded on one channel of the tape recorder.

Discrimination score

Speech sound discrimination was investigated in seven subjects with unilateral total stapedius muscle paralysis (group III), during the acute stage of the paralysis and after the recovery of the stapedius muscle function (Borg & Zakrisson, 1973). Monosyllabic phonetically-balanced nonsense words (consonant-vowel-consonant, in a carrier phrase) were presented at several intensity levels from 35 to 127 dB SPL. Measurements were usually made with one list of 50 monosyllables in one ear at each intensity level during masking of contralateral ear. The levels of the monosyllables, measured by a noise level meter (fast setting), were defined as the maximum amplitude of the signal recorded on the level recorder.

In four of the subjects of group III the speech phrases were also mixed with low-pass filtered noise (cut-off at 700 Hz (−3 dB), attenuation rate at high frequency of 24 dB/octave). These measurements were also made at several intensity levels of the speech phrase. For each level the signal-to-noise ratio was adjusted to give a discrimination score of approximately 80% in the non-affected ear. The same ratio was thereafter used for the ear with the paralysed stapedius muscle (affected ear). The difference in discrimination score between the two ears was determined.

Measurement of masking

Masking produced by 0·5 kHz narrow-band noise (bandwidth at −3 dB, 0·3 kHz, Brüel & Kjaer type 1024 noise generator) was measured in the three subjects of group IV (Borg & Zakrisson,

1974). Hearing thresholds in the frequency range 1·0 to 8·0 kHz were traced by pulsed tone Békésy audiometry at fixed frequency before, during and after exposure to the masking noise at various levels 85 to 120 dB SPL of the masker. In two of the subjects, who later recovered normal stapedius muscle function, the measurements were repeated after recovery.

RESULTS

Stapedius muscle activity during vocalization

In the experimental situation with the subject in a relaxed state, insignificant or no spontaneous electromyographic activity was recorded from the stapedius muscle. During the vocalization of [*a*:] motor unit activity was obtained in almost all recordings at the lowest vocal intensity. The electromyogram and the speech signal occurred largely simultaneously, yet there was no close correlation between the amplitude of the integrated electromyogram and the speech signal as a function of time.

The threshold for stapedius muscle activation during vocalization of [*a*:] was at or just above the lowest vocal intensity level which the subject could produce. The average threshold value, defined as the vocal intensity giving an integrated electromyogram amplitude that was 10% of the maximum obtained, was 67 dB SPL as measured 20 cm in front of the mouth (four subjects of group I). Movement of the stapedius tendon during vocalization and acoustic stimulation could also be observed by an operating microscope.

Four subjects in which comparisons could be made by this method, gave threshold values up to about 10 dB higher than the electromyogram. During vocalization the stapedius muscle threshold by visual observation was 73 dB SPL in seven subjects.

The relation of the amplitude of the integrated electromyogram to the vocal intensity of [*a*:] in two subjects of group I is shown in Fig. 1. The maximum amplitude of each integrated electromyogram, visually smoothed over approximately 0·2 sec, was used as the measure of stapedius muscle activity.

Figure 1 reveals that the stapedius muscle activity generally increased with increased vocal intensity, though the data show a considerable spread. This variance is most likely due both to an inherent variability in the stapedius muscle activity during vocalization and to uncertainty in the measurements of the amplitudes of the integrated electromyogram and the vocal signals due to their variable time course. The threshold values in these cases were

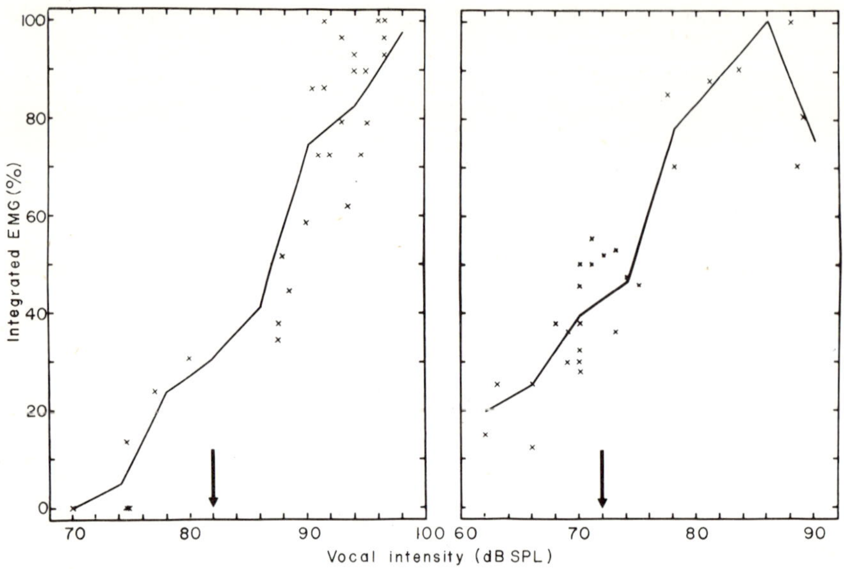

FIG. 1. Amplitude of integrated electromyogram (EMG) as a function of the vocal intensity of [*a*:] measured 20 cm in front of the mouth in subjects GO (left graph) and BS (right graph). Crosses represent individual values expressed as percentages of the maximum amplitude obtained during vocalization. The continuous lines connect mean values in 4 dB intervals. Arrows indicate the lowest vocal intensity giving a microscopically visual movement of the stapedius tendon.

75 dB SPL (left graph) and about 60 dB SPL (right graph, extrapolated). At a normal voice effort, which corresponds to approximately 80 dB SPL 20 cm in front of the mouth (Fletcher, 1953), the stapedius muscle activity was 50 to 60% of maximum, on the average.

The electromyogram was also obtained in response to contralateral stimulation with a 0·5 kHz pure tone (close to the fundamental frequency of [*a*:]). The threshold for this stimulus was, on the average, 36 dB above threshold during vocalization. In one subject with normal bilateral hearing the difference in threshold was 65 dB.

It should be mentioned that the sound pressure of [*a*:], 20 cm in front of the mouth, is close to that at the external ear (Sundberg, personal communication).

The large difference between the thresholds for acoustic activation and activation during vocalization might indicate that stapedius muscle activity is part of the vocalization process and not simply an acoustic reflex. Conclusive evidence might be obtained

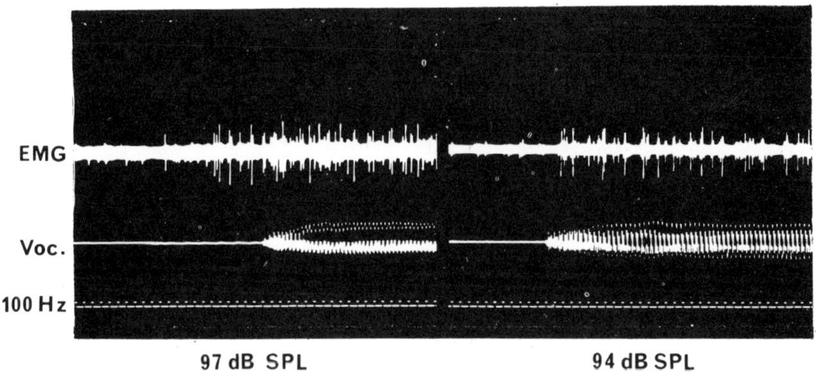

FIG. 2. Temporal relation between electromyogram (EMG) and speech signal at two vocal intensity levels for subject GO.

from measures of the temporal relation between the start of the electromyographic signal and the speech sound. For 55 recordings, the onset of the electromyogram and the speech signals were so distinct that definite measurements could be made. In 30 of these, the electromyogram preceded the speech by up to 150 msec. There was, however, considerable variability in the temporal relation between the two signals. This did not seem to be dependent on vocal intensity. In all subjects the electromyogram preceded the vocal sound in at least some recordings. Figure 2 shows the temporal relation between the electromyogram and the speech signal at two nearly equal intensities (results from one subject). The electromyogram precedes the speech sound in one of the two recordings and is delayed in the other.

The motor unit potentials recorded before the start of the speech were analysed with a faster time base and compared with those obtained during acoustic stimulation in the same subjects. They were found to be closely identical in shape and duration and differed markedly from the broad potentials recorded e.g. during teeth clenching and facial movements.

On the basis of these results it seems justified to conclude that the stapedius muscle is active during most vocalizations, that it is activated as part of the vocalization process, and that it is not only activated as an acoustic reflex.

Stapedius muscle activity when listening to speech

The speech phrases used to determine discrimination score were also used to elicit the stapedius reflex by contralateral

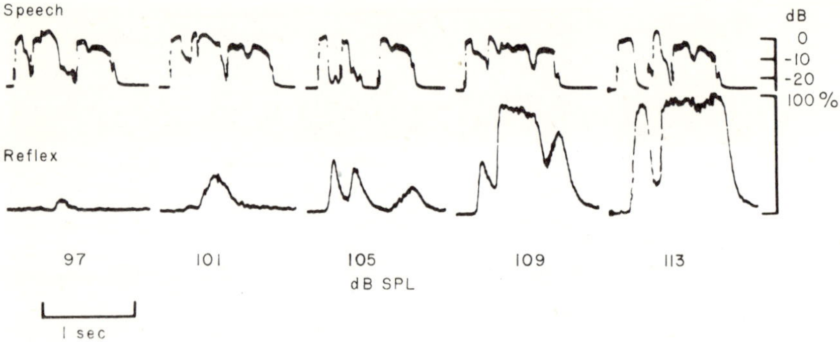

FIG. 3. Upper row: rectified, low-pass filtered (cut-off frequency 50 Hz) trace of speech phrases used as reflex stimuli. Scale to the right: sound pressure levels re maximum value of the speech phrases. Lower row: acoustic impedance changes in response to the speech phrases, the maximum intensities of which are given by the figures below the corresponding recordings. Scale to the right: impedance change in percentages of the maximum.

stimulation in group II. Figure 3 shows the rectified low-pass filtered speech signal plotted after logarithmic transformation (upper row) and the corresponding contralateral impedance change (lower row) at five sound pressure levels. Speech sound elicits an impedance change at 97 dB SPL. At higher levels the impedance change, somewhat delayed, is a smoothed "copy" of the speech envelope. The average threshold (defined as 10% of the maximum) for the six subjects tested was 97 dB SPL with contralateral stimulation. Ipsilateral recordings could not be made with speech sound stimulation owing to the interaction of the speech signal with the 800 Hz impedance measuring tone. However, by using pure tone at 0·5 kHz, ipsilateral and contralateral reflex responses could be recorded simultaneously. The difference in threshold sensitivity between the ipsi- and contralateral reflexes was then calculated to give an average difference of 5 dB. Thus, it would seem evident that the stapedius muscle will contract in the ear exposed to speech sounds, the maximum level of which is above about 92 dB SPL. This most likely means that the stapedius muscle is not active in an ordinary listening situation (at 1 m distance from a speaker talking with a normal voice effort, Fletcher, 1953).

Discrimination score

It has been seen that the stapedius muscle is active during almost all vocalization and while listening to speech, the intensity of which is above about 90 dB SPL. The significance of the stapedius muscle for understanding speech was also investigated in subjects

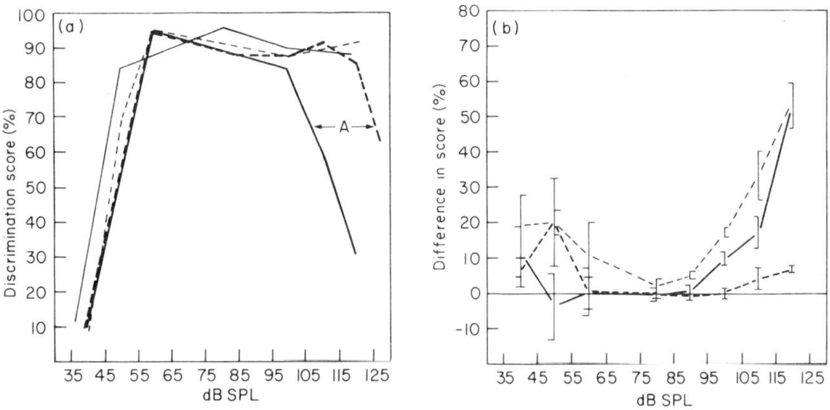

FIG. 4. (a) Discrimination scores of nonsense monosyllables (percentage) as a function of their maximal speech level (dB re 2×10^{-5} N/m^2) in one subject with normal Békésy thresholds. ——— during stapedius muscle paralysis; ——— the same ear after complete recovery of the stapedius muscle function; – – – – unaffected ear during paralysis of the contralateral side; – – – – unaffected ear after recovery from paralysis. "A" shows the influence of the stapedius muscle in terms of equivalent change of sound pressure level. (b) Average difference in discrimination scores as a function of sound pressure level of nonsense monosyllables (abscissa). ——— scores for the unaffected minus the affected ear. – – – – scores for affected ear after recovery from stapedius muscle paralysis minus scores during paralysis; – – – – scores in unaffected ear after recovery from paralysis minus scores during paralysis (reproducibility);]: standard errors of each mean.

with total unilateral stapedius muscle paralysis (group III). Figure 4 (left graph) shows discrimination scores as a function of speech sound intensity in one representative subject with normal hearing up to 8 kHz. He was tested with lists of monosyllables over a wide intensity range in both ears, both during unilateral stapedius paralysis and after recovery.

All curves agree closely except for the ear with acute stapedius paralysis above 90–100 dB SPL. Here discrimination score in the absence of a stapedius reflex was considerably lowered. Also, the intensity of the monosyllables has to be raised "A" dB in the non-affected ear to obtain the same discrimination score as in the affected ear at high intensity. The value of "A" was at maximum about 20 dB, which corresponds to the maximum attenuating effect provided by the stapedius reflex (Borg & Zakrisson, 1974).

The influence of the stapedius reflex on discrimination score on all the material is presented in the right graph of Fig. 4. The improvement in score with a normal stapedius reflex is shown as a function of the intensity of the monosyllables. The heavy continuous line shows the difference between the non-affected and affected ears during the acute stage of the palsy. The thin broken line shows the difference in scores in the affected ear during and after paralysis. The heavy broken line shows the reproducibility of the measurements. It is obtained as the difference in scores in the non-affected ear during and after paralysis. With a normal stapedius reflex the scores were improved significantly at and above a level of 100 dB SPL. The reproducibility measurements (heavy broken line) suggest that there may have been a slight learning effect.

Thus speech sounds up to 90 dB are adequately analysed even in the absence of stapedius reflex. Also, in subjects with normal stapedius function there is no activity in the stapedius muscle below 90 dB SPL. Only at levels above the estimated stapedius reflex threshold (see p. 58) does discrimination decline in the ear with the paralysed muscle.

Speech intelligibility in noise

Since the stapedius reflex attenuates low frequency sound (Borg & Zakrisson, 1974) it could be expected to influence the

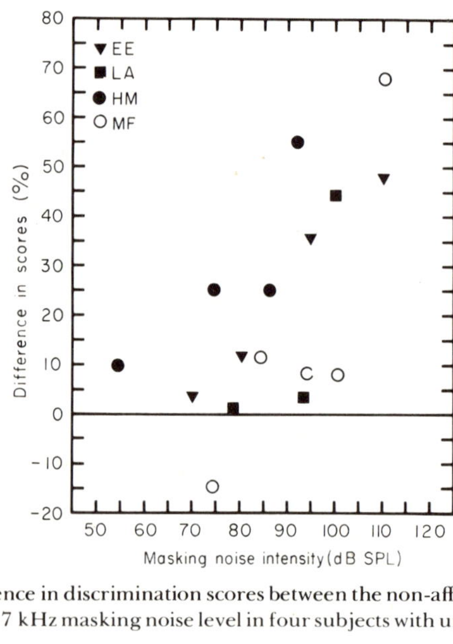

FIG. 5. Difference in discrimination scores between the non-affected and the affected ear as a function of 0·7 kHz masking noise level in four subjects with unilateral stapedius muscle paralysis.

masking produced by such sounds on speech. It was of course found that low frequency noise could completely mask speech sounds, both in ears with normal stapedius function and in ears with stapedius paralysis. The signal-to-noise ratio was hence adjusted for each subject and each speech sound intensity level so that the discrimination score in the non-affected ear was about 80%. The difference in scores was determined when presenting monosyllables masked in the same way in the affected ear. Figure 5 shows the difference in scores between the two ears as a function of 700 Hz masking noise level, when the stapedius muscle was paralysed unilaterally.

Although the data are limited it is clear that discrimination is better in the non-affected than in the affected ear. This difference is seen at levels within the range of noise intensities occurring in industry and in traffic.

Masking produced by low frequency noise

A possible mechanism behind the decline of discrimination score in the absence of stapedius muscle function is an increase of masking effects of low frequency speech sounds on speech sound of high frequency. To test this, masking produced by 0·5 kHz-centred narrow-band noise was investigated in subjects with unilateral stapedius paralysis (group IV).

In both ears it was found that hearing thresholds for pure tones were elevated over a wide frequency range during exposure to ipsilateral noise. The magnitude of the increase was dependent on the masking noise level. Figure 6 shows masking produced by 0·5 kHz-centred noise (heavy broken line) at two intensity levels in one subject with unilateral stapedius muscle paralysis. The shaded area shows a difference in masking between the affected and the non-affected ear during stapedius paralysis.

For noise levels below reflex threshold, the masking effect was equal throughout, both in the affected ear during stapedius paralysis and after recovery, and in the non-affected ear. At high noise levels (115 dB SPL) masking was considerably greater (up to about 50 dB) in the ear without a stapedius reflex as shown by the shaded area. The increase in masking was greatest at 4–8 kHz.

Masking increased more rapidly as a function of noise level when the stapedius muscle was paralysed. The curves relating masking to noise level (not shown in Fig. 6) shifted to higher levels and their slopes were increased. The shift of the curves was equal to the decrease in the noise level in the ear with the paralysed

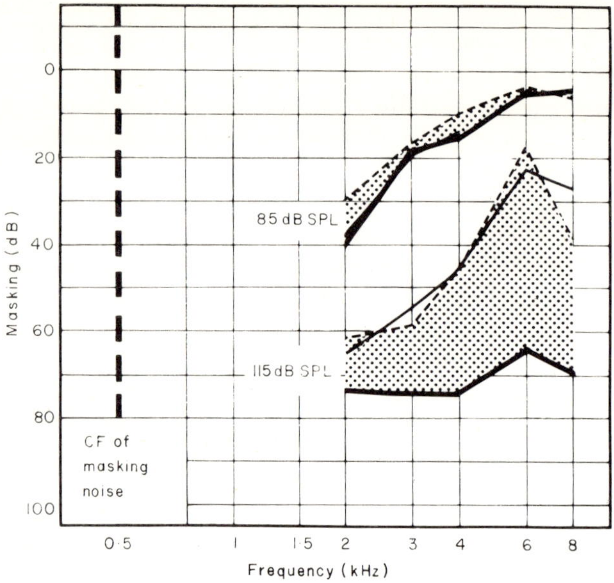

FIG. 6. Masking as a function of test frequency at two masking noise levels (85 and 115 dB). ——— affected ear during stapedius paralysis; ——— the same ear after recovery; — — — — non-affected ear; dotted areas show the difference in masking during unilateral stapedius muscle paralysis; CF: centre frequency.

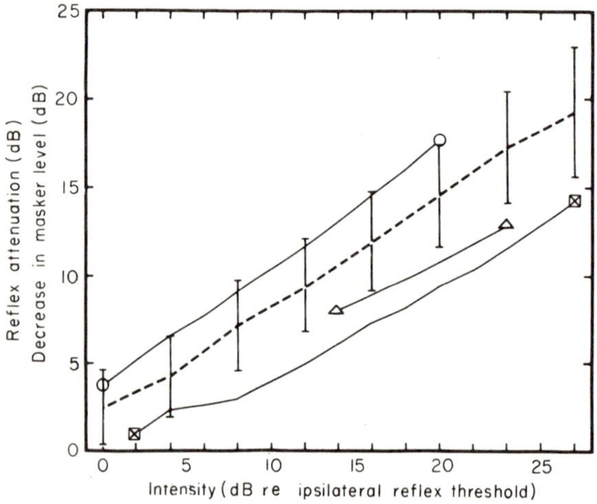

FIG. 7. The relation between attenuation of pure tones at 0·5 kHz and the equivalent decrease in masker level. The interrupted line shows the mean and standard deviation of the attenuation provided at 0·5 kHz by the stapedius reflex (group V). Abscissa for the interrupted line: stimulus intensity above ipsilateral reflex threshold for 0·5 kHz pure tones. Continuous lines show average decrease in masker level that is necessary to obtain the same masking at the frequencies 3·0, 4·0, 6·0 and 8·0 kHz in the affected ear as in the normal ear. Abscissa of continuous lines: masker intensity above estimated individual stapedius reflex thresholds for the 0·5 kHz masking noise. Each continuous line represents one subject.

stapedius muscle necessary to obtain the same masking as in the non-affected ear.

The relation between the attenuation of sound at 0·5 kHz (broken line) determined in group V as described by Borg (1968) and the decrease of masker level needed to obtain equal masking in the three subjects of group IV (continuous lines with symbols) is shown in Fig. 7. Here the decrease necessary to obtain equal masking was about the same as the attenuation provided by the stapedius reflex.

The stapedius reflex evidently improves hearing for pure tones at high frequency by up to about 50 dB because of the attenuation of low frequency masking sounds. The antimasking effect of the stapedius reflex is thus large and is likely to be significant for the discrimination of high level speech sounds and speech in noise.

DISCUSSION

Mechanisms for stapedius muscle activation during vocalization

The results showed that the electromyogram of the stapedius muscle during vocalization often appeared before the speech sound (group I). This result is in agreement with Henson's (1965) findings in some bat species and with the work of Carmel & Starr (1963) in the cat. Thus, it is evident that the stapedius muscle can be activated non-acoustically both in animals and in man as a part of the vocalization process. Furthermore, in the studies in man presented here it was found that the stapedius muscle is active throughout almost the whole range of vocal intensities.

The voice can be raised to levels well above 100 dB SPL when measured 20 cm in front of the mouth (one subject of group I reached 122 dB SPL). For [*a*:] the sound pressure at the external ear is similar to that 20 cm in front of the mouth (Sundberg, personal communication). It may be concluded that an acoustic reflex is evoked at high voice levels, but at low and medium voice levels the situation is more ambiguous. It was found that the stapedius muscle was activated during vocalization at a sound level (20 cm in front of the mouth) about 36 dB below the acoustic reflex threshold for contralateral sound stimulation. One subject with a bilateral conductive hearing impairment had the lowest hearing threshold of 40 dB SPL (2·0 kHz). This value was only 20 dB below the threshold for stapedius muscle activation during the subject's vocalization. The conductive hearing impairment did not seem to influence the stapedius muscle activity during the subject's own speech but greatly influenced the stapedius muscle threshold for sound stimulation. For another subject with air-conduction thresholds in both ears within 15 dB (ISO Standard, 1964) in the range 0·5 to 2·0 kHz, thresholds for stapedius muscle activation during vocalization were 65 dB and upon acoustic stimulation 130 dB SPL. Since the acoustic reflex thresholds were far above the thresholds for stapedius activation during vocalization it seems that airborne sound plays no or very little part in activating the stapedius muscle during vocalization, at least at low and medium vocal intensities.

It is difficult to evaluate the role of bone conduction in activating the stapedius muscle during vocalization on the basis of the present results and the current literature. According to von Békésy (1960: 187) the bone-conducted voice energy that reaches the inner ear is equal to the air-conducted energy. Djupesland, Flottorp, Sundby & Szalay (1973) claim that the stapedius muscle is activated by bone-conducted sound at levels 5 to 25 dB below that for air-conducted sound. In any case, the great difference in threshold for stapedius muscle activation can hardly be explained by bone conduction, since the bone-conducted voice energy would have to be at least 40 dB above the air-conducted energy, even in the subject with bilateral normal air-conduction thresholds (see above). The most likely explanation for stapedius muscle activation at low and moderate vocal intensity is that the stapedius muscle is an integral part of the mechanisms of vocalization.

Mechanisms for the influence of the stapedius muscle on speech sound intelligibility

The subjects with unilateral stapedius muscle paralysis (group III) had equally good discrimination scores, close to 100% for both ears, up to 90 dB SPL (maximum level of the monosyllables). This means that the ear without stapedius reflex functions satisfactorily for unmasked speech up to this sound level. When speech sounds exceed 90 dB, however, they can only be analysed adequately when the stapedius muscle functions normally.

The masking at high intensity, mainly by low frequency sound on sound of high frequency, becomes very marked. This asymmetry is relevant for speech discrimination, since speech sound has maximum energy at low frequency (400–500 Hz, Fletcher, 1953). These low frequencies then would have a masking effect on the weaker high frequency components which are more important for the intelligibility of speech (Licklider & Miller, 1951). According to the present study the stapedius reflex reduces masking by low frequency noise on high frequency sound by up to 50 dB. If direct masking is responsible for decline in discrimination score, the vowels ought to be the most affected since they contain both strong low frequency and weaker high frequency components. If backward and forward masking (see Kryter, 1970: 39) is important, consonants ought also to be affected.

In a preliminary analysis of the errors made in the discrimination measurements it was found that the majority concerned vowels (Borg & Sjögren, in preparation). Consonant errors were small. This specificity supports the assumption that direct masking has a much greater role than forward and backward masking in the decline of discrimination for speech sounds in absence of stapedius reflex.

The role of centrifugal control of middle ear sound transmission during vocalization

In some bat species the stapedius muscle contracts shortly before the start of the vocalization and continues to contract throughout the duration of the vocal sound. The cochlear microphonic potentials are thereby considerably depressed (Henson, 1965, 1970) and the sensitivity of the ear is centrifugally controlled by stapedius muscle activity. The importance of this mechanism is most likely to protect and maintain the auditory system in a sensitive state for perception of faint echoes (Henson, 1965).

The stapedius muscle activity during vocalization in man is unlikely to be significant for perception of echoes of the voice. However, there are signals analogous to the bat's echo. Verbal messages from other speakers and other acoustic signals must often be perceived at the same time as the person receiving the message is talking himself. The low frequency components of a person's own speech sounds are effectively transmitted to his ear, their intensity at the entrance of the auditory canal being closely similar to that of 15–20 cm in front of the mouth (Fletcher, 1953; Sundberg, personal communication). The intensity of these low frequency components is high and presumably has a considerable masking effect on the perception of speech sounds, even those of high frequency, which are very important for intelligibility. The stapedius muscle contraction during vocalization attenuates these low frequency components by up to about 20 dB (see Fig. 7) and thereby reduces their masking effect on the sound received by up to 50 dB. This is analogous to the present finding that a low frequency noise masked speech sounds much more effectively when the stapedius muscle was paralysed than under normal conditions. On the basis of the present observations it seems that the stapedius muscle activity during vocalization controls the sound transmission to the inner ear in such a way that the masking effect of the subject's own voice on the simultaneously perceived speech sounds from other sources decreases, and thereby the intelligibility is improved.

An interesting correlation to this centrifugal control function of the stapedius muscle is found in the lateral line organ of fish which is sensitive to displacement of water. Roberts & Russell (1972) showed that the efferent nerve fibres to the hair cells were activated during the animal's own movements. They thereby blocked the afferent neuronal activity from the sensory receptor cells. The sense organ would be prevented from overstimulation and could maintain its sensitivity. These centrifugal functions thus seem analogous: they decrease the effect of the activity of the organism itself, i.e. those vocalizations or movements that otherwise may reduce the receptivity of the sensory organ.

SUMMARY

The stapedius muscle was investigated with respect to its activation during vocalization and listening to speech and with respect to its role for perception of speech sounds. It was found that the muscle was activated over almost the whole range of vocal intensities that subjects could produce, and while they listened to speech sound at high level. The intelligibility of speech sounds with maximum levels at and above 100 dB SPL was significantly improved when stapedius muscle function was normal compared with conditions of stapedius muscle paralysis. It was also found that masking produced by low frequency sound on sound of high frequency was reduced by up to 50 dB. It was concluded that the antimasking effect of the stapedius muscle was of great importance for the intelligibility of high intensity speech, of speech in noise and especially of speech and other acoustic signals while the subject is talking himself.

ACKNOWLEDGEMENTS

This study was supported by grants from the Swedish Medical

Research Council, the Medical Faculty, University of Umeå, P. Frenckners fund, B. Fromms fund and Karolinska Institutets funds. We are indebted to Assistant Professor Aage R. Møller for valuable suggestions and criticism. We thank Assistant Professor Henry Anderson and Dr Lars Gleisner of the Department of Audiology, Karolinska sjukhuset, Stockholm, for providing the Bell's palsy cases and electrical engineer Bertil Johansson and his colleagues of the Department of Technical Audiology, Karolinska sjukhuset, for placing their measurement facilities at our disposal. Acta Otolaryngologica (Stockholm) is acknowledged for permission to publish Figs 1, 2, 6 and 7 and the Journal of the Acoustical Society of America for Figs 3 and 4.

References

von Békésy, G. (1960). *Experiments in hearing*. New York: McGraw-Hill.

Borg, E. (1968). A quantitative study of the effect of the acoustic stapedius reflex on sound transmission through the middle ear of man. *Acta otolaryng. (Stockh.)* **66**: 461–472.

Borg, E. & Sjögren, H. (in prep.) *A matrix analysis of speech discrimination data in stapedius paralysis*.

Borg, E. & Zakrisson, J. E. (1973). Stapedius reflex and speech features. *J. acoust. Soc. Am.* **54**: 525–527.

Borg, E. & Zakrisson, J. E. (1974). Stapedius reflex and monaural masking. *Acta otolaryng. (Stockh.)* **78**: 155–161.

Borg, E. & Zakrisson, J. E. (in press). The activity of the stapedius muscle in man during vocalization. *Acta otolaryng. (Stockh.)*.

Carmel, P. W. & Starr, A. (1963). Acoustic and nonacoustic factors modifying middle ear muscle activity in waking cats. *J. Neurophysiol.* **26**: 598–616.

Djupesland, G., Flottorp, G., Sundby, A. & Szalay, M. (1973). A comparison between middle ear muscle reflex thresholds for bone- and air-conducted pure tones. *Acta otolarying. (Stockh.)* **75**: 178–183.

Fletcher, H. (1953). *Speech and hearing in communication*. New York: D. Van Nostrand.

Henson, O. W., Jr (1965). The activity and function of the middle ear muscles in echo-locating bats. *J. Physiol. Lond.* **180**: 871–887.

Henson, O. W., Jr (1970). The ear and audition. In *Biology of bats*. **2**: 181–263. Wimsatt, W. A. (ed.). New York and London: Academic Press.

Jepsen, O. (1963). Middle ear muscle reflexes in man. In *Modern developments in audiology*: 193–239. Jerger, J. (ed.). New York and London: Academic Press.

Kryter, K. D. (1970). *The effects of noise on man*. New York and London: Academic Press.

Licklider, J. C. R. & Miller, G. A. (1951). Perception of speech. In *Handbook of experimental psychology*: 1040–1074. Stevens, S. S. (ed.). New York: Wiley.

Møller, A. R. (1960). Improved technique for detailed measurements of the middle ear impedance. *J. acoust. Soc. Am.* **32**: 250–257.

Møller, A. R. (1961). Bilateral contraction of the tympanic muscles in man, examined by measuring acoustic impedance change. *Ann. Otol. Rhinol. Laryngol.* **70**: 735–753.

Møller, A. R. (1972). The middle ear. In *Foundations of modern auditory theory*: 193–239. Tobias, J. V. (ed.). New York and London: Academic Press.

Roberts, B. L. & Russell, I. J. (1972). The activity of lateral-line efferent neurons in stationary and swimming dogfish. *J. exp. Biol.* **57**: 435–448.

Zakrisson, J. E., Borg, E. & Blom, S. (1974). The acoustic impedance change as a measure of stapedius muscle activity in man. A methodological study with electromyography. *Acta otolaryng. (Stockh.)* **78**: 357–364.

Symp. zool. Soc. Lond. (1975) No. 37, 69–95.

PERIPHERAL ASPECTS OF THE PHYSIOLOGY OF MIDDLE EAR MUSCLES

S. D. ANDERSON

Institute of Laryngology and Otology, London, England

SYNOPSIS

An investigation has been carried out into the types of muscle fibres found in the middle ear muscles of various mammals, in order to identify tonic and phasic motor units (as indicated in the literature). Silver impregnation and teasing of cotton rat and guinea-pig tensor tympani muscles failed to reveal multiterminal motor endings.

Histochemical fibre typing techniques were therefore applied to distinguish whether or not tonic and phasic focally innervated skeletal muscle fibres were present in the middle ear muscles.

Classification of fibres has been based upon the standard methods for staining the enzymes myofibrillar ATPase (Padykula & Herman, 1955) and reduced coenzyme I diaphorase (Novikoff, Shin & Drucker, 1961), supplemented by the modified method of staining for myofibrillar ATPase involving acid and alkaline pre-incubation of sections (Guth & Samaha, 1969). On the basis of findings with these methods in the hind limb and middle ear muscles of the species studied (cotton rat, guinea-pig and rabbit), a *histochemical* classification based on a concept of *transformation* between fibre categories was proposed to enable a coherent description of the results for the middle ear muscles. The middle ear muscles were shown to be heterogeneous in their fibre composition, but there are differences in the types of fibres found in different species. There are also differences in the arrangement of the fibre types within the muscles. These interspecies variations in the middle ear muscles are discussed in relation to the dynamic characteristics of contraction that can be inferred.

INTRODUCTION

In all mammals except certain fossorial forms, the middle ear contains two muscles, the stapedius and the tensor tympani. These muscles are attached to the ossicular chain and constitute the effectors of an input control system of the ear.

There are considerable interspecies variations in gross morphology of the middle ear muscles, in terms of shape and size, (e.g. Kobayashi, 1956a,b; Hinchcliffe & Pye, 1969). Similarly, the threshold for middle ear muscle contraction in response to sound, in an acoustic reflex, has been shown to be different in man from that in commonly studied laboratory animals such as the rabbit and the cat (e.g. Borg, 1972a). It is now apparent from the present study that there are interspecies differences in the types of muscle fibres and the proportions in which they occur in the middle ear muscles. Differences in the fibre composition of muscles reflect differences in the dynamic characteristics of contraction (Edgerton

& Simpson, 1971). This in turn suggests that the requirements for optimal middle ear muscle contraction are different in different animals.

Previously the main measure of heterogeneity within the middle ear muscles has been the description of slow and fast motor units. Electromyographic studies of the tensor tympani muscles of the cat (Okamoto, Sato & Kirikae, 1954) distinguish two types of motor units: those with a tonic activity, low contraction threshold and small action potentials, and a second type with phasic properties, high contraction threshold, and larger action potentials. This distinction has been substantiated by myographic recording of the contraction characteristics of individual cat tensor tympani motor units in response to threshold stimulation of their motor neurones (Teig, 1972b).

The question posed here is: what is the structural basis for the different types of motor units in the middle ear muscles? Do the tonic motor units of the middle ear muscles reflect the presence of multiterminally innervated slow fibres (Erulkar, Shelanski, Whitsel & Ogle, 1964; Fernand & Hess, 1969), or the presence of tonic, focally innervated twitch fibres? Evidence will be presented from histochemical staining of middle ear muscle sections that the middle ear muscles of the cotton rat (*Sigmodon hispidus*), guinea-pig (*Cavia porcellus*), fruit bat (*Rousettus aegyptiacus*) and rabbit (*Oryctolagus cuniculus*), are heterogeneous in their muscle fibre composition. The staining properties of the majority of fibres are comparable with those of the "twitch" muscle fibre types identified in limb muscles (Close, 1972). However, it was found necessary to propose a classification of fibres into six categories. It was also found that the types of muscle fibres in the middle ear muscles differ in both type and/or relative numbers between the different species studied.

METHODS

Silver impregnation

The middle ear muscles from six cotton rats and the tensor tympani from four guinea-pigs were removed and impregnated with silver by the method described by Barker & Ip (1963). The muscle fibres were then teased apart in glycerol and examined for nerve endings under a microscope.

Histochemistry

Animals

Initially, histochemical techniques were applied to the middle ear muscles of the cotton rat (35 adult animals). The refined techniques were then applied to middle ear muscles of the fruit bat (seven adult animals kindly provided by Dr Ade Pye) and the albino guinea-pig (11 young adult animals weighing 670–800 g). A preliminary study has recently been made of the muscles of the Dutch rabbit (two young adult animals each weighing about 1 kg).

Dissection procedure

All animals were anaesthetized by intraperitoneal injections of 40 mg pentobarbitone sodium per kilogram of body weight or allobarbital and urethan. The middle ear muscles were removed for histochemical procedures in one of two ways. Either (a) the middle ear cavity of the live animal was exposed and the muscles carefully dissected free from their insertions, or (b) the animal was decapitated, the auditory bullae immediately exposed and fractured, and the muscles again carefully dissected free.

Specimen mounting

Because of the small size of the specimens it was found necessary to mount them in tissue blocks, prior to freezing in liquid nitrogen and subsequent sectioning. These tissue blocks were taken from the same animal during dissection of the middle ear. Cotton rat specimens were mounted in several types of tissue, (kidney, brain, liver or hind limb skeletal muscle). These tissues served as control tissues for the staining methods. Although liver was found to be the most satisfactory tissue for ease of cutting of the frozen sections and because freezing artefacts rarely occurred, hind limb skeletal muscle was used in half of the cotton rat specimens and with all specimens of the guinea-pig and rabbit. The use of skeletal muscle ensured that comparisons could be made between the staining intensities in the limb muscle fibre types and the fibres of the middle ear muscles. Similarly, on some occasions, the tensor tympani and stapedius were mounted in the same tissue block to enable direct comparison of staining intensity between the two muscles.

The tissue blocks containing the middle ear muscles were placed on pieces of cork, and immediately frozen in liquid nitrogen. Serial sections 10 μm thick were cut in a cryostat at −22°C.

Staining methods

The following staining methods were applied. (a) The standard method for myofibrillar adenosine triphosphatase (ATPase) at pH 9·4 (Padykula & Herman, 1955). (b) The modified method for myofibrillar ATPase involving comparison of sections pre-incubated in acid (pH 4·35), and sections pre-incubated in alkali (pH 10·4) (Guth & Samaha, 1969). (c) The standard method for reduced coenzyme I (NADH)-diaphorase (Novikoff *et al.*, 1961).

The staining properties of the different muscle fibres were studied by comparison of photographs taken on a Zeiss "Ultraphot" microscope of consecutive cross-sections stained by the different methods. Muscle fibre diameter was calculated by taking the square root of the product of the major axis and that at right angles to it, measured from photographic enlargements. The cross-sectional area of a muscle was taken as the sum of the areas of the individual fibres and thereby did not include supporting tissues. The relation between fibre type and range of fibre diameters will be discussed more fully elsewhere.

CLASSIFICATION OF FIBRES

Dubowitz & Pearse (1960) distinguished two types of muscle fibres: one with high oxidative activity and low glycolytic activity, and another type with high glycolytic and low oxidative activity. It was subsequently shown that these were type I fibres with low myofibrillar ATPase activity and type II fibres with high myofibrillar ATPase activity (Engel, 1962). However, more recent investigations using histochemical methods and electron microscopical techniques have concluded that there are three general categories of focally innervated skeletal muscle fibres (Close, 1972). A very simplified representation of their properties is given in Table I (the type I fibres of Engel are equivalent to the intermediate fibres).

The intensity of oxidative enzyme staining may be influenced by exercise and other environmental variables. For this reason the classification of Stein & Padykula (1962), which is based on qualitative differences in the cytochemical distribution of the mitochondria within the fibres, is preferable to a distinction on the basis of intensity of staining of oxidative enzyme activity. Unfortunately, the small size of the fibres makes assessment of such differences in mitochondrial distribution extremely difficult for all but the largest middle ear muscle fibres. There is a similar problem

TABLE I

Histochemical types of fibre in mammalian skeletal muscle

Classifications:			
Padykula & Gauthier (1967)	White	Intermediate	Red
Romanul (1964)	I	III	II
Barnard, Edgerton, Furukawa & Peter (1971)	Fast-twitch	Slow-twitch	Fast-twitch
Edström & Kugelberg (1968a,b)	Rapid fatigue	No evidence of fatigue	Slow fatigue
Morphological Properties:			
Mitochondrial content	Small	Intermediate	Large
Neuromuscular junction	Large and complex	Intermediate	Small and simple
Fibre diameter (not always)	Large	Intermediate	Small
Histochemical Properties:			
Distribution of succinic dehydrogenase	Even network	Even network	Predominantly subsarcolemmal
Oxidative enzyme activities	Low	High or intermediate	Intermediate or high
Myofibrillar ATPase at pH 9·4	High	Low	High
pH sensitivity of myofibrillar ATPase	Acid labile	Alkali labile	Acid labile
Formaldehyde sensitivity of myofibrillar ATPase	Sensitive	—	Stable
Glycolytic activities	High	Low or variable	Intermediate or low

Modified after Close (1972).

with fibres having the atypical myofibrillar arrangement as described, for example, by Malan (1934) and Candiollo (1965).

Consequently, a second classification (Guth, Samaha & Albers, 1970; Samaha, Guth & Albers, 1970) which is based on the myofibrillar ATPase reaction, and in which the fibres are divided according to qualitative differences in the pH lability of the enzyme, has been applied to the middle ear muscles. This technique is valuable because the reaction again reflects relatively stable qualitative differences between cell types. Also, a reversal of staining intensity is obtained for two types of fibres in sections pre-incubated in acid compared with those pre-incubated in alkali (Fig. 1).

Intermediate fibres possess myofibrillar ATPase that is acid-stable and alkali-labile (β), and thus appear to stain heavily after acid pre-incubation, but hardly at all after alkali pre-incubation.

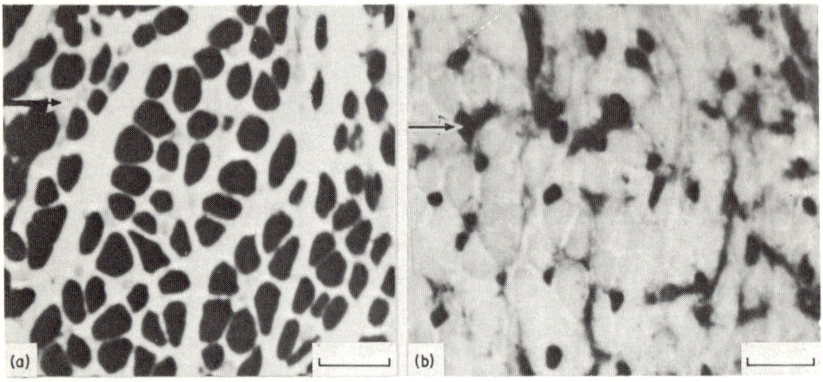

FIG. 1. Reversal of staining intensity for myofibrillar ATPase in fibres of the guinea-pig tensor tympani, when pre-incubating in (a) alkali, pH 10·4, (b) acid, pH 4·35. An arrow indicates the position of an intermediate fibre. Staining and pre-incubation times are such as to accentuate the reversal of staining intensity in these fibres. Note that there is heavy staining of blood vessels in (b). Fine blood vessels were distinguished from intermediate fibres by staining for alkaline phosphatase. Scale = 45 μm.

Conversely, the red fibres are generally the opposite, possessing acid-labile myofibrillar ATPase (α). In most muscles, white fibres represent a hybrid or intermediate position ($\alpha\beta$). However, Yellin & Guth (1970) showed that the pH lability of myofibrillar ATPase of white and red fibres is reversed in the tibialis muscle of the cat compared with the same muscle in the rat. In the cat tibialis muscle the red fibres, as indicated by the mitochondrial distribution, have hybrid characteristics ($\alpha\beta$) and the white fibres acid-labile (α) characteristics.

Because of the species dependence of this second method of classification, a comparison was made in this study between the mitochondrial distribution and the intensity of oxidative enzyme staining in the hind limb muscle as demonstrated by NADH-diaphorase activity, and the pH lability of myofibrillar ATPase activity. Thus, this muscle was used both as a control and as supporting tissue. The results of the two methods were found to be incompatible with the hypothesis that there are only three categories of fibres.

Many authors have reported that some fibres appear to be intermediate in enzymatic activity between the three general categories. Thus, when multiple staining techniques are used, classification into eight or more categories becomes possible (Romanul, 1964; Yellin, 1967). What seems particularly significant is that transformations between the three general categories have now been demonstrated (Guth, 1968; Barnard, Edgerton & Peter, 1970; Zíka, Lojda & Kuĉera, 1973). On the basis of findings using

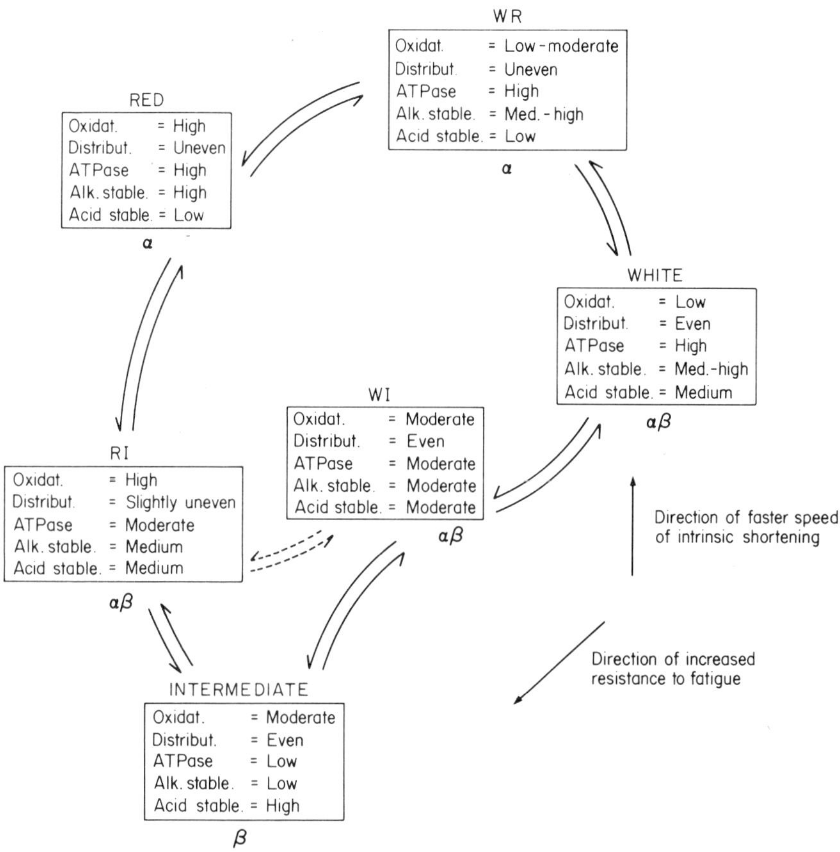

FIG. 2. Proposed classification of skeletal muscle fibres on the basis of mitochondrial distribution and pH stability of myofibrillar ATPase. Oxidat.: intensity of staining for oxidative enzyme activity; Distribut.: distribution of mitochondria within the fibre (uneven signifies high concentration at the sarcolemma); ATPase: intensity of staining for myofibrillar ATPase with standard method; Alk. stable.: intensity of staining for myofibrillar ATPase after alkali pre-incubation; Acid stable.: intensity of staining for myofibrillar ATPase after acid pre-incubation. Examples: WR: "white fibres" in cat tibialis (Yellin & Guth, 1970). RI: "red.fibres" in cat tibialis (Yellin & Guth, 1970). WI: guinea-pig tensor tympani.

both methods in the hind limb and middle ear muscles of the species studied (cotton rat, guinea-pig and rabbit), the following *histochemical* classification was proposed to enable a coherent description of the results in the middle ear muscles (Fig. 2).

The three general categories were retained, but with three more proposed. These are (a) category WR: with fibres somewhat intermediate in properties between white and red fibres (cf. Barnard, Edgerton & Peter, 1970; and others); (b) category WI: intermediate between white and intermediate fibres and (c) category RI: intermediate between red and intermediate fibres. The schematic representation of the classification is an attempt to visualize these three additions within a concept of *transformation*. In other words, there are fibres with characteristics equivalent to those that would be observed if these fibres were undergoing an adaptive transformation from one main category to another. This is not meant to exclude the existence of these fibres without adaptive transformation. Rather, the characteristics of the fibres are envisaged as being determined by the properties and "demands" of the neural elements of the motor control system.

It is postulated that there is a complete spectrum of fibres between the three main categories, depending upon the optimal contractile ability required from a fibre. It must be stressed that any adaptive transformation that may occur in a muscle does not necessarily mean that morphological properties associated with the three main categories will also be modified. This concept is considered useful for the purpose of classification, whether or not all muscles are capable of adaptive transformation. However, it is not intended as a biochemical classification. The histochemical methods can only accurately indicate the relative extent of enzyme activity between fibres in the same tissue block. For confirmation of the extent of the enzyme activity, biochemical analysis is required.

It is assumed that a medium to high oxidative enzyme activity, evenly distributed throughout the fibre cross-section, in conjunction with a base-labile myofibrillar ATPase (β) activity, signifies a maintained and slow contraction ("intermediate"). Similarly, a high activity of acid-labile myofibrillar ATPase (α) is assumed to signify a potentially fast speed of shortening. Also, the greater this activity, the more any oxidative enzyme activity is heavily concentrated in the periphery of the fibre (i.e. uneven distribution of mitochondria). In the fibres which possess both acid- and alkali-labile myofibrillar ATPase activity (hybrid condition $\alpha\beta$), it is assumed that the greater the activity of α-ATPase, the greater the concentration of mitochondria at the periphery of the fibres.

The relative intensity of oxidative enzyme activity, and thus the indicated extent of oxidative metabolic activity within the muscle fibres, may be a somewhat independent variable when compared with the two factors used in this classification system. However, the extent of oxidative metabolism in skeletal muscle is reciprocally related to the capacity for glycolytic metabolism (Pette & Bücher, 1963; Dawson & Romanul, 1964; Jöbsis & Meijer, 1973). This in turn has been shown to be related to the *speed* of contraction in cross reinnervation experiments (Romanul & Van Der Meulen, 1967). Muscles appeared to become *slower* when there was a decrease in the extent of glycolytic metabolism and a concomitant increase in oxidative metabolism. Since these are processes that resynthesize adenosine triphosphate (ATP), it is hardly surprising that the time course of active contraction was affected, nor that the activity of oxidative enzymes is directly related to resistance to fatigue (Kugelberg & Edström, 1968; Edström & Kugelberg, 1968b; Edström & Lindquist, 1973). However, other factors are also important in resistance to fatigue, such as the extent of glycogen stores in a particular fibre type. Further, in contrast to the normal reciprocal relationship mentioned above, both oxidative enzyme activity and glycolytic activity increases in tonically stressed skeletal muscle (Zíka *et al.*, 1973).

Since intensity of oxidative enzyme staining was a noted parameter in this investigation and is related to resistance to fatigue, it was included in this classification. This is not intended to represent a comprehensive division of fibre types on metabolic grounds. Fibres with moderate to high α-ATPase activity are not divided into three groups (those with low, moderate or high oxidative enzyme activity). Those with low and moderate oxidative enzyme activity are represented by WR fibres, and those with high activity by the red category. Similarly, in the cotton rat tensor tympani most of the fibres were identified as white or RI fibres. Those fibres with a moderate oxidative enzyme activity in this muscle are presumably intermediate between the white and RI fibres in properties. For this reason it could be argued that the category WI, into which these fibres were placed, should be divided into two blocks to maintain the concept of transformation between the fibre types. Instead, for simplicity, this finding is represented by the dotted lines in Fig. 2.

RESULTS

Silver impregnation

Despite the demonstration of multiterminally innervated fibres in control extraocular muscle, no such endings could be demonstrated in the tensor tympani of the cotton rat or guinea-pig. Typical single endplates were demonstrated.

In one cotton rat tensor tympani, a very fine complex of nerve endings in association with a nerve fibre spiralling around a very thin muscle fibre (2 μm) was observed. This was thought to be a proprioceptive structure.

Histochemical techniques

General histochemical characteristics of the middle ear muscles

Myofibrillar ATPase. In the middle ear muscles of *Rousettus*, the guinea-pig and the rabbit, fibres with a low activity of myofibrillar ATPase (type I) and fibres with a high activity (type II) were identified (e.g. Fig. 3). In the cotton rat middle ear muscles there

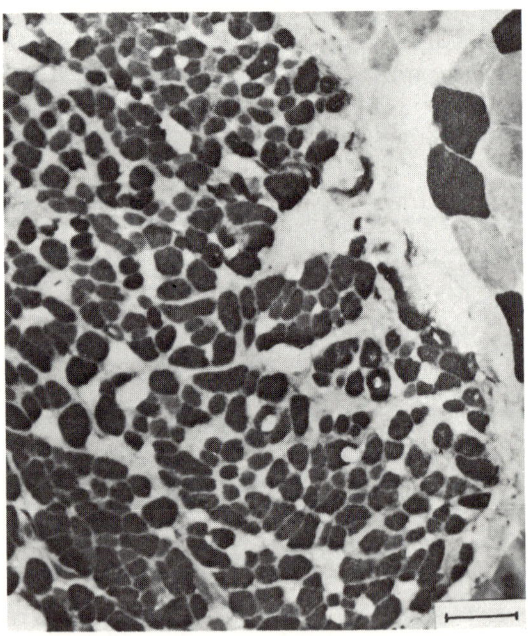

FIG. 3. Staining of *Rousettus* tensor tympani fibres for myofibrillar ATPase at pH 9·4. Type I (light) and Type II (dark) fibres can be distinguished. Large diameter fibres to the right of the photograph are those of hind limb skeletal muscle. Scale = 58 μm.

were characteristically only type II fibres. Typically all the middle ear muscle fibres stained more intensely than comparable limb muscle fibres.

Very precise pre-incubation and incubation times were required when demonstrating the pH lability of myofibrillar ATPase in the middle ear muscle fibre types.

NADH-diaphorase. In the tensor tympani muscles of all the animals studied, a range of intensity, from light to heavy staining, was observed for this enzyme. In the stapedius muscles there was generally more enzyme activity and intensities ranged from medium to heavy. Both muscles appear to show more activity than surrounding limb muscle. This was particularly pronounced in the guinea-pig muscles.

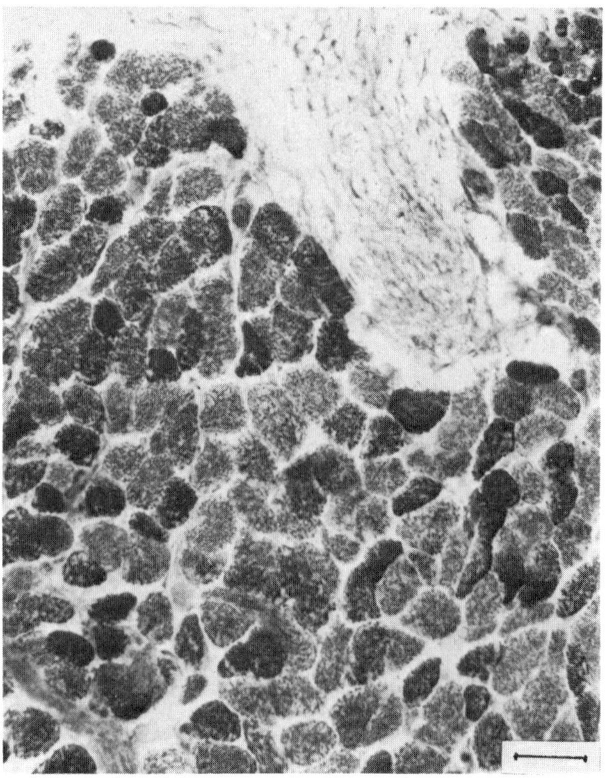

FIG. 4. Cotton rat tensor tympani muscle fibres stained for NADH-diaphorase. Fibres show a range of intensity of staining. Scale = 41 μm.

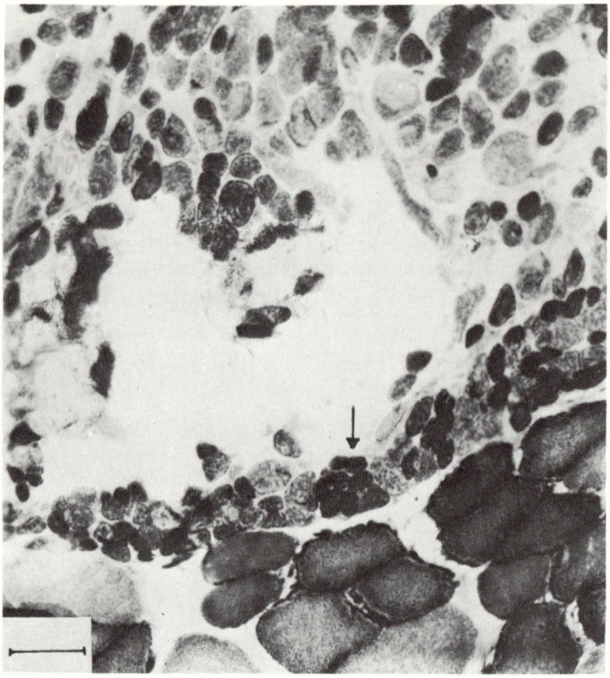

FIG. 5. Peripheral portion of the cotton rat tensor tympani stained for NADH-diaphorase and showing small groups of RI fibres (arrow) and larger white fibres. Very large muscle fibres are part of the mounting tissue block. Scale=71 μm.

Fibre types in individual muscles: Table II

Cotton rat

Tensor tympani. No intermediate (type I) fibres were found in the tensor tympani muscle. A range of intensity of oxidative enzyme staining was observed (Fig. 4). All of the fibres had the hybrid ($\alpha\beta$) type of myofibril ATPase activity. Some 47% of the fibres were identified as white, and 33% as RI (heavy oxidative, medium "$\alpha\beta$" staining). The remaining 20% were moderately stained for oxidative enzyme activity and again identified "$\alpha\beta$" (WI). RI fibres were typically smaller than white fibres, particularly in peripheral portions of the muscle (Fig. 5). A very small number of fibres with larger diameters stained very intensely for myofibrillar ATPase.

Stapedius. Typically, all fibres in the stapedius muscle had a high myofibrillar ATPase activity and exhibited "$\alpha\beta$" staining. The muscle as a whole stained more intensely for NADH-diaphorase than the tensor tympani. This was especially noticeable for the

TABLE II

Distribution of fibre types (% total number of fibres) identified in the middle ear muscles

Muscle	White	WR	Red	RI	WI	Intermediate
Cotton rat						
Tensor tympani	47	—	—	33	20	—
Stapedius	—	—	—	43	57	—
Guinea-pig						
Tensor tympani	—	15·5	—	—	56·5	28
Stapedius	—	11	—	—	44	45
Fruit bat						
Tensor tympani	53*	—	—	—	22*	25
Stapedius	35*	—	—	—	42*	23
Rabbit†						
Tensor tympani	—	—	Yes	Yes	—	Yes
Stapedius	—	—	Yes	Yes	—	Yes

* Determined by NADH-diaphorase staining only.
† Insufficient data.

fibres with the lowest intensity staining which were classified as WI fibres (57%) (see Table II). The largest fibres were all WI fibres. The remaining RI fibres (43%), although smaller, still represented a considerable proportion of the cross-sectional area of the muscle.

A small number of fibres stained lightly for myofibrillar ATP-ase after either acid or alkali pre-incubation.

Guinea-pig

Tensor tympani. A small percentage of the fibres in the tensor tympani muscle are intermediate (28%), staining intensely for myofibrillar ATPase after acid pre-incubation (Fig. 6). These fibres were small compared to the majority of fibres and only represented a small proportion of the total cross-sectional area of the contractile elements (Table III). The majority of these fibres stained heavily for NADH-diaphorase.

The type II fibres stained either lightly or moderately for NADH-diaphorase activity. The oxidative enzyme activity could be seen to be evenly distributed within the larger fibres, and the greater the oxidative activity, the more individual fibres showed acid-stable myofibrillar ATPase activity. Such fibres were classified as WI fibres and represented about 83·5% of the cross-sectional

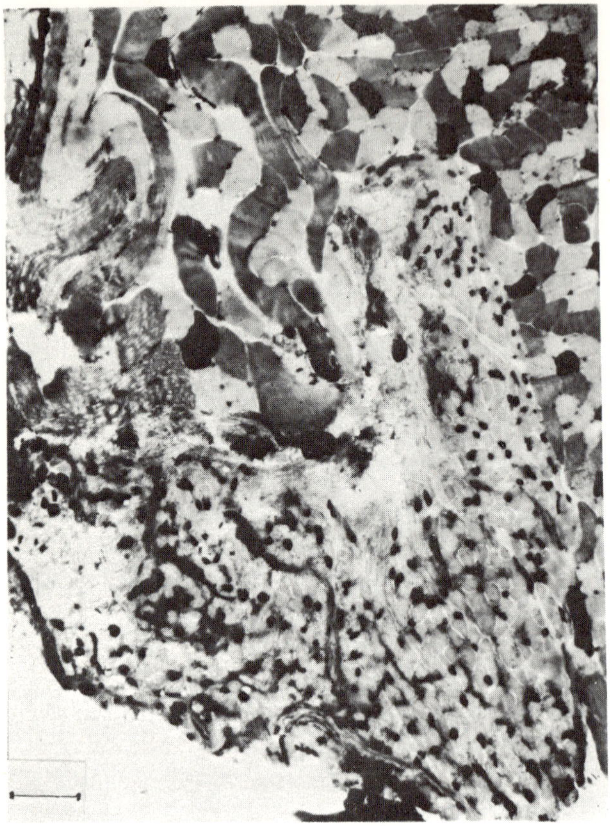

FIG. 6. Whole section of guinea-pig tensor tympani mounted in hind limb muscle and stained for myofibrillar ATPase after acid pre-incubation. Darkly staining intermediate fibres have a relatively even distribution throughout the section. Three types of fibres can be distinguished in both the limb and tensor tympani muscle and also dark staining capillaries in the limb muscle. Scale = 93 μm.

area. The remaining fibres had acid-labile myofibrillar ATPase (α) activity and were classified as WR (14%).

Stapedius. A large percentage of the fibres in the stapedius muscle had low myofibrillar ATPase activity (45%). These intermediate fibres stained heavily for acid-stable myofibrillar ATPase (β), and in contrast to the tensor tympani, represented some 35% of the cross-sectional area (Table III).

Half of the cross-sectional area (53%) was provided by WI fibres ($\alpha\beta$). A few WI fibres were low in alkali-stable myofibrillar ATPase (α), and a few fibres were classified as WR because of the low

TABLE III

Frequency, size of fibres, and percentage of cross-sectional area of the fibre types in the guinea-pig middle ear muscles

Muscle	% of fibres			Mean diameter (μm)			% cross-sectional area		
	WR	WI	I	WR	WI	I	WR	WI	I
Tensor tympani	15·5	56·5	28	15·5	21	10	14	83·5	2·5
Stapedius	11	44	45	6	9	7·5	12	53	35

acid-stable (β) activity. All fibres showed a moderate to high oxidative enzyme staining intensity.

Even though the range in fibre diameter was smaller for the stapedius than the tensor tympani, the intermediate fibres had a mean diameter smaller than that of the WI fibres (Table III). The diameter of the WR fibre may have been underestimated because of the level of the section at which diameters were measured.

The distribution of fibre diameters of the whole muscle and of the intermediate fibres is plotted in Fig. 7.

Rabbit

Tensor tympani. In the rabbit tensor tympani the majority of fibres stained heavily for oxidative enzymes. Fibres were present in large numbers which stained particularly intensely for acid-stable

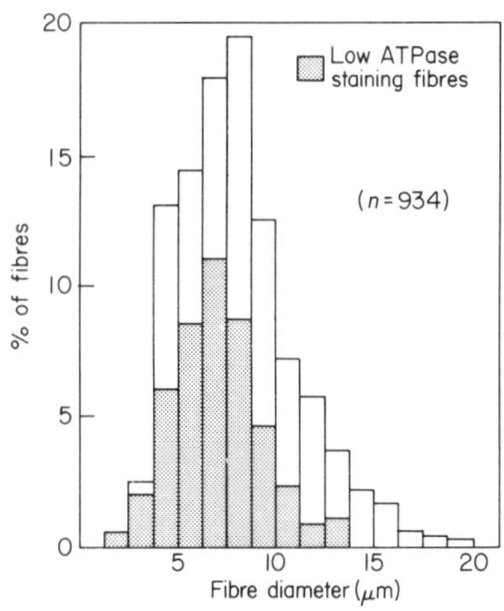

FIG. 7. Histogram of the percentage distribution of fibre diameters in microns, of the muscle fibres of the guinea-pig stapedius. The distribution of intermediate fibres (low ATPase staining) is represented by the cross-hatched area.

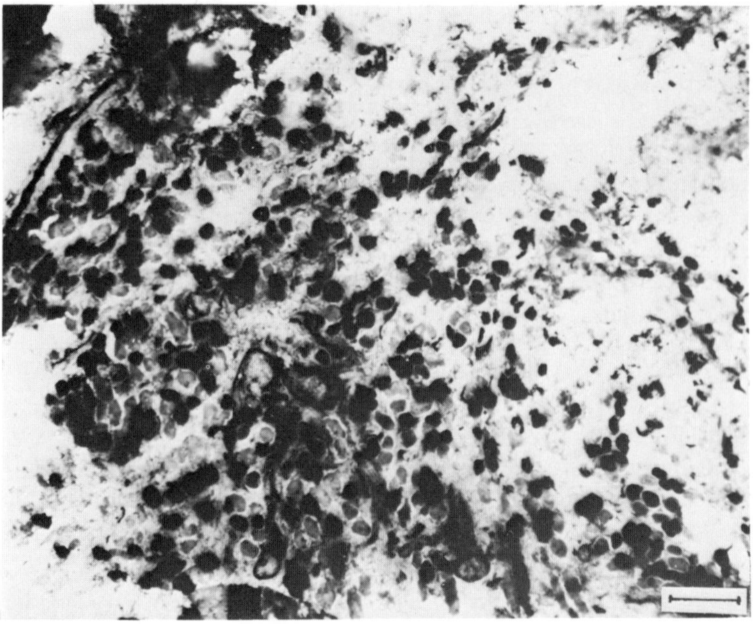

FIG. 8. Rabbit tensor tympani. Part of a section stained for myofibrillar ATPase after acid pre-incubation. Note large numbers of heavily staining fibres. Scale = 88 μm.

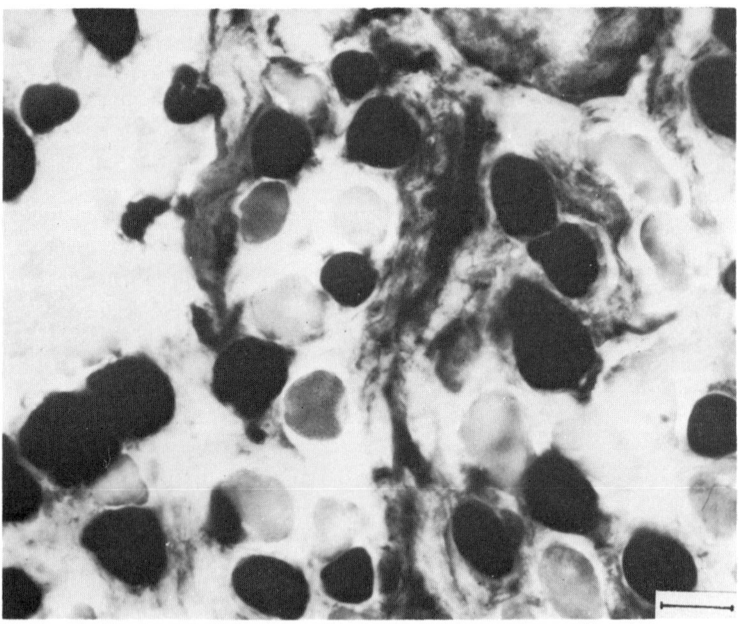

FIG. 9. Higher magnification of rabbit tensor tympani fibres to show three types of fibre with three staining intensities, dark, medium and light, for myofibrillar ATPase (acid pre-incubation). Scale = 22 μm.

myofibrillar ATPase (β) and NADH-diaphorase activity (Fig. 8). Some of these fibres were intermediate, their staining intensity reversing after alkaline pre-incubation. However, many of the target fibres showed no reversal of staining intensity. This "dual" ($\alpha\beta$) heavy staining characteristic requires further study. At present these fibres are classified as RI. A smaller number of fibres possessed only acid-labile myofibrillar ATPase and were classified as red fibres. The three types of fibres, distinguished by acid pre-incubation, can be seen in Fig. 9.

Stapedius. In the stapedius muscle acid pre-incubation revealed similar fibre types to those found in the tensor tympani. Once again intense staining for NADH-diaphorase was produced. No accurate data on fibre proportions are available, although the intermediate fibres did appear to be of a relatively large size compared with the type II fibres as in the tensor tympani.

Rousettus

Tensor tympani. Some 25% of the fibres of the tensor tympani muscle had low myofibrillar ATPase activity and were classified as intermediate. A complete range of oxidative enzyme activity from low to high was demonstrated. It was found to be evenly distributed within each fibre, indicating that those fibres with a high myofibrillar ATPase activity should be divided into white and WI fibres.

Stapedius. The percentage of intermediate fibres in the stapedius muscle was similar to that of the tensor tympani, but the number of white fibres was somewhat less (35%), and the number of WI fibres was almost double (42%) that of the latter muscle. In other words, the number of fibres with moderate or heavy staining for oxidative activity, was greater in the stapedius than the tensor tympani.

Summary. Slow twitch, intermediate fibres were found in quite large numbers in the middle ear muscles of the guinea-pig and *Rousettus*. This was especially so in the guinea-pig stapedius, and appeared to be so in the middle ear muscles of the rabbit. No such fibres were found in the middle ear muscles of the cotton rat; nor were there any fibres with acid-labile myofibrillar ATPase (α) activity only. Fibres with only β-myofibrillar ATPase (red and WR) were not found in this study except for a small number in the guinea-pig and rabbit middle ear muscles. Thus the majority of type II fibres were divided into types on the basis of intensity of oxidative staining, since they have $\alpha\beta$-ATPase staining characteristics (white, RI and WI). This was also done with the middle ear muscles of *Rousettus* where the pH lability of fibre myofibrillar ATPase was not studied.

Larger numbers of WI and RI fibres in the stapedius muscles relative to the number in the tensor tympani muscles of each animal reflect the greater oxidative enzyme activity.

Distribution of fibre types within the muscles

Some interesting and characteristic patterns of fibre type distribution within individual muscles were obtained. In general, intermediate fibres, when present, were quite evenly distributed throughout the cross-sections of both tensor tympani and stapedius muscles (e.g. guinea-pig, Fig. 10). Conversely, there was a general tendency for fibres low in oxidative enzyme activity and high in myofibrillar ATPase activity, to occupy the central region, while fibres high in oxidative activity were concentrated around the periphery of the muscles.

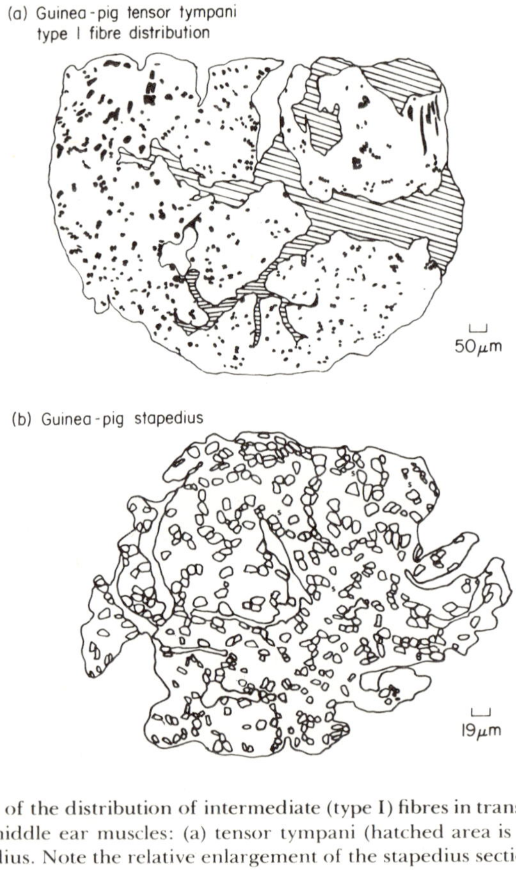

FIG. 10. Plots of the distribution of intermediate (type I) fibres in transverse sections of the guinea-pig middle ear muscles: (a) tensor tympani (hatched area is nerve and other tissue); (b) stapedius. Note the relative enlargement of the stapedius section.

For example, in the cotton rat tensor tympani a central core of large diameter white fibres is clearly evident. At the base of the muscle, the general tendency of the RI fibres to be concentrated

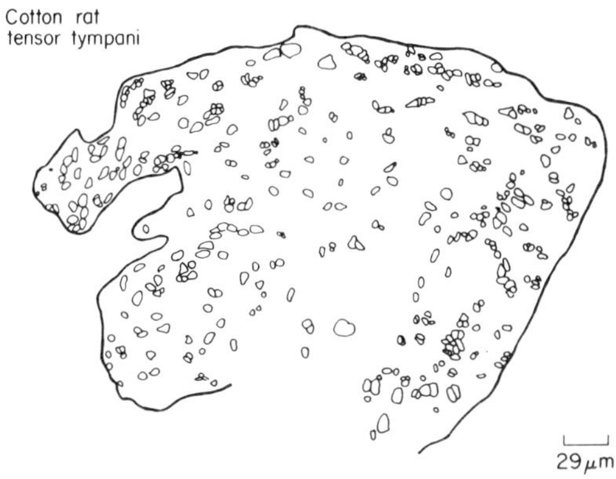

FIG. 11. Plot of the distribution of RI fibres in a section of the cotton rat tensor tympani. The section is taken towards the basal or insertion end of the muscle. Note the tendency of the fibres to be grouped near the periphery. Scale = 29 μm.

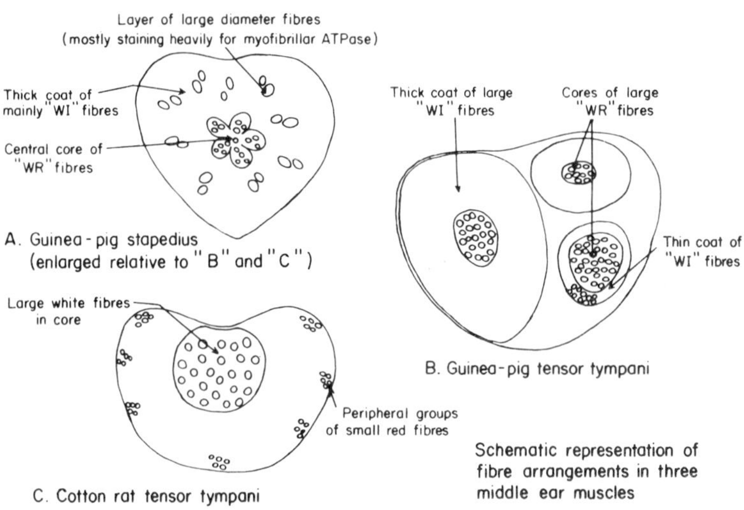

FIG. 12. Schematic representation of fibre arrangements in three middle ear muscles. A: guinea-pig stapedius; B: guinea-pig tensor tympani; C: cotton rat tensor tympani. Not to scale.

towards the periphery may be seen (Fig. 11). Proceeding towards the ossicular insertion of the muscle these fibres may be distinguished in characteristic small groups of six of seven fibres (Fig. 5). This is represented schematically in Fig. 12.

Similarly, the guinea-pig tensor tympani is also represented in Fig. 12. Here, there is more than one core, represented by three main groups of fibres. Each core is predominantly composed of WR fibres (relatively lower oxidative enzyme activity). One group has a relatively thin peripheral coat of WI fibres with moderate oxidative enzyme activity. However, the major part of the muscle has a thick peripheral band of WI fibres, many of the WI fibres having a similar diameter to those of the WR fibres in the muscle cores.

Finally, in the guinea-pig stapedius a central core of WR fibres is present, but it is quite small in comparison with a thick coat of WI fibres (Fig. 12). The largest diameter fibres are in the middle of this coat and the majority of them are WI fibres with a moderate content of alkali-stable myofibrillar ATPase (α) (Fig. 12).

Fibre diameters

The muscle fibres of both the tensor tympani and stapedius muscles are very small in comparison with limb muscle fibres (e.g. Fig. 3 and Table III). Fibre size is often related to the histochemical characteristics of the fibre. For example, the intermediate fibres in the guinea-pig tensor tympani are very small in comparison with the WI and WR fibres. Also, the RI fibres of the cotton rat tensor tympani are smaller than the majority of white fibres.

Even in the very small guinea-pig stapedius, those fibres with a low activity of myofibrillar ATPase have a smaller mean and range of diameter than the high ATPase activity fibres.

The distribution of fibre diameters for the whole muscle is plotted in Fig. 7 with the percentage of intermediate fibres also indicated.

It can be seen from this histogram that the general distribution shows a characteristic skew to the left towards smaller diameter fibres.

DISCUSSION

The histochemical properties of the middle ear muscles studied varied in certain respects in the different species. However, certain similarities were apparent, such as the higher intensity of staining for the oxidative enzyme in the stapedius muscles as compared with the tensor tympani muscles. Correlations can be made between the histochemical and contraction properties of limb muscle. In order to predict contraction characteristics from the results of this study, it must first be established that such correlations will also apply to the middle ear muscles, as part of the cranial musculature.

The number of type II fibres has been correlated with contraction time for cat jaw muscles (Bosley, Cody & Taylor, 1972) and intrinsic laryngeal muscles (Edström, Lindquist & Mårtensson, in press). A direct comparison between histochemical findings in middle ear muscles and contraction characteristics may be made for the cat. Teig (1972b) recorded the contraction of individual motor units myographically in response to threshold stimulation of their motor neurones in both the stapedius and tensor tympani. In the tensor tympani, slow and fast twitch types of motor units were distinguished. This correlates with Teig & Dahl's (1972) finding of two fibre populations on the basis of staining for myofibrillar ATPase (i.e. low and high enzyme activity) in the cat's tensor tympani. Although myofibrillar ATPase staining again distinguished two fibre populations in the stapedius, no slow twitch motor units were distinguished myographically. This might be explained by a bias in favour of stimulating faster units with the method of threshold stimulation used. Another possibility is that the slow units could be served by motor neurones in the lateral group of the facial nucleus, which Teig did not stimulate. Borg (1973) observed retrograde changes in this region of the facial nucleus of the rabbit after removal of the stapedius.

It is important to note that at present only general inferences concerning the speed of contraction can be made on the basis of histochemical results for myofibrillar ATPase alone. For example, among the facial muscles of the cat, the orbicularis oculi and the orbicularis oris have contraction times of 8·5 to 33 msec respectively (Lindquist, 1973), but the percentage of type II fibres in the two muscles does not differ very much (Edström & Lindquist, 1973). The staining intensity for myofibrillar ATPase should reflect the rate by which myosin may hydrolyse ATP, and it has been suggested that this rate determines the speed of intrinsic shortening of the contractile material. Dissimilarities in twitch contraction times between striated muscles then seem mainly dependent upon differences in the intrinsic speed of shortening (Close, 1965). However, variations in the time course of contraction in the active state and in the properties of the series-elastic elements must also be taken into account. Recently, Close & Luff (1974) have shown that the relation between intrinsic speed of shortening and duration of the active state of the contractile material is not the same for rat extraocular muscles and hind limb muscles.

Even so, the discovery of slow twitch intermediate fibres in the tensor tympani of the rabbit correlates with the finding of a slow component in contraction studies of this muscle (Teig, 1972a). A similar slow component is indicated in the middle ear muscles of *Rousettus* and the stapedius muscle of the guinea-pig. The percentage of cross-sectional area represented by the intermediate fibres in the guinea-pig tensor tympani is relatively low and would therefore indicate a very small slow component. Conversely, all the middle ear muscles studied have a large percentage of fibres staining intensely for myofibrillar ATPase, indicating that as a group they are fast muscles. This must be especially so for the cotton rat middle ear muscles.

The relatively higher intensity of oxidative enzyme staining in the stapedius muscles in comparison with the tensor tympani muscles suggests a more tonic or repetitive activity and a greater resistance to fatigue in the former. In the guinea-pig stapedius muscle, the substantial intermediate fibre population and the heavy oxidative staining of the majority of other fibres, suggests both tonic and repetitive activity, while, in the cotton rat stapedius muscle, fast repetitive fibre contractions would seem to be clearly indicated.

Although muscular fatigue is related to exhaustion of energy stores at low frequencies of activation (Edström & Lindquist, 1973), a significant initial effect at high frequencies of activation (above 20–30 Hz) may be caused by failing neuromuscular transmission (Del Pozo, 1942). In this respect, on the basis of Teig's motor unit contraction studies in the cat middle ear muscles, a maximal firing frequency of more than 50 Hz is to be expected for the fast motor units. It is possible that the very high ratio of area of motor endplate to muscle fibre volume in the middle ear muscles (Mayr & Mayr, personal communication) also relates to a degree of resistance to neuromuscular fatigue. The ratio is always higher in the stapedius than in the tensor tympani, and is greatest in certain fibres of the cat stapedius and in the fibres of the guinea-pig stapedius muscle. On the other hand white muscle fibres fatigue rapidly, yet characteristically have large and complex neuromuscular junctions.

The distribution of the different types of twitch fibres of individual middle ear muscles is also of interest. The arrangement of a central core of large fibres with a peripheral coat of small fibres, as in the cotton rat tensor tympani, is very similar to that of some extraocular muscles, where finely controlled and finely graded contractions are performed. In the middle ear muscles the core fibres seem to be lower in oxidative enzyme activity but high in myofibrillar ATPase activity (α), suggesting a strong, phasic, but rapidly fatiguing, contraction. Conversely, the higher oxidative enzyme activity in the muscle coat suggests more maintained or repetitive contractions. One possible mode of action for this arrangement in the tensor tympani may be that the central core fibres contract first, rapidly and strongly, to pull the malleus into position and tauten the tympanic membrane. Then the weaker but less phasic outer fibres may contract to maintain muscle tension and ossicle position as the contractile force of the white fibres weakens. At other times, perhaps both groups contract separately to affect the ossicular chain in different ways. The relatively even distribution of intermediate fibres, when present, suggests that they participate in all contractions. A clearer understanding of these fibre arrangements may prove useful for anticipating the effects of middle ear muscle contraction in different animals.

The histochemical evidence suggests that focally innervated twitch fibres are the structural basis for the tonic motor units in middle ear muscles. The failure to find multiterminal motor innervation on extrafusal fibres with the silver impregnation technique is in accordance with the findings of Barker (1967) and of Mayr & Mayr (personal communication). The latter authors stained middle ear muscle preparations for cholinesterase, but failed to find any multiterminally innervated fibres despite their presence in extraocular muscle used as a control tissue. Cholinesterase positive sites in myomyous junctions were observed and these may bear some relation to the "elongated" nerve terminals reported by Fernand & Hess (1969). Alternatively, these latter endings may be concerned with a proprioceptive efferent innervation. Simple spindle-like structures have been occasionally observed in this study. The low myofibrillar ATPase fibres observed in the cotton rat stapedius may perhaps be involved in proprioception.

In conclusion, the middle ear muscles studied are heterogeneous in fibre composition, and their histochemical properties differ from species to species. It is interesting that even in such small muscles the histochemical fibre types are still related to fibre size. The tendency in distribution towards smaller diameter fibres in the guinea-pig stapedius muscle (Fig. 7) is comparable to the preponderance of motor units producing smaller tensions found by Teig (1972b) in cat middle ear muscles. This simply reflects the fact that the middle ear muscles have a preponderance of small weak motor units enabling a very fine gradation of contraction. The relative distribution of different fibre types becomes particularly significant when two points are considered. First, small tension changes in the middle ear muscles of the cat have been shown to be highly frequency selective in their effect on the transmission of sound through the middle ear (Teig, 1973), weak contractions being more selective than strong contractions. Second, variations in the number and type of motor units (high/low threshold, phasic/tonic contraction) can be expected to affect profoundly the speed of response and stability of a feedback system such as the middle ear muscles (Borg, 1973). The variation in fibre composition of the tensor tympani and stapedius of *Rousettus, Sigmodon, Cavia* and *Oryctolagus* reflects differences in the dynamic characteristics of contraction of these muscles. This in turn suggests that the requirements for optimal middle ear muscle contraction are different in different animals. The trend towards larger numbers of fibres with a high intensity of staining of oxidative enzyme activity, particularly in the stapedius muscles, indicates a tendency for repetitive or maintained contractions. Frequent or tonic activity of the middle ear muscles is a prerequisite for the performance of many of the perceptual functions proposed for these muscles by Simmons (1964).

Acknowledgements

The author would like to thank Mr D. Ellames for his valuable assistance in this work. The author is also indebted to Dr R. Hinchcliffe and Dr A. Pye for their advice and encouragement.

The help of Miss S. Damps, Mrs J. Anderson, Mrs M. Wheatley and Mrs M. Bramwell in the preparation of the manuscript, and of Mr J. D. Connolly and Mr D. Froude in the preparation of the illustrations, is gratefully acknowledged.

References

Barker, D. (1967). Comment. In CIBA Foundation Symposium *Myotatic, kinesthetic, and vestibular mechanisms*: 38–39. De Reuck. A. V. S., & Knight, J. (eds). London: Churchill Ltd.

Barker, D. & Ip, M. C. (1963). A silver method for demonstrating the innervation of mammalian muscle in teased preparations. *J. Physiol., Lond.* **169**: 73–74P.

Barnard, R. J., Edgerton, V. R., Furukawa, T. & Peter, J. B. (1971). Histochemical, biochemical and contractile properties of red, white and intermediate fibres. *Am. J. Physiol.* **220**: 410–414.

Barnard, R. J., Edgerton, V. R. & Peter, J. B. (1970). Effect of exercise on skeletal muscle. I. Biochemical and histochemical properties. *J. appl. Physiol.* **28**: 762–766.

Borg, E. (1972a). Acoustic middle ear reflexes: a sensory control system. *Acta otolaryngol.* Supplement **304**.

Borg, E. (1972b). On the neuronal organisation of the acoustic middle ear reflex. A physiological and anatomical study. *Brain Res.* **49**:101–123.

Borg, E. (1973). Nonlinear dynamic properties of a somatomotor reflex system. A model study. *Acta physiol. Scand.* **87**: 15–26.

Bosley, M. A., Cody, F. W. J. & Taylor, A. (1972). The correlation of histochemistry and speed of contraction in cat jaw muscles. *J. Physiol., Lond.* **224**: 92P.

Candiollo, L. (1965). Osservazioni e considerazioni sulle miofibrille a decorso atipico nel m.tensore del timpano di Ratto (*Epimys norv.*, var., *albina*, Erxl.). *Atti Accad. Sci. Torino.* **99**: 517–541.

Close, R. I. (1965). The relation between intrinsic speed of shortening and duration of the active state of muscle. *J. Physiol., Lond.* **180**: 542–559.

Close, R. I. (1972). Dynamic properties of mammalian skeletal muscles. *Physiol. Rev.* **52**: 129–197.

Close, R. I. & Luff, A. R. (1974). Dynamic properties of inferior rectus muscle of the rat. *J. Physiol., Lond.* **236**: 259–270.

Dawson, D. M. & Romanul, F. C. A. (1964). Enzymes in muscle. II. Histochemical and quantitative studies. *Arch. Neurol.* **11**: 369–378.

Del Pozo, E. C. (1942). Transmission fatigue and contraction fatigue *Am. J. Physiol.* **135**: 763–771.

Dubowitz, V. & Pearse, A. G. E. (1960). A comparative histochemical study of oxidative enzyme and phosphorylase activity in skeletal muscle. *Histochemie* **2**: 105–117.

Edgerton, V. R. & Simpson, D. R. (1971). Dynamic and metabolic relationships in the rat extensor digitorum longus muscle. *Expl Neurol.* **30**: 374–376.

Edström, L. & Kugelberg, E. (1968a). Properties of motor units in the rat anterior tibial muscle. *Acta physiol. Scand.* **73**: 543–544.

Edström, L. & Kugelberg, E. (1968b). Histochemical composition, distribution of fibres and fatiguability of single motor units. Anterior tibial muscle of the rat. *J. Neurol, Neurosurg. Psychiat.* **31**: 424–433.

Edström, L. & Lindquist, C. (1973). Histochemical fibre composition of some facial muscles in the cat in relation to their contraction properties. *Acta physiol. Scand.* **89**: 491–503.

Edström, L., Lindquist, C. & Mårtensson, A. (in press). Correlation between functional and histochemical properties of the intrinsic laryngeal muscles in the cat. *International Symposium on ventilatory and phonatory control systems.* Wyke, B. (ed.). London: Oxford University Press.

Engel, W. K. (1962). The essentiality of histo- and cytochemical studies of skeletal muscle in the investigation of neuromuscular disease. *Neurology* **12**: 778–784.

Erulkar, S. D., Shelanski, M. L., Whitsel, B. L. & Ogle, P. (1964). Studies of muscle fibres of the tensor tympani of the cat. *Anat. Rec.* **149**: 279–298.

Fernand, V. S. V. & Hess, A. (1969). The occurrence, structure and innervation of slow and twitch muscle fibres in the tensor tympani and stapedius of the cat. *J. Physiol., Lond.* **200**: 547–554.

Guth, L. (1968). "Trophic" influences of nerve on muscle. *Physiol. Rev.* **48**: 645–687.

Guth, L. & Samaha, F. J. (1969). Qualitative differences between actomyosin ATPase of slow and fast mammalian muscle. *Expl Neurol.* **25**: 138–152.

Guth, L., Samaha, F. J. & Albers, R. W. (1970). The neural regulation of some phenotypic differences between the fibre types of mammalian skeletal muscle. *Expl Neurol.* **26**: 126–135.

Hinchcliffe, R. & Pye, A. (1969). Variations in the middle ear of the Mammalia. *J. Zool., Lond.* **157**: 277–288.

Jöbsis, A. C. & Meijer, A. E. F. H. (1973). Evaluation of enzyme histochemical observations for metabolic studies. I. *Histochemie* **36**: 51–61.

Kobayashi, M. (1956a). The comparative anatomical study of the stapedial muscles of the various kinds of mammalian animals. *Hiroshima J. Med. Sci.* **5**: 63–84.

Kobayashi, M. (1956b). Comparative anatomical studies of the morphology of the tensor tympani muscles of various mammals. *Hiroshima J. Med. Sci.* **5**: 85–108.

Kugelberg, E. & Edström, L. (1968). Differential histochemical effects of muscle contractions on phosphorylase and glycogen in various types of fibres: relation to fatigue. *J. Neurol. Neurosurg. Psychiat.* **31**: 415–423.

Lindquist, C. (1973). Contraction properties of cat facial muscles. *Acta physiol. Scand.* **89**: 482–490.

Malan, E. (1934). Etude d'histologie comparée sur quelques modifications particulières des fibres du tensor tympani dues a la senescence. *Arch. Biol. (Liège)* **45**: 355–375.

Novikoff, A. B., Shin, W. & Drucker, J. (1961). Mitochondrial localisation of oxidative enzymes: staining results with two tetrazolium salts. *J. Biophys. Biochem. Cytol.* **9**: 47–61.

Okamoto, M., Sato, M. & Kirikae, I. (1954). Studies of the acoustic reflex. Part II. Experimental studies on the function of the tensor tympani muscle. *Ann. Otol.* **63**: 950–959.

Padykula, H. A. & Gauthier, G. F. (1967). Morphological and cytochemical characteristics of fibre types in normal mammalian skeletal muscle. In *Exploratory concepts in muscular dystrophy and related disorders*: 117–128. Milhorat, A. T. (ed.). Amsterdam: Excerpta Medical Foundation.

Padykula, H. A. & Herman, E. (1955). The specificity of the histochemical method for adenosine triphosphate. *J. Histochem. Cytochem.* **3**: 170–195.

Pette, D. & Bücher, T. (1963). Proportionskonstante Gruppen in Beziehung zur Differenzierung der Enzymactivit ätsmuster von Skelettmuskeln des Kaninchens. *Hoppe-Seylers Z. Physiol. Chem.* **331**: 180–195.

Romanul, F. C. A. (1964). Enzymes in muscle I. Histochemical studies of enzymes in individual muscle fibres. *Arch. Neurol.* **11**: 355–369.

Romanul, F. C. A. & Van Der Meulen, J. P. (1967). Slow and fast muscles after cross innervation. Enzymatic and physiological changes. *Arch. Neurol.* **17**: 387–402.

Samaha, F. J., Guth, L. & Albers, R. W. (1970). Phenotypic differences between the actomyosin ATPase of the three fibre types of mammalian skeletal muscle. *Expl Neurol.* **26**: 120–125.

Simmons, F. B. (1964). Perceptual theories of middle ear muscle function. *Ann. Otol. Rhinol. Laryngol.* **73**: 724–729.

Stein, J. M. & Padykula, H. A. (1962). Histochemical classification of individual skeletal muscle fibres of the rat. *Am. J. Anat.* **110**: 103–124.

Teig, E. (1972a). Force and contraction velocity of the middle ear muscles in the cat and the rabbit. *Acta physiol. Scand.* **84**: 1–10.

Teig, E. (1972b). Tension and contraction time of motor units of the middle ear muscles in the cat. *Acta physiol. Scand.* **84**: 11–21.

Teig, E. (1973). Differential effect of graded contraction of middle ear muscles on the sound transmission of the ear. *Acta physiol. Scand.* **88**: 382–391.

Teig, E. & Dahl, H. A. (1972). Actomyosin ATPase activity of middle ear muscles in the cat. *Histochemie* **29**: 1–7.

Yellin, H. (1967). Neural regulation of enzymes in muscle fibres of red and white muscle. *Expl Neurol.* **19**: 92–103.

Yellin, H. & Guth, L. (1970). The histochemical classification of muscle fibres. *Expl Neurol.* **26**: 424–432.

Zíka, K., Lojda, Z. & Kuĉera, M. (1973). Activities of some oxidative and hydrolytic enzymes in musculus biceps brachii of rats after tonic stress. *Histochemie* **35**: 153–164.

Symp. zool. Soc. Lond. (1975) No. 37, 97–105.

COMPARISON OF HISTOLOGICAL CHANGES IN THE COCHLEA OF THE GUINEA-PIG AND THE BAT *ROUSETTUS* AFTER ACOUSTIC TRAUMA

ADE PYE

Institute of Laryngology and Otology, London, England

SYNOPSIS

Damage to the sensory hair cells of the cochlea has been produced in two species of mammals after exposure to intense (120–126 dB SPL) pure tones of frequencies between 2 kHz and 40 kHz. Assessment of damage was carried out by using the microdissection and surface specimen technique.

A restricted area of damage was induced in the guinea-pig *Cavia* after an exposure duration of 30 min at frequencies from 2 kHz to 20 kHz, but for the highest frequencies 7 h of exposure were required to produce any damage. The damage produced was never greater in extent than one turn of the four and a quarter turns of the cochlea.

For *Rousettus*, the only echolocating megachiroptean fruit bat, the frequencies used varied from 4 kHz to 20 kHz. Minimal damage occured in some specimens at 4 kHz after long exposure durations, but none was obtained at 20 kHz, even after 9 h of exposure. However, they appear to be very sensitive to frequencies of about 10–12 kHz, where even a 3-min exposure may cause complete destruction to the sensory hair cells of the whole of the two turns of the cochlea.

The results are discussed in relation to the hearing sensitivity and the importance of sound to the two species of mammals studied.

INTRODUCTION

Exposure of mammalian ears to high intensity sound can cause temporary or permanent damage to the sensory hair cells of the cochlea. This can be a serious hazard to human beings working in noisy environments (e.g. factories, airfields, artillery ranges and demolition sites). To study this damaging effect experimentally, animals other than man have to be used, especially for demonstrating permanent damage.

The study of experimental trauma was begun more than two decades ago, for example by Davis *et al.* (1953), Eldredge & Covell (1958) and Beagley (1965), but most of these workers relied on conventional histological serial sections for assessment of damage. However, in 1966, Engström, Ades & Andersson developed a more accurate, more reproducible and quicker method, which was first used by Retzius as long ago as 1884.

Acoustic trauma studies using pure tones have been carried out by Stockwell, Ades & Engström (1969) and other workers, although in the case of the guinea-pig most research workers have concentrated on lower frequencies, usually only up to 4 kHz. The present study used frequencies from 2 kHz to 40 kHz (Pye, 1971, 1973) as it has been ascertained that cochlear microphonic responses have been obtained for such high frequencies in the guinea-pig (Pestalozza & Davis, 1956).

It is well known that lower frequencies are analysed in the cochlea at the apical part and higher frequencies towards the base, although not on a linear scale. Higher frequencies tend to be compressed towards the base, but this varies a great deal in the different mammalian species, depending on the frequency ranges of hearing.

MATERIAL

Two species of mammals were used for this study: the laboratory guinea-pig *Cavia porcellus* and the fruit bat *Rousettus aegyptiacus* from East Africa. The latter produces echolocation cries by clicking its tongue. The frequency components of the pulse vary from 4 kHz to 70 kHz, a significant part being in the audible range (Pye, J. D., personal communication). This echolocation system enables this fruit bat to live in deep caves. Pulses are produced mainly in darkness, and the bats' vision is considered to be very good (Suthers, 1970). Early acoustic trauma experiments on this species were begun at the Nuffield Unit of Tropical Animal Ecology, Queen Elizabeth Park, Uganda, but the rest of the animals were kindly collected and sent to London by Dr S. K. Eltringham.

The ages of the guinea-pigs at the beginning of the experiments varied between six and eight weeks. For the *Rousettus* specimens the ages ranged down to six months but many were older than this.

METHODS

Apparatus

The pure tones were produced by a sonic-ultrasonic generator (built at the Institute of Laryngology and Otology) and an Ionophone 601 loudspeaker. A small 2·5 cm square horn replaced the large exponential horn of the original Ionophone, so as to reproduce the higher frequencies at high local intensities. The intensities varied between 120–126 dB SPL, depending on the frequencies used, and were measured by a Bruel and Kjaer 6·4 mm microphone type 4135 and a Bruel and Kjaer type 2604 microphone amplifier reading in dB SPL. The experiments were performed in a silent room.

Procedure

Only one ear of the experimental animal was exposed, the other being used as a control. The loudspeaker was placed 1 cm from the left ear in all experiments.

The guinea-pigs were restricted in a crush-cage and held firmly in place by foam rubber pads, so that no movement of the head was possible. The bats had to be anaesthetized with pentabarbitone sodium and kept warm during anaesthesia and exposure.

Assessment of damage

The damage obtained was assessed by the microdissection and surface specimen technique of Engström *et al.* (1966). The animals were sacrificed 3–4 weeks after exposure, so that destruction of the sensory cells would be well advanced. The bullae were opened and partially dissected, and a number of small holes were made in the bony wall of the cochleae, which were then fixed in a 1·3% buffered solution of osmic acid for 2–3 h. They were thoroughly washed in water and dehydrated up to 70% alcohol, the processing taking place in a refrigerator. Each cochlea was then dissected from the apex downwards, a careful note being kept of each piece and its order of removal. The pieces were mounted in glycerine and observed under phase contrast microscopy. This shows the three rows of outer sensory cells in surface view, which are arranged in a regular, rectangular pattern, so that any damaged cells can immediately be recognized.

Since previous results had shown that the extent of damage was never less than one-eighth of *any* turn, it was considered sufficient to note the damage by the divisions of the turns, rather than by counting every sensory cell.

RESULTS

General

On either side of the main area of damage there occurred a small area of partial damage. If more than every third cell was found to be destroyed in any row, this was considered to represent the beginning of the main area of damage. Usually, in a severely damaged area, all the outer three rows of sensory cells would be destroyed. In partially damaged areas, the innermost row of outer sensory cells would be destroyed first, followed by the middle row and then the outermost row. No damage was ever observed in guinea-pigs to the inner row of sensory cells, although this occurred in *Rousettus.*

Caviá

The guinea-pig has four and a quarter turns to the cochlea and the height from the base to the apex is 5·5 mm. At first, longer exposure durations were used (Pye, 1971) but later it was discovered that except for the highest frequencies (25 and 40 kHz) 30-min exposures to sounds at intensities of 120–126 dB SPL were sufficient always to cause damage. This gave a restricted area of damage, which was desirable for mapping out the histological damage caused by the different frequencies along the spiral organ.

Six animals were used for each series (Fig. 1), except for exposures at 40 kHz, where only five animals were used. For 2 kHz the median area of damage falls at about one and three-quarter turns from the base. The location of damage was most variable at 4 kHz, ranging from a point three-quarters of a turn from the base to nearly two turns from the base, but for any animal the area of damage did not exceed three-quarters of a turn. These results are comparable with those of Stockwell *et al.* (1969) who found the major area of damage for 4 kHz from half a turn from the base to the second turn, although minor damage extended further. This may have been due to the fact that these workers used higher intensities (130–150 dB SPL) and longer exposure durations (1–4 h). They also obtained damage to the inner row of sensory cells, which never occurred in the present experiments.

As can be seen in Fig. 1 the areas of damage in individual animals vary not only in location along the spiral organ, but also in extent, indicating that some animals are more sensitive to sound damage than others.

The extent of damage generally decreases towards the basal part of the cochlea and reaches the base at frequencies of 12 kHz and above. It is interesting to note that no damage was obtained in two animals at 25 kHz and only one animal showed damage at 40 kHz.

Rousettus

The cochlea of *Rousettus* has only two complete turns and the height from the base to the apex is 2·8 mm. This small size caused

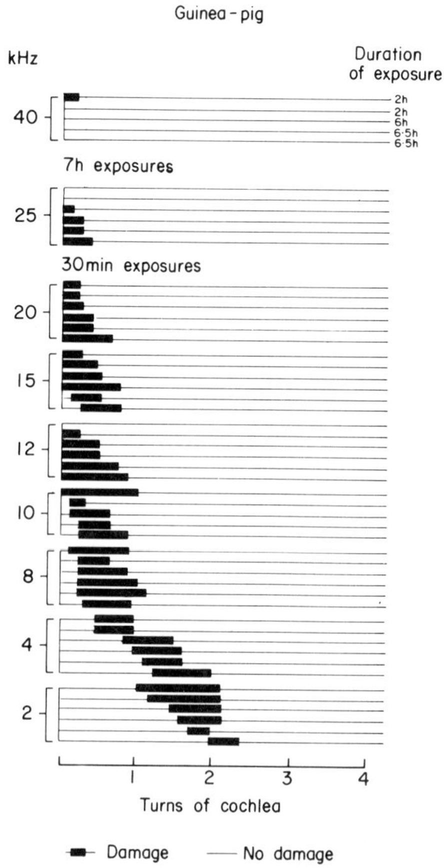

FIG. 1. Showing the areas of damage in the cochlea of the guinea-pig after exposure to intense pure tones of frequencies ranging from 2 kHz to 40 kHz.

some difficulties in microdissection, but these were overcome with patience and practice.

The number of animals used for each frequency varied because of the difficulty of obtaining adequate numbers. The preliminary experiments were conducted in Uganda and because of shortage of time, only two frequencies were chosen: 4 kHz and 20 kHz. Even after 9 h of exposure, no damage could be obtained at first, but later two individuals showed damage at 4 kHz (Pye, 1971).

Experiments were continued in London, where the animals were in short supply. Because of this and the results from the preliminary experiments, long exposure durations were employed for the other frequencies: 8, 10, 12 and 15 kHz. It was astounding to find total destruction of all the cochlear sensory cells, generally after only 2 h of exposure to sound. The inner row of sensory cells was also destroyed. Exposure durations were then reduced and it was even more surprising to find that at 10 kHz even a 3-min exposure could destroy the sensory cells of the whole cochlea (Fig. 2). By this time cochlear microphonic responses had been obtained by Pye (personal communication) and later by Brown (1973), who showed that a peak of sensitivity existed between 10 and 12 kHz. The cochlea was found to be more sensitive to histologically assessed damage at 10 kHz than at 12 kHz, since the exposure time at 12 kHz needed to be at least 30 min to produce damage. The cochlea was found to be equally sensitive to 8 kHz and 15 kHz. No damage occurred at 20 kHz.

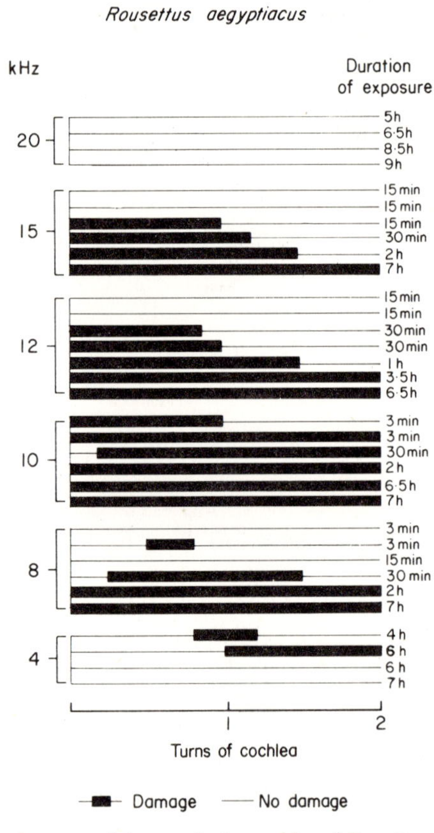

FIG. 2. Showing the areas of damage in the cochlea of *Rousettus aegyptiacus* after different durations of exposure to intense pure tones of frequencies ranging from 4 kHz to 20 kHz.

Brown (1973) performed cochlear microphonic experiments on some of the exposed animals before they were sacrificed and confirmed that no physiological responses could be obtained at any of the frequencies which were found to have caused total histological damage to the cochlea.

Grinnell & Hagiwara (1972), in recording collicular responses from *R. amplexicaudatus*, found a broad peak at 50 kHz. Unfortunately, at this time no further live specimens of *R. aegyptiacus* were available to try exposure experiments at frequencies between 40 and 50 kHz, although Brown (1973) failed to find a sensitive peak in the cochlear microphonic responses at these higher frequencies in the latter species.

No damage was ever obtained in the control ears of any of the experimental animals, all of which received the same assessment as the exposed ears. This may be accounted for by the fact that high frequencies produce pronounced sound shadows and in these experiments the exposed ear was very close to the loudspeaker.

DISCUSSION

The guinea-pig *C. porcellus* and the bat *R. aegyptiacus* show different patterns of histological damage to the cochlea after exposure to intense pure tones at similar intensities and frequencies.

In the case of the guinea-pig, whereas there are individual variations, the extent and location of damage along the spiral organ shows a definite trend from the lowest frequency used at about the second turn, to the higher frequencies located towards the base of the cochlea. The sensitivity appears to be about the same from 2 kHz to nearly 20 kHz, but decreases above this frequency, i.e. longer exposure durations are required to produce damage. One must not forget, however, that the frequency range of hearing in the guinea-pig starts at about 125 Hz, the lower frequencies having already been investigated by Stockwell *et al.*

In *Rousettus* a different picture emerges. At 4 kHz only two of the four specimens showed damage. At 8 kHz, one specimen showed damage after a 3-min exposure, but otherwise 30 min were required to produce damage. Ten kHz was found to be the most effective frequency, partial or even complete damage to the whole spiral organ occurring after only a 3-min exposure. The sensitivity began to fall off at 15 kHz and no damage was obtained at 20 kHz.

CONCLUSIONS AND SUMMARY

For the guinea-pig the greatest extent of damage for 30-min sound exposures was never more than one complete turn of the cochlea, the smallest extent being one-eighth of a turn. For *Rousettus* complete damage was caused to the whole cochlea at 10 kHz and to half of the cochlea at 8, 12 and 15 kHz for the same duration of exposure.

Cochlear microphonic responses can be obtained from the guinea-pig from frequencies of 125 Hz to above 40 kHz, although histological damage after acoustic trauma at the latter frequency has been difficult to demonstrate. The main sensitivity lies in the audible range of frequencies, although the social cries of guinea-pigs almost certainly contain ultrasonic components. As far as is known, no purely ultrasonic communication takes place among guinea-pigs of any ages. This is in contrast to many other rodents, which often communicate by pure ultrasound.

For *Rousettus* a peak of sensitivity occurs at about 10–12 kHz and this is not surprising, since it coincides not only with the cochlear microphonic results but also with a significant audible component of its echolocation pulse. These pulses are of great importance to *Rousettus* in enabling it to find its way in the dark. Because of this ability it is the only fruit bat able to live in deep caves. The most surprising fact is that at certain frequencies, an exposure of only 3 min can cause complete damage to all sensory cells of the whole cochlea.

Unfortunately, most other bats are difficult to obtain and keep as experimental animals, but it is hoped in the future to investigate some echolocating microchiropteran bats, which should also yield interesting results in sound exposure experiments.

ACKNOWLEDGEMENTS

Thanks are due to Dr S. K. Eltringham for the capture and sending of live *Rousettus* to London and Mr D. J. Connolly of the Department of Clinical Photography, Institute of Laryngology and Otology, for the preparation of the illustrations. This research was sponsored by the Wellcome Trust.

REFERENCES

Beagley, H. A. (1965). Acoustic trauma in the guinea-pig. 1. Electrophysiology and histology. *Acta otolar.* **60**: 437–451.

Brown, A. M. (1973). An investigation of the cochlear microphonic response of two species of echolocating bats: *Rousettus aegyptiacus* (Geoffroy) and *Pipistrellus pipistrellus* (Schreber). *J. comp. Physiol.* **83**: 407–413.

Davis, H., Benson, R. W., Covell, W. P., Fernandez, C., Goldstein, R., Katsuki, Y., Legouix, J. P., McAulisse, D. R. & Tasaki, I. (1953). Acoustic trauma in the guinea-pig. *J. acoust. Soc. Am.* **25**: 1180–1189.

Eldredge, D. H. & Covell, W. P. (1958). A laboratory method for the study of acoustic trauma in the guinea-pig. *Laryngoscope* **68**: 465–477.

Engström, H., Ades, H. W. & Andersson, A. (1966). *The structural pattern of the organ of Corti.* Stockholm: Almqvist and Wiksell.

Grinnell, A. D. & Hagiwara, S. (1972). Studies of auditory neurophysiology in non-echolocating bats, and adaptations for echolocation in one genus *Rousettus. Z. vergl. Physiol.* **76**: 82–96.

Pestalozza, G. & Davis, H. (1956). Electric responses of the guinea-pig ear to high audio frequencies. *Am. J. Physiol.* **185**: 595–600.

Pye, A. (1971). Effect of exposure to intense pure tones on the hearing organ of animals. *Revta Acustica* **2**: 199–203.

Pye, A. (1973). The destructive effect of intense pure tones on the cochleae of mammals. In *Disorders of auditory function*: 89–96. Taylor, W. (ed.). London and New York: Academic Press.

Retzius, G. (1884). *Das Gehörorgan der Wirbelthiere. Das Gehörorgan der Reptilien, der Vögel und der Säugethiere.* Stockholm: Die Centraldruckerei.

Stockwell, C. W., Ades, H. W. & Engström, H. (1969). Patterns of hair cell damage after intense auditory stimulation. *Ann. Otol. Rhinol. Laryngol.* **78**: 1144–1169.

Suthers, R. A. (1970). Vision, olfaction, taste. In *The biology of bats.* **2**: 265–318. Wimsatt, W. A. (ed.). New York and London: Academic Press.

Symp. zool. Soc. Lond. (1975) No. 37, 107–120.

RELATIONSHIPS BETWEEN NOISE-INDUCED THRESHOLD SHIFTS, MORPHOLOGICAL CHANGE AND SOCIAL HANDICAP

R. R. A. COLES

Operational Acoustics and Audiology Group, Institute of Sound and Vibration Research, University of Southampton, Southampton, England

SYNOPSIS

Noise exposure causes both temporary (TTS) and permanent (PTS) threshold shifts, and it is often associated with tinnitus, discrimination loss and other phenomena. After an initial period of about two minutes, both growth and recovery of TTS are proportional to log time, depending on the intensity of noise and amount of TTS_2 respectively. But if TTS_2 is above about 40 dB or acquired slowly from an intermittent noise, recovery may be very slow or even incomplete.

An apparent equality (but not necessarily a causal relationship) between $TTS_{2\ min}$ and risk of $PTS_{10\ years}$ from a given noise has been used as an experimental tool for derivation of damage risk criteria (DRC). Likewise, an equal-energy relationship between noise exposure and PTS has been postulated, but it would seem that neither model fits all circumstances, particularly with intermittent or prolonged noise exposures.

The relationships between cochlear hair cell damage and threshold shift have been studied, in human beings pre- and post-mortem and in animal experimentation. Again, there are wide variations; quite extensive outer hair cell damage may be accompanied by no significant PTS, and absence of inner hair cell damage has been found in presence of 80 dB of PTS. This leads to speculation as to the function of the hair cells, and to anxiety in connection with the ethics of TTS experimentation.

Whilst there is a general correlation between hearing loss and auditory handicap, there are again wide individual variations. Correlation coefficients between handicap questionnaire responses and audiometric measurements in the order of 0·6–0·7 have been reported, and it seems likely that neither are very good measurements of actual social handicap. This depends on many other factors including adaptation, amelioration by lip reading and hearing aid, interaction with environment and other handicaps, all of which are themselves influenced by attitude and intelligence.

TEMPORARY AND PERMANENT THRESHOLD SHIFTS

Temporary symptoms such as dullness of hearing, tinnitus and distortion are often experienced after noise exposure. All or most of these disappear quite soon afterwards, as does the threshold elevation which can be measured with an audiometer, the so-called temporary threshold shift (TTS). Noise exposure sufficiently great to cause a TTS will cause a permanent threshold shift (PTS) if repeated many times or if, on a single occasion, it is severe enough. The relationship between TTS and PTS is not clear, but it is of considerable importance for both our fundamental understanding of noise-induced hearing loss and for the practical applications to which we put TTS measurements.

Growth and recovery of threshold shifts

The amount of TTS resulting from a noise exposure depends both on the degree to which the sound pressure level (SPL) of the noise exceeds a critical level and on the logarithm of the exposure time (Ward, Glorig & Sklar, 1958). PTS behaves similarly with respect to the sound pressure level (Glorig, Ward & Nixon, 1961) and the duration of exposure (Taylor, Pearson, Mair & Burns, 1965). Recovery from TTS is more complex but, over much of its time course, it follows somewhat similar laws. Hirsh & Ward (1952) describe an initial R-1 phase of recovery, which occupies the first two minutes after cessation of noise exposure. During this phase recovery curves of TTS against time show an irregular shape, which they refer to as a "bounce" effect. It then stabilizes and follows a more defined course of recovery (R-2 phase); hence, TTS_2, the TTS measured 2 min after cessation of noise exposure, has come to be the conventional measure of TTS. In the R-2 phase, provided the TTS_2 does not exceed about 40 dB, its recovery is linear with the logarithm of recovery time, and the rate of recovery is proportional to the amount of TTS_2 (Ward *et al.*, 1958). However, for higher values of TTS_2 the R-2 phase proceeds at a much slower rate during the first few hours, though still as a function of the logarithm of time. Later, a more rapid, R-3 phase, recovery process sets in which is linear in time (Ward, 1960).

Applications of TTS data

$TTS_{2m} \simeq PTS_{10y}$

The series of TTS studies by Ward and his colleagues since 1958 have received widespread acceptance, but their practical importance and application depended on the demonstration (Glorig *et al.*, 1961) of a remarkable degree of agreement between the magnitude of PTS at 4 kHz resulting from 10 years of habitual exposure to certain industrial noises and the amount of TTS_2 calculated from the equations put forward by Ward and his colleagues. The considerable body of evidence in support of this apparent quantitative relationship between TTS_2 and PTS_{10} was considered carefully by the members of the CHABA (Committee on Hearing, Bioacoustics and Biomechanics, of the U.S. National Academy of Science—National Research Council) Working Group

46 and accepted as the basis for drawing up damage risk criteria (Kryter, Ward, Miller & Eldredge, 1966). It has also been used as the basis for defining damage risk criteria for impulse noise (Coles, Garinther, Hodge & Rice, 1968) and has been accepted for this purpose by CHABA Working Group 57.

Possible mechanisms of growth of PTS

It is interesting to speculate on the possible relationships between TTS and PTS. It seems quite likely, but uncertain, that there are several mechanisms of TTS corresponding to the R-1, R-2 and R-3 phases of recovery, and certainly the slow variety of the R-2 phase which occurs with high values of TTS_2 attracts suspicion as to whether it may represent a link between recoverable and non-recoverable threshold shifts.

However, this could not account for the PTS which still develops where TTS_2 arising from continuous noise exposure throughout the working day is distinctly less than 40 dB. The gradual growth of PTS, when studied in groups exposed to progressively longer durations of noise (Taylor *et al.*, 1965), is reminiscent of the successive layers and skins of an onion. From such consideration, it is easy to derive a concept of growth of PTS_{10} whereby there is an accumulation of 5 (days) × 48 (weeks) × 10 (years) = 2400 layers of PTS. In turn, this implies that in every TTS there is a small component of PTS. This alone is not large enough to be measurable by audiometric means. So, when a TTS recovers to within 5 dB of the pre-exposure threshold, recovery is regarded to be complete. But perhaps this is not so. Indeed, histological evidence (*vide infra*) now lends some support for supposing that there may be components of morphological damage which are not measurable functionally.

The above theory provides a convenient model by which the apparent quantitative relationship between TTS_2 and PTS_{10} may be explained. But what about further growth of PTS after the tenth year of exposure? According to some authorities, growth of PTS ceases after 10 years (Glorig *et al.*, 1961); others (Taylor *et al.*, 1965; Burns & Robinson, 1970) say that growth slows up, but does not cease. Indeed, one would not expect it to cease when considering the amount of TTS measurable in presence of a PTS. Ward (1973) has published data on this which show, for example, that a noise capable of producing an average TTS at 4 kHz of 26 dB in normal-hearing persons will still produce an average of about 15 dB in persons having a PTS of 26 dB. Thus, if TTS is causally related to PTS in some way, further increments of PTS would be expected to occur with further noise exposure leading to perhaps 26 plus 15 dB of PTS after 20 years of exposure.

Another theory is that of "chronic TTS". If recovery from TTS does proceed as a function of the logarithm of time, there must be some residual, though usually unmeasurable, TTS on a Tuesday morning some 16 h after the end of noise exposure on a Monday, and perhaps also on a Monday morning some 64 h after the end of noise exposure on the previous Friday. It seems reasonable to argue that a threshold shift that is never allowed to recover completely may gradually become persistent and then permanent: this could explain the small amount of recovery that often takes place over a period of up to 6 months after the end of habitual noise exposure. On the other hand, Coles & Knight (1966) have demonstrated apparent complete recovery from what they termed the "background noise effect" which results from continuous exposure to background noise, e.g. ventilation and machinery in ships, for 24 h per day for months or years. However, this may not be the same as a chronic or unrecovered component of relatively large TTSs. Whilst the correspondence between TTS_2 and PTS_{10} was not considered to be as precise as desirable (Kryter *et al.*, 1966), there was reasonable agreement. It is interesting, though, and probably significant, that the study conducted by Martin (1970) which showed a considerable discrepancy between TTS_2 and PTS_{10}, was concerned with intermittent noises.

Effects of intermittent noise

This discrepancy, studied by Martin, corresponds to a later series of studies by Ward (1970) who showed that the amount of TTS_2 was not the sole factor governing its rate of recovery. Contrary to his original assertions that it did not matter how a TTS had been caused, he showed that the recovery from TTS of less than 40 dB, but derived from a long period of intermittent noise, might exhibit the slow R-2 and linear R-3 phases characteristic of recovery from high-level TTS. By contrast, in the same subjects, a TTS of the same magnitude but derived from a short continuous exposure to the same type and level of noise, followed the pattern of recovery he and his colleagues originally described in 1958. Thus, it does matter how the TTS was caused, and it is reasonable to postulate that the slow-recovering TTS_2 from intermittent noise is more hazardous in terms of risk of a PTS component than is the fast-recovering TTS_2 of the same magnitude. Ward went on to question whether TTS_{30} or TTS_{1000} might be a better indicator of potential hazard than TTS_2, on which so many of the American criteria and allowances for intermittency are based.

Individual susceptibility

Another aspect in which there is doubt about the role of TTS in the development of PTS, either as a cause or part of a sequence of events, concerns the wide range of individual variability in the amount of TTS and PTS that results from noise exposure. It might be expected that there would be a close link between them, measurement of the amount of TTS being an indication of the susceptibility of the individual to PTS, at least from that particular type of noise exposure. Unfortunately, susceptibility is a multifaceted phenomenon (Ward, 1965) and one that varies not only between individuals but also, with respect to TTS at least, within individuals from one occasion to another, and between right and left ears of the same individual.

Correlation between TTS and PTS cannot be studied experimentally in man for obvious reasons, and adequately controlled data from industry are scanty: some studies, and in particular the recent work of Burns, Stead & Penney (1970), have shown correlations between TTS and PTS which are significant, but not sufficiently good to form the basis of a sufficiently reliable or simple test of susceptibility for use in industry. In animal experiments, however, it would appear that TTS and PTS are not closely related (Ward, 1969). Thus there is no conclusive evidence for or against the hypothesis that TTS is a phenomenon closely linked, together with other factors, with the development of PTS.

The energy principle

We come now to another mathematical relationship between noise exposure and PTS, namely the energy principle. Its first major application, when still very much a hypothesis, was in 1956 when it was adopted by the U.S. Air Force in its much publicized and influential damage risk criterion. Since then, more positive evidence for it has come from the Medical Research Council/ National Physical Laboratory studies of Burns & Robinson (1970). It postulates that the magnitude of PTS depends on the amount of energy which has fallen upon the ear in the period of time concerned, the so-called "noise immission". This has been shown to be true with large samples of industrial workers exposed to continuous noise for quite a wide range of exposure durations.

Since then, there is evidence that the principle applies also to intermittent impactive noises such as drop forging (Atherley & Martin, 1971). When extrapolated to high intensities, Rice & Martin (1973) have shown it to agree remarkably well with the criteria for gunfire-type noises put forward by Coles, Garinther *et al.* in 1968 (which, incidentally, were based mainly on TTS studies). Most authorities do, however, agree that it would probably be risky to use the principle in such a way as to permit exposures on one or several days that are above criterion limits, on the grounds that there is to be, or was, no exposure at all on other days. This condition is particularly necessary for high-intensity impulse noises where substantial PTS can result from a single exposure. Such caution makes sense when one recalls that when noise exposures are either severe enough to cause a TTS_2 of over 40 dB, or are intermittent (as is usually the case with gunfire noise exposures), the R-2 phase of recovery may be slow and prolonged, which may in turn be associated with a greater risk of PTS.

On the other hand, field studies of PTS arising from exposure to intermittent noises in America and Sweden indicate that sometimes this kind of noise gives less hearing loss than predicted by the energy principle. The causes of this discrepancy are uncertain. One possibility is that it is due to activity of the acoustic reflex, whereby contraction of the stapedius muscle within the middle ear attenuates the stimulating noise, particularly its lower frequency parts, and thereby produces some sort of protective effect. There is some doubt, though, about how long this could be maintained in an industrial situation of continuous noise (Brasher, Coles, Elwood & Ferres, 1969), but certainly in short term experiments it is protective. With intermittent, fluctuating or semi-impulsive noises the adaptation may be less, owing to frequent reactivation of the reflex, which would then reduce substantially the noise immission; that is, the noise would be reduced in passing through the middle ear to the inner ear.

Another possible explanation for the PTS being smaller than expected is that it has something to do with the fact that there is a recovery period between each burst of noise as the Americans postulate. This permits some recovery to take place, and thereby leads to a smaller TTS and presumably, therefore, to a smaller PTS. This idea is less acceptable now as a result of Ward's work (1970), showing that the TTS from an intermittent noise may have a much longer recovery time than the same amount of TTS from a continuous noise. This longer recovery time may well be associated with greater risk of PTS.

From some other evidence, however, the present author believes that recovery periods are important to safety. The evidence concerns a group of sailors who were noise-exposed continuously for periods of up to four days. The level of noise was not far above accepted levels for industrial noise, but because of the unusual pattern of exposure the hearing of the ship's company was monitored closely. A number of quite substantial PTSs occurred from what, by ordinary energy principles, had been a trivial noise exposure. It was felt that the reason for the PTSs was the fact that the men's hearing was not given time to recover in the way that happens overnight with daily exposures to industrial noise.

Thus, there are doubts as to whether the energy principle can apply to intermittent noises with respect to PTS. The same reservation holds for TTS_2 arising from intermittent noise. Ward (1970) showed how different amounts of energy can produce the same TTS_2. It is not difficult also to postulate situations in which the same amount of energy can be distributed differently in time throughout the day and produce different TTS_2: an example of this has been given by the author (Coles, Rice & Martin, 1973) as evidence of TTS_2 data not being compatible with the energy concept.

MORPHOLOGICAL CHANGE AND AUDITORY IMPAIRMENT

The most commonly considered impairment of function is the elevation of the threshold, so-called "hearing loss", although there are other impairments which must be taken into account. Studies of the relationships between threshold shifts and morphological change fall into two main categories. First there is animal work, mainly carried out with the chinchilla in the U.S.A. (Eldredge & Miller, 1969; Miller, Rothenberg & Eldredge, 1971; Ward & Duvall, 1971; Henderson, Hamernik & Sitler, 1972). In various experiments, the chinchillas were trained by conditioning techniques to perform threshold audiometry over a range of frequencies. They were exposed to noise, their TTS was measured initially and during recovery and finally any PTS present was measured. They were then sacrificed and the histology of the inner ear was studied.

On the human side, there are the studies of Bredberg (1968) in Sweden, who obtained audiograms of 16 persons shortly before their death (from non-otological causes) and subsequently examined their temporal bones.

In both types of study there were remarkably wide individual variations between the amount of damage to hair cells and other structures and the auditory threshold, although, in general, the greater the hearing loss the greater the damage. There was also redundancy, both of function and of structure. In all the chinchilla work, there were instances of quite substantial outer hair cell damage with no auditory threshold elevation in the corresponding parts of the frequency range. The most obvious discrepancies were in the work of Ward & Duvall (1971) who found the complete loss of outer hair cells in the first one and a half to two turns of the cochlea in an animal with normal hearing at all frequencies. They also found no striking morphological differences between inner hair cells in a normal ear and those in an ear suffering from more than 80 dB TTS. Furthermore, they found that from exposures in which the total energy was constant but derived from various combinations of intensity and duration, the structural damage was not always the same. These findings are worrying in two contexts. With respect to ideas on cochlear function, it would appear that inner hair cells alone can mediate normal threshold hearing, yet with a 80 dB hearing loss the inner hair cells may appear virtually normal. As severe damage can occur to outer hair cells while thresholds remain normal, this raises ethical problems in doing TTS experimentation in human beings; the return to pre-exposure threshold has been believed to represent a return to normal without any residual damage.

What then can be the function of the outer hair cells, without which the threshold seems to be unaltered? The body does have considerable redundancies of structure, but it is difficult to imagine that these cells have no function. Ward & Duvall (1971) suggest that possible disorders of loudness, of critical bandwidth or of temporal integration are associated with such damage; chinchillas will have to be trained to more complex tasks in order to find out! Another possibility is a disturbance of frequency discrimination. Such damage may relate to the functional impairments that (in humans with cochlear disorders) we call recruitment of loudness, diplacusis, tone perversion, and which in turn lead to loss of clarity, distortion and speech discrimination loss.

Another common symptom in cochlear disorders is tinnitus. We know even less about the nature of the morphological change and functional impairment that produces the tinnitus than we do about threshold shift and disorders of loudness and frequency discrimination. We do, however, know that there is great individual variation—because only some of the people with a given amount of noise exposure and/or similar audiograms experience tinnitus; others do not. There is much latency in this symptom: years later, sometimes as much as 30 years later, they develop tinnitus, which when matched for pitch is found to be at or near that of their maximum hearing loss. Presumably the onset of tinnitus is associated with further deterioration due to ageing, but its association with long-past noise exposure, an audiogram characteristic of noise-induced hearing loss and an associated pitch match is so strong and frequent, that it seems highly likely that such tinnitus is often a long-term result of the noise exposure. One imagines that here too a temporary noise-induced tinnitus has largely recovered, but left a residual latent component of permanent type. Perhaps damaged outer hair cells are associated with latent tinnitus too.

FUNCTIONAL IMPAIRMENT AND SOCIAL HANDICAP

Finally, we move from the area of relatively soundly based data and closely reasoned argument to the area of handicap, which is less well understood. We seek a measurement of social handicap for several purposes. The aim of hearing conservation is to prevent social handicap. The target of our rehabilitational efforts is to reduce social handicap. And when we seek to "compensate" financially for the effects of noise exposure or other injury to the cochlea, we should be thinking in terms of social handicap.

The relationships between impairment and handicap are well illustrated by the studies of Acton (personal communication in 1973). In a group of 120 persons with noise-induced hearing loss, he performed a range of measurements of impairment. These included pure-tone audiometry, speech audiometry by earphone and speech audiometry where the speech signals came from a loudspeaker in front of the subject and competing noise was presented to him from loudspeakers on either side. The subjects also described their auditory handicap with the help of a questionnaire to standardize the data into a format from which a handicap rating could be drawn (Kell, Pearson, Acton & Taylor, 1971). The correlations between the wide variety of audiometric data and the described-social-handicap rating were analysed. Surprisingly perhaps, the correlations between handicap and speech audiometry were not notably better than between handicap and pure-tone audiometry.

One of the best correlations, with a coefficient of 0·643, was that using the 1–2–3 kHz average hearing level which, incidentally, is the formula recommended by the Industrial Injuries Advisory Council for measurement of auditory impairment when occupational deafness becomes a prescribed industrial disease and compensatable under the National Insurance (Industrial Injuries) Act, 1965. The individual data are shown in Fig. 1.

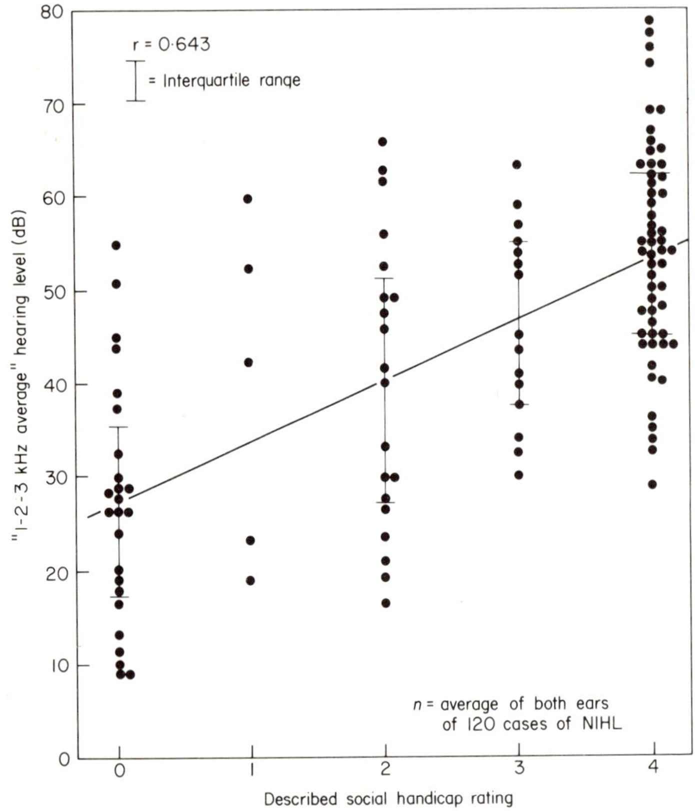

FIG. 1. A relationship between functional impairment and described social handicap. After Acton (personal communication in 1973).

On the figure is plotted the regression line of hearing level as a function of handicap rating, and also the interquartile range of hearing levels for each rating. It can be seen that there is a disturbingly wide scatter. The way in which the figure is drawn suggests also that the principal source of variance is in the measurement of hearing level. Indeed, it is almost traditional amongst otologists to regard clinical audiometric data with suspicion when they do not accord with the patient's handicap, but this self-critical attitude may be out of accord with reality. This is illustrated by the plan of inter-relationships shown in Fig. 2.

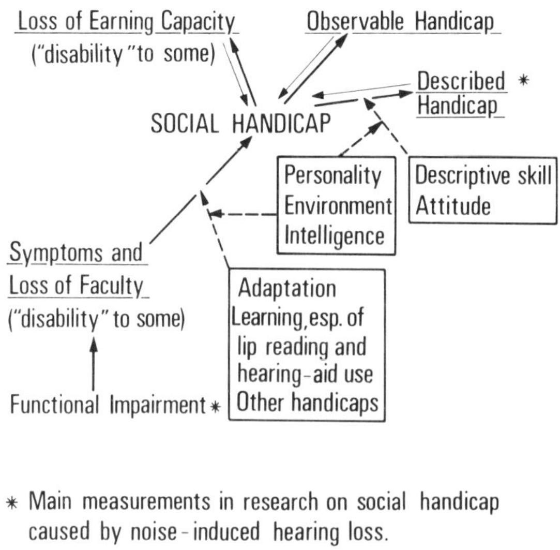

FIG. 2. Factors causing, influencing and indicating social handicap.

It is at once apparent that the lack of correlation which is criticized is hardly surprising. Many factors influence the social handicap and a close correlation with the degree of impairment is not therefore to be expected. There are further factors of equal, if not still greater, importance which influence the manner in which a person describes his handicap. This is what Acton assessed for his research and what the otologist assesses in his clinic. It could be that the measured impairment is in fact a better guide to the true social handicap than the patient's description of his hardships. It is therefore worthwhile to consider these inter-relationships and the many factors which influence them.

Noise-induced hearing loss produces symptoms such as have already been discussed. These, together with the impairment, are often described as "loss of faculty" and sometimes as "disability". The latter word is best avoided as to some other people it has a different meaning—associated with loss of earning capacity.

Compensation under the National Insurance Act concerns itself with loss of faculty only but, under common-law, loss of earning capacity can be an important additional factor. Prevention is, of course, concerned with both; prevention of the former being a means of avoiding the latter. In rehabilitation, the restoration of earning capacity is one aim, although certainly not the only one.

There are a number of factors which influence the relationship between the symptoms and the social handicap of noise-induced hearing loss. First, there is adaptation to the slowly developing impairment which reduces its immediate social impact, but carries hazards in encouragement of bravado or the "it can't happen to me" attitude so common in man when it comes to taking safety precautions; in this case the wearing of hearing protectors. Then, there is the degree to which the person learns to take advantage of lip reading and hearing aids. Finally, there are interactions with other handicaps, especially visual ones. Each of these factors is itself influenced by the individual's personality, environment and intelligence. The actual social handicap is further influenced by the following. It is influenced detrimentally by any loss or reduction of earning capacity and by the social rejection which often occurs. It is also influenced detrimentally by the way in which a man reacts to his handicap, which is closely related to the extent to which he lets his handicap lead him into social withdrawal.

It is often influenced beneficially when the social handicap is recognized and a helping hand is given. The conspicuous nature of the Medresco OL 56–58 series of NHS hearing aids is perhaps an insufficiently appreciated asset.

The handicap described depends to some extent on the man's descriptive skill, and honesty, but the key factor is his attitude to his handicap. These factors are themselves influenced by the factors of personality, environment and intelligence already mentioned.

CONCLUSIONS

The various inter-relationships described reveal how slender is our real knowledge, in spite of the large amount of research effort expended. It is right, though, that we continue to use our existing knowledge as the basis for prevention, rehabilitation and financial "compensation", provided that in so doing the incompleteness of our knowledge is not forgotten, that further research is not inhibited thereby, and that we are prepared to alter and adjust our beliefs and procedures in the light of that research.

References

Atherley, G. R. C. & Martin, A. M. (1971). Equivalent-continuous noise level as a measure of injury from impact and impulse noise. *Ann. occup. Hyg.* **14**: 11–28.

Brasher, P. F., Coles, R. R. A., Elwood, M. A. & Ferres, H. M. (1969). Middle ear muscle activity and temporary threshold shift. *Int. Audiol.* **8**: 579–584.

Bredberg, G. (1968). Cellular pattern and nerve supply of the human organ of Corti. *Acta otolar.* Suppl. **236**.

Burns, W. & Robinson, D. W. (eds) (1970). *Hearing and noise in industry.* London: H.M. Stationery Office.

Burns, W., Stead, J. C. & Penney, H. W. (1970). The relations between temporary threshold shift and occupational hearing loss. In *Hearing and noise in industry*: Appendix 13. Burns, W. & Robinson, D. W. (eds). London: H.M. Stationery Office.

Coles, R. R. A., Garinther, G. R., Hodge, D. C. & Rice, C. G. (1968). Hazardous exposure to impulse noise. *J. acoust. Soc. Am.* **43**: 336–343.

Coles, R. R. A. & Knight, J. J. (1966). The "background-noise effect" and auditory threshold. *Jl. R. nav. med. Serv.* **52**: 25–30.

Coles, R. R. A., Rice, C. G. & Martin, A. M. (1973). Noise-induced hearing loss from impulse noise: present status. *Proceedings of the International Congress on Noise as a Public Health Problem:* 211–218. May 13–18, 1973 at Dubrovnik, Yugoslavia, Washington D.C. 20460: U.S. Environmental Protection Agency. Report 550/9–73–008.

Eldredge, D. H. & Miller, J. D. (1969). Acceptable noise exposures—damage risk criteria. In *Noise as a public health hazard*: 110–120. Ward, W. D. & Fricke, J. E. (eds). Washington D.C.: American Speech and Hearing Association.

Glorig, A., Ward, W. D. & Nixon, J. (1961). Damage risk criteria and noise-induced hearing loss. *Archs Otolar.* **74**: 413–423.

Henderson, D., Hamernik, R. P. & Sitler, W. (1972). Comparisons between hair cell losses and permanent threshold shifts produced by three levels of impulse noise. Abstract of *Paper J8, 83rd Meeting of the Acoustical Society of America, Buffalo, 1972; and personal communication.*

Hirsh, I. J. & Ward, W. D. (1952). Recovery of the auditory threshold after strong acoustic stimulation. *J. acoust. Soc. Am.* **24**: 131–141.

Kell, R. L., Pearson, J. C. G., Acton, W. I. & Taylor, W. (1971). Social effects of hearing loss due to weaving noise. In *Occupational hearing loss*: 179–191. Taylor, W. (ed.). London and New York: Academic Press.

Kryter, K. D., Ward, W. D., Miller, J. D. & Eldredge, D. H. (1966). Hazardous exposure to intermittent and steady-state noise. *J. acoust. Soc. Am.* **39**: 451–464.

Martin, A. M. (1970). *Industrial impact noise and hearing.* Ph.D. Thesis: University of Salford.

Miller, J. D., Rothenberg, S. J. & Eldredge, D. H. (1971). Preliminary observations on the effects of exposure to noise for seven days on the hearing and inner ear of the chinchilla. *J. acoust. Soc. Am.* **50**: 1199–1203.

Rice, C. G. & Martin, A. M. (1973). Impulse noise damage risk criteria. *J. Sound Vibrat.* **28**: 359–367.

Taylor, W., Pearson, J. C. G., Mair, A. & Burns, W. (1965). Study of noise and hearing in jute weaving. *J. acoust. Soc. Am.* **38**: 113–120.

United States Air Force (1956). Hazardous noise exposure. *Air Force Regulation* No. 160–3. Washington D.C.: Department of the Air Force.

Ward, W. D. (1960). Recovery from high values of temporary threshold shift. *J. acoust. Soc. Am.* **32**: 497–500.

Ward, W. D. (1965). The concept of susceptibility to hearing loss. *J. occup. Med.* **7**: 595–607.

Ward, W. D. (1969). Effects of noise on hearing thresholds. In *Noise as a public health hazard*: 40–48. Ward, W. D. & Fricke, J. E. (eds). Washington D.C.: American Speech and Hearing Association.

Ward, W. D. (1970). Temporary threshold shift and damage risk criteria for intermittent noise exposures. *J. acoust. Soc. Am.* **48**: 561–574.

Ward, W. D. (1973). Auditory fatigue and masking. In *Modern developments in audiology* (second edn.): 301–344. Jerger, J. (ed.). New York and London: Academic Press.

Ward, W. D. & Duvall, A. J. (1971). Behavioural and ultrastructural correlates of acoustic trauma. *Ann. Otol. Rhinol. Lar.* **80**: 881–896.

Ward, W. D., Glorig, A. & Sklar, D. L. (1958). Dependence of temporary threshold shift at 4 kc on intensity and time. *J. acoust. Soc. Am.* **30**: 944–954.

Symp. zool. Soc. Lond. (1975) No. 37, 121–132.

THE PATHOLOGY OF DEAFNESS DUE TO TRAUMA

L. MICHAELS

The Institute of Laryngology and Otology, London, England

SYNOPSIS

The pathological changes in the cochlea damaged by stimulation, i.e. direct injury, blast and gunshot and sound waves, and also by chemical effects, are reviewed with reference to the findings both in human cochleae and in experimental studies with animals. Transverse fractures of the human temporal bone tend to involve the cochlea. Direct injury to the skull is also associated with microscopical damage to the hair cells, particularly the external hair cells of the basal coil. This is associated with hearing losses of frequency 3000—8000 Hz. Similar lesions are found after trauma by blast and gunshot and sound waves. Internal hair cells, supporting cells and nerve fibres seem to suffer damage as the intensity of the insult increases. The pathways for the entrance of ototoxic substances into the cochlea are reviewed. These substances, the most damaging of which are a number of therapeutically useful antibiotics and diuretic drugs, usually enter the cochlea from the bloodstream. We have not been able to induce ototoxic damage by the application of the antibiotic gentamycin to the middle ear of guinea-pigs. The damage produced by ototoxic substances, notably ototoxic antibiotics, is similar to that produced by stimulation trauma and predominantly located in the external hair cells of the basal coil. Possible factors in the production of this selective vulnerability are discussed.

Although the Shorter Oxford English Dictionary defines trauma as "wound or external bodily injury" the term will be used in a broader sense as meaning any type of damage to the cochlea. The "wound" aspect of cochlear trauma will be discussed and, in the context of this labyrinthine structure within a dense bone, this is a question of fractures of the petrous bone involving the cochlea. The microscopical effects of direct injury and also of blast and gunshot injury and of intense sound waves are examples of trauma due to excessive stimulation of cochlear structures. The damaging effects of ototoxic substances will then be discussed and it is of interest that the microscopical and functional effects of this very different form of injury are similar.

FRACTURES INVOLVING THE COCHLEA

Trauma caused by automobile accidents, until the recent reduction of the speed limit provoked by the oil shortage, was by far the commonest type of trauma in the Western world. It was the chief cause of death between the ages of one year and 34 years and in the U.S.A. 10 000 serious injuries were said to take place as a result of car accidents last year.

It is possible, and to be hoped, that enforced major changes in driving patterns will alter this epidemic of trauma, but in the meantime its effects are still with us. Three-quarters of automobile injuries were stated to involve the head, and it is my personal experience, as a pathologist who was engaged in the performance of numerous post-mortem examinations on subjects involved in car accidents, that fractures of the skull are extremely common in cases of trauma. Fractures take place in the vault and/or the base of the skull. The vault fractures are usually of little significance unless, and this is rare, they are depressed onto the underlying brain. The base-of-skull fractures are very serious and an indication of the great severity of the trauma. I have observed in the majority of patients with fractures of the base of the skull, that the rigid and more brittle wedge of bone occupying about one-third of the skull base—the temporal bone—is the seat of a fracture. Although such post-mortem accident material has provided a vast potential source for research into the mechanism of cochlear damage by gross trauma, unfortunately this has not been used, and what knowledge there is has been largely derived from clinical and radiological observations.

On the basis of such studies it has been found that fractures of the petrous temporal bone seem to fall within two anatomical groups: those in which the fracture line passes along the length of the petrous temporal bone—longitudinal fractures—and those in which the fracture line passes at right angles to the length of the petrous temporal bone—transverse fractures.

The effects on the cochlea of these two types of lesion are quite different. Longitudinal fractures arise as a result of direct blows to the temporal and parietal areas of the head. The fracture line starts in the squamous portion of the temporal bone and usually involves the external auditory canal, the tympanic membrane and one or more of the ossicles of the middle ear, and finishes up in the region of the foramen lacerum near the apex of the petrous temporal bone. The cochlea is not involved (Fig. 1). The transverse fractures are caused by blows to the front or back of the skull producing a sideways tearing effect. The fracture line in these cases passes from the dural membrane on the posteromedial aspect of the petrous temporal, often through the internal auditory meatus to involve the seventh and eighth cranial nerves and then into the cochlea in the region of the basal turn at its posterolateral side. The adjacent vestibule, round and oval windows are also frequently involved. (Fig. 2). Thus, in the case of the longitudinal fractures, the hearing loss is usually a mild or moderate one of the conductive type which may be helped by surgery. The hearing loss caused by a transverse fracture, involving as it does both the sensory organ and the afferent nerve derived from it, produces a severe sensorineural deafness from which little improvement is to be expected. A further serious side effect of transverse fractures is the result of their establishing a communication between the meninges and the middle ear, so that there is a leak of cerebrospinal fluid which invariably leads to infection spreading to the meninges; and this meningitis may cause death if it is not controlled by antibiotic therapy. Another form of fluid leakage in transverse fractures, that of perilymph, may be associated with the symptoms of Menière's disease. The pathological changes in these cases come about from the lowering of perilymph pressure with secondary endolymphatic hydrops. Reissner's membrane is seen to be grossly distended throughout the cochlear duct and the saccule may be dilated.

The effect of a fracture in other parts of the body, e.g. in a long bone, is to give rise to a local haemorrhage between the bone ends. This is soon invaded by granulation tissue which then gives way to a primitive type of bone called callus, which acts as a splint to secure the two fractured ends. In the temporal bone, however, callus does not seem to form; moreover, the union of the two fractured portions is by fibrous tissue, not bone. This type of fracture healing seems to be present in skull fractures in general and may be related to the immobility of these fractures. Small fractures joined up by this type of fibrous union are often found in sections of temporal bone at post-mortem in cases with no history of trauma and no symptoms related to the bone damage which usually is of insignificant degree. These healed fractures are found between the vestibule or cochlea and the middle ear. Their pathogenesis is unknown.

MICROSCOPIC COCHLEAR DAMAGE IN HEAD INJURY

After a head injury a sensorineural type of deafness may develop without any detectable macroscopic damage to the cochlea. The hearing loss is in the higher frequency ranges from 3000 to 8000 Hz. The pathological changes giving rise to this functional loss have not been explored in human material. Experimental studies have revealed microscopic cochlear damage following direct head injury. Schuknecht, Neff & Perlman (1951) trained ten cats to respond to auditory stimuli by walking forward in a rotating cage. The anaesthetized animals were then given a blow to the exposed skull in the midline, via a 2 lb metal rod 1 in. in diameter, by a 1 lb mallet. The animals were allowed to recover and then tested audiometrically until they were killed some weeks after trauma had been inflicted. All animals had hearing losses at least between 3000 and 8000 Hz, and some animals had an even more

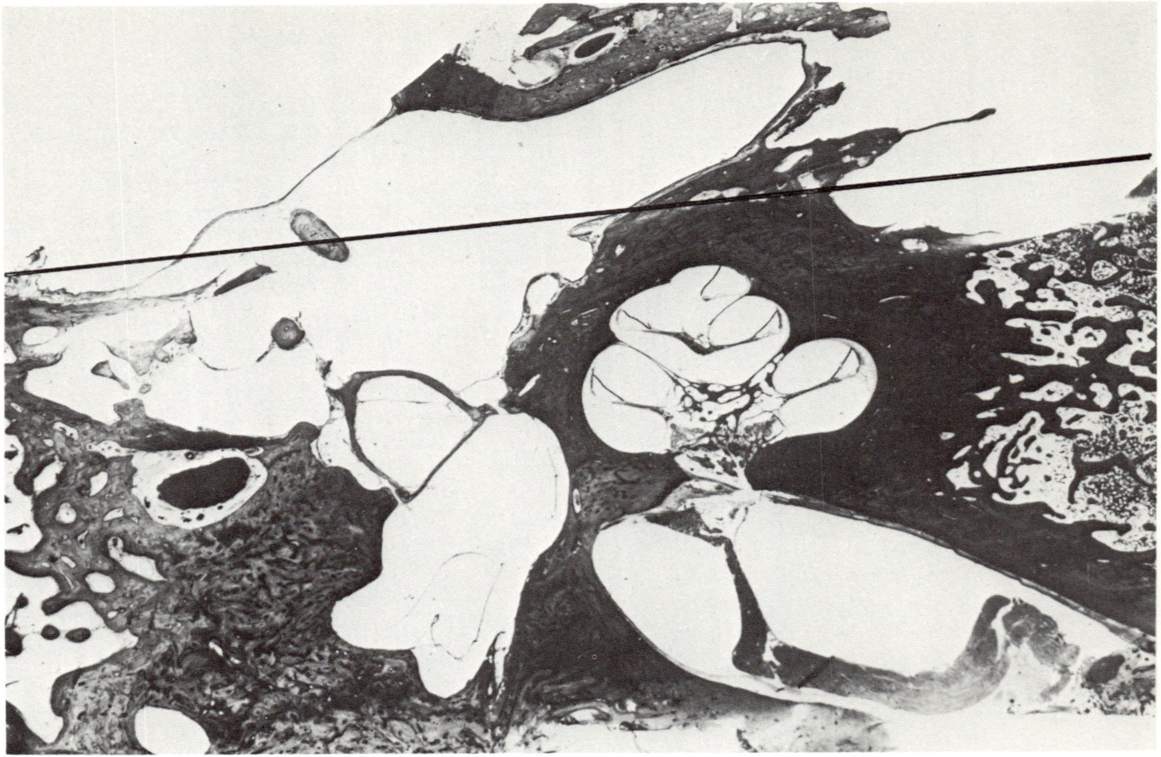

FIG. 1. Photomicrograph of a transverse section of the normal human temporal bone. Note normal degree of thinness of the footplate of the stapes, the only structure separating middle ear from scala vestibuli at this point. The black line represents the pathway of a frequent variety of longitudinal fractures. Middle ear structures are involved. Haematoxylin and eosin × 6·5.

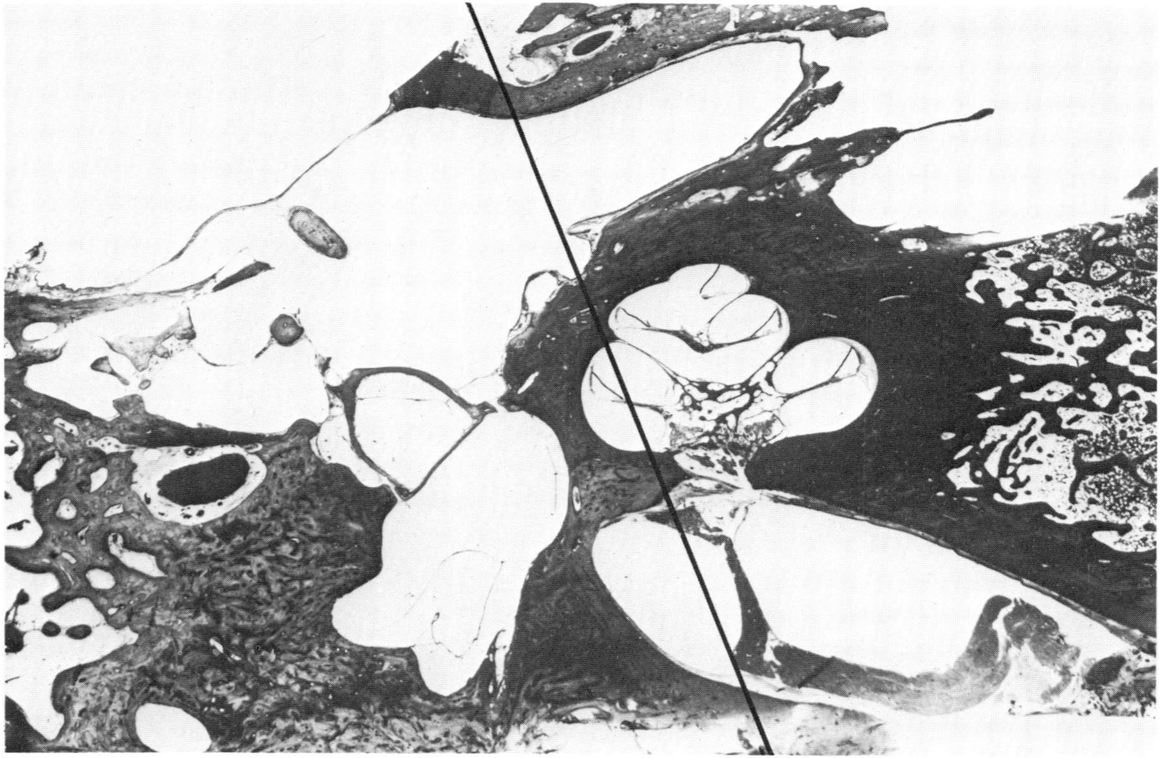

FIG. 2. Transverse section of normal human temporal bone. The black line represents the pathway of a frequent variety of transverse fractures. Auditory nerve and cochlea, as well as middle ear, are involved. Haematoxylin and eosin × 6·5.

widespread damage. The cochleae of all the animals were examined histologically by serial section. A constant change was found to be a loss of external hair cells in the upper basal coil region. In some animals damage was more marked and in a few there was a complete disappearance of internal and external hair cells in some areas. Nerve fibres and ganglion cells were correspondingly reduced in these cases. Microscopic haemorrhage was also frequently detected in cochlear sections. It was not provoking an inflammatory reaction and it was thus not felt to be of any significance in producing the hearing loss.

BLAST AND GUNSHOT INJURY

Peripheral damage to the ear, particularly rupture of the tympanic membrane, is the most striking result of explosive blast and gunshot injury and vestibular damage frequently takes place; however, permanent damage to the internal ear and its hearing mechanism is also a likely sequel. In the experimental study of injury from this source and other types of injury to the hair cells of the cochlea, a particularly valuable technique has been the examination of surface preparations of the spiral organ. By this method the exact distribution of hair cell damage can be mapped out easily and correlated with functional studies of sound reception. The effects of blast and gunshot injury are initially on the external hair cells of the basal coil. When this type of trauma is particularly severe, outer hair cells in long stretches of the cochlea may be entirely damaged and the supporting cells in these areas may become disrupted (Engström, Ades & Bredberg, 1970).

SOUND WAVE INJURY

As described by other speakers in this Symposium, sound waves of high intensity damage the cochlea. Again the pathological changes are those of microscopical destruction of hair cells. The earliest changes affect isolated external hair cells. As the intensity of the sound is increased, large groups of external hair cells die and their supporting cells are lost with them. The basal coil is once more the main centre for external hair cell loss. These findings have been obtained mainly in animal experiments but there is a little material from human pathology indicating that the same pattern of injury occurs.

CYTOPATHOLOGY OF HAIR CELLS IN CONDITIONS OF "STIMULATION" DEAFNESS

The changes in the inner ear following the trauma of direct blows to the head, of explosive blasts and of high-intensity sound waves, which together produce a condition which may be categorized as "stimulation" deafness (Paparella & Melnick, 1967), all seem to be directed particularly to the external hair cells of the basal coil. Internal hair cells and higher cochlear coils may be affected by greater degrees of these insults to the inner ear. Light microscopical studies, whether by examination of stained sections or by surface preparations of osmic acid-fixed cochleae, show little more than loss of stereocilia and necrotic features of the cells, followed by disappearance of the hair cells. Attempts have been made to identify the underlying nature of the damage produced to the hair cells by "stimulation" trauma using the electron microscope. The changes of increased lysosome formation beneath the basal bodies, the appearance of myelin figures and vacuoles in the cytoplasm and the swelling of the hair cell nuclei would seem to be features of any degenerating cell. The development of "Hensen's bodies" in the cytoplasm—areas of organelle-free cytoplasm surrounded by mitochondria and endoplasmic reticulin—have not been proved to be of any definite significance (Engström *et al.*, 1970). Studies of histochemical changes in external hair cells under stimulation trauma have likewise lacked evidence of specificity, and have indicated only the metabolic slowing down that would be expected prior to death of a cell. Such studies have shown a decrease in acetylcholinesterase, and alkaline phosphatase and mitochondrial oxidative enzyme activity, and the total protein level is stated to be lowered (Vinnikov, Ya, A.; Titova, L. K. 1964, quoted by Paparella & Melnick, 1967).

OTOTOXIC INJURY TO COCHLEA

The examples of stimulation trauma to the cochlea considered above may be regarded as forms of mechanical injury. Ototoxic injury is produced by the effects of molecular or chemical change. Although many therapeutic substances have been incriminated as causes of ototoxic injury, the most important ones fall into two main groups: certain antibiotic substances and certain diuretic substances. The usual target effect of the ototoxic antibiotics in therapeutics is on 70 *S*-ribosomes of bacteria leading to a paralysis of the reproductive capacity of the organism. The two diuretics frusemide and ethacrynic acid are used in medicine for their inhibition of certain enzyme activities of the renal tubules, thereby diminishing water absorption. It is unfortunate that these two groups of drugs are among the most important in the modern pharmacopoeia; infection and heart failure with oedema, the conditions treated by the two types of drugs respectively, are among the most prevalent serious illnesses.

Possible means of transport of any of these ototoxic substances to the cochlea are shown in Fig. 3. The commonest route is through

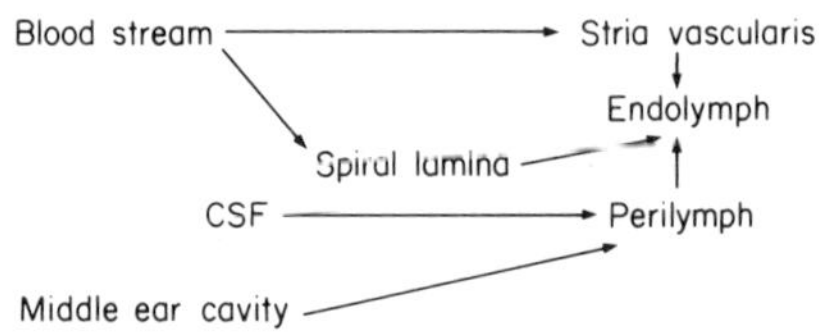

FIG. 3. Routes of access of ototoxic substances to the cochlea.

the bloodstream after administration by mouth or by injection. The damage is done to the hair cells, but the exact means by which these structures are reached by the ototoxic substances is not known. It is possible that they enter the endolymph by diffusing from blood vessels in the stria vascularis or even the spiral lamina. There seems to be no microcirculation among the outer hair cells so that ototoxic substances—indeed all substances, nutritive as well as toxic—must diffuse from surrounding fluids.

During the early years of the treatment of tuberculous meningitis with streptomycin until about the middle 1950s, it was felt that daily injections of antibiotics into the subarachnoid space were necessary in order to achieve adequate contact between antibiotic and tubercle bacillus. Such treatments were noted to be particularly dangerous for the development of the cochlear damage, suggesting that high concentrations of antibiotics in the cerebrospinal fluid would more easily pass to the cochlea than similar concentrations in the bloodstream. There is a pathway from the subarachnoid space into the perilymph at the scala tympani by way of the cochlear aqueduct. It is possibly this route that is taken by ototoxic substances injected intrathecally.

Several antibiotic substances are given therapeutically into the middle ear for the direct treatment of otitis externa and otitis media. At the Royal National Throat, Nose and Ear Hospital, London, four preparations are in frequent use and are presented in Table I in order of the quantity used. It will be noticed that all four of these preparations contain markedly ototoxic substances. These ear drops are used by insertion into the ear canal, but many of the patients on whom this form of treatment is used have

TABLE I

Antibiotic ear drops used at The Royal National Throat, Nose and Ear Hospital, London 1974

Preparation	Substances contained
Sofradex	Gramicidin, framycetin* and dexamethasone
Otosporin	Polymixin, neomycin* and hydrocortisone
Gentacin	Gentamycin*
Neocortef	Neomycin* and hydrocortisone

In order of the quantity used.
* Known severe ototoxic effect.

perforations of the tympanic membrane, so that the fluid passes into the middle ear. The middle ear cavity is separated from the perilymph space in most places by a layer of mucosa and the dense petrous temporal bone. In two places, however, only a thin fibrous membrane lies between the mucosa of the middle ear and the perilymph. One of these is the annulus fibrosus surrounding the stapes footplate which is the only barrier between the middle ear and scala vestibuli (Fig. 1), and the other is the round ligament separating the middle ear from scala tympani (Fig. 4). In addition, the stapes footplate itself is a very thin lamina of bone separating the middle ear from the scala vestibuli. It is therefore possible that antibiotic substances placed into the middle ear would diffuse through into the cochlea at these sites.

There is no definite clinical evidence that this does in fact happen during local treatment with antibiotics. Sensorineural deafness is a naturally occurring complication of otitis media well known to clinicians and usually ascribed to labyrinthitis—a direct involvement of the cochlea by the inflammatory process. Thus, the process of ototoxic damage by locally instilled antibiotic substances could easily be missed in clinical practice.

Animal experiments provide only little support for the idea of transmission of substances from the middle ear to the cochlea. Webster, Carroll, Benitez & McGee (1971) applied gentamycin

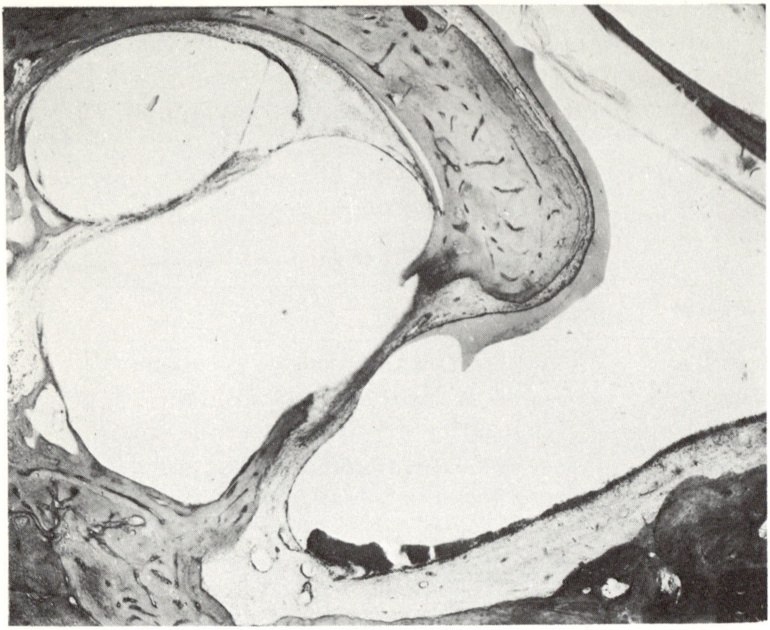

FIG. 4. Transverse section of human temporal bone at the level of the round window. Note that only the thin round ligament separates the perilymph space of the scala tympani (to left of centre) from the middle ear space (to right of centre). Haematoxylin and eosin × 35.

concentrations of up to 10%, adsorbed on gelfoam, into the marsupialized bullae of cats and showed that the cochlear hair cells were damaged. These concentrations are heavy compared with the 1% levels of gentamycin which are used clinically. My colleagues, Dr Ade Pye and Dr J. McLaughlin, and I, in an extended series of experiments in which we applied various concentrations of gentamycin through the tympanic membrane into the middle ear of guinea-pigs, were unable to demonstrate definite hair cell damage in concentrations up to 1%. It is possible that under conditions of suppuration in human disease the antibiotic might maintain a longer contact with the mucosa near the perilymph and so be more likely to diffuse into it.

The pathological basis for the sensorineural deafness induced by excessive amounts of ototoxic drugs is curiously similar to that of stimulation trauma. The lesion produced in the early stages is one of degeneration and death of the external hair cells principally of the basal coil of the cochlea. In the later stages internal hair cells, supporting cells and nerve elements may be involved. The examination of surface preparations of the spiral organ is particularly valuable in the experimental study of the effect of ototoxic drugs on the cochlea (Kohonen, 1965). In ototoxicity by some antibiotics, sensory cells in other parts of the membranous labyrinth may be involved, notably the maculae of the utricle and the saccule. The surface preparation method may be adapted to the study of cellular damage in these situations.

Electron microscopical studies of the spiral organ in experimental animals, which had been given an excess of ototoxic drugs, showed apical lysosome formation, mitochondrial degeneration and vacuole formation in the external hair cells (Lundquist & Wersall, 1967). These are similar to changes described following stimulation trauma, suggesting that these lesions are non-specific manifestations of cell disturbance prior to death of the cell. Indeed, both types of insult probably work on the hair cells through a final common path.

CONCLUSIONS

Thus it would seem that the microscopic damage produced on the one hand by stimulation trauma blows, blast and noise, and on the other hand by ototoxic chemical damage, are similar and take the form of necrosis of external hair cells predominantly in the upper portion of the basal coil of the cochlea. A selective vulnerability of external hair cells in this area would appear to be present towards injurious agents of quite different kinds. Possible reasons for this are at present highly speculative. The vascular supply to the external hair cells, particularly in the basal coil, may be more delicately balanced than elsewhere. We know from electron microscopical studies that there is no microcirculation to the spiral organ. It is possible that this area in particular has a barely sufficient blood supply, which could easily be rendered insufficient by the increased metabolic demands stimulated by the injurious agents described above. The nerve supply to the outer hair cells is different from that to the inner; that to the outer is derived from a much smaller number of ganglion cells, predominantly of the olivocochlear efferent system, than that to the inner, which communicate with more numerous afferent ganglion cells. This difference in nerve supply may possibly be related to some fundamental differences, at present unknown, in the metabolic activity of outer and inner hair cells, which may influence the susceptibility of the cells to toxic influences. The reason for the particular involvement of the basal coil may be related to the fact that the basal coil of the cochlea lies in the path of vibratory motion between the round and oval windows (Hinchcliffe, 1967). Thus, sounds of all frequencies could affect this area particularly.

References

Engström, H., Ades, H. W. & Bredberg, G. (1970). Normal structure of the organ of Corti and the effect of noise induced cochlear damage. In *Sensorineural hearing loss*: 127–152. Wolstenholme, G. E. W. & Knight, J. (eds). London: J. and A. Churchill.

Hinchcliffe, R. (1967). Occupational deafness. *Proc. R. Soc. Med.* **60**: 1111–1117.

Kohonen, A. (1965). Effect of some ototoxic drugs upon the pattern and innervation of cochlear sensory cells in the guinea pig. *Acta otolar.* Suppl. **208**.

Lundquist, P. & Wersall, J. (1967). The ototoxic effect of gentamycin. An electron microscopical study. "Gentamycin". *First International Symposium, Paris:* 26–46.

Paparella, M. M. & Melnick, W. (1967). Stimulation deafness. In *Sensorineural hearing processes and disorders*: 427–443. Graham, A. B. (ed.). London: J. and A. Churchill.

Schuknecht, H. F., Neff, W. O. & Perlman, H. B. (1951). An experimental study of auditory damage following blows to the head. *Ann. Otol. Rhinol. Lar.* **60**: 273–289.

Webster, J. C., Carroll, R., Benitez, J. T. & McGee, T. M. (1971). Ototoxicity of topical gentamycin in the cat. *J. infect. Dis.* **124**: Suppl. 138–144.

Symp. zool. Soc. Lond. (1975) No. 37, 133–165.

NORMAL AND ABNORMAL FUNCTIONING OF THE COCHLEAR NERVE

E. F. EVANS

Department of Communication, University of Keele, Staffordshire, England

SYNOPSIS

Studies of individual fibres from the normal cochleae of cats and guinea-pigs indicate that the cochlear nerve is relatively homogeneous in terms of its constituent fibre threshold and filtering properties. The fibres form a population of elements with filter characteristics of an overlapping filter bank, where the minimum thresholds correspond to the behavioural threshold audiogram of the animal. These filtering properties are up to an order of magnitude sharper than those of the basilar membrane and appear to be linear and not to depend on lateral inhibitory mechanisms. The filter characteristics of the cochlear nerve filters can account for the frequency selectivity observed by various psychophysical techniques in man.

In cochleae rendered abnormal through the action of anoxia, ototoxic agents or mechanical damage, the affected fibres have greatly raised thresholds and broadly tuned properties, approximating to those of the basilar membrane. These effects can be produced transiently and reversibly by brief periods of hypoxia and the transient application of antimetabolic or ototoxic agents.

These and other data suggest that the normal, low threshold, sharply tuned properties of the cochlear nerve are determined by the action of a "second filter" process subsequent to that of the basilar membrane; this second filter is particularly vulnerable to noxious influences. It is suggested that it is damage of a similar nature which is the essential lesion in human deafness of cochlear origin, and which in turn accounts for the accompanying widening of the critical band, for loudness recruitment and (partly at least) for the degradation of speech perception, as well as for the changes observed in the electrocochleogram (ECochG). Finally, there is some evidence that the normally sharply tuned characteristics of the cochlear nerve may depend on interaction between the inner and outer hair cell systems and/or their innervations.

INTRODUCTION

Considerable attention is being focused at the present time on the peripheral auditory system by auditory neurophysiologists and clinical audiologists alike. Recent data, much of it quantitative, on the activity and properties of single fibres in the cochlear nerve of animals under normal conditions, are assuming a particular relevance for our understanding of certain psychophysical phenomena, on the one hand, and the interpretation of results obtained by clinical electrocochleography, on the other. In turn, data obtained under abnormal conditions of the cochlea are giving us insights into the mechanisms underlying the normal functioning of the auditory periphery.

The pioneering work in this direction is that of Kiang, Moxon & Levine (1970) who described in detail the differences observed in cochlear nerve activity between normal cats and those poisoned with an ototoxic antibiotic, kanamycin. The emphasis of the work presented here, however, has been more or less restricted to the problems of how the frequency selectivity of the cochlear nerve is determined, how this may be affected by a variety of conditions of the cochlea, and what in turn is the possible significance of these measurements in animals for our understanding of human frequency selectivity and of cochlear deafness. (A detailed survey of the properties of single fibres in the normal cochlear nerve will be found elsewhere, Evans, 1975.)

By frequency selectivity, we mean the ability of the ear to resolve or separate out the individual frequency components of a complex signal. This frequency filtering property of the auditory system must be distinguished from its ability for frequency discrimination, some 30 times finer, although the underlying mechanisms of the two may well be related (e.g. Siebert, 1970; Zwicker, 1970; Evans, 1975).

Details of the experimental methods are given elsewhere (Evans, 1972b). Briefly, broad-band stimuli and tone bursts, the latter shaped to avoid click artefacts, are presented at a repetition rate of 4–5/sec into a closed sound system calibrated at the tympanic membrane by probe tube microphone (as in Figs 2 and 3). Single-fibre activity in pentobarbitone anaesthetized cats and guinea-pigs is recorded by ultrafine micropipettes filled with 3 M potassium chloride solution and inserted into the cochlear nerve under direct vision in the internal auditory meatus. Fibres were positively identified as such on the grounds of spike polarity, latency and histology. The tuning curves of Figs 4–7 and 9 were determined automatically and rapidly by on-line computer control of the frequency and intensity of the tone bursts, so that they covered the frequency-intensity space of the figures in an even but random manner (as in Fig. 4). The computer also counted the number of discharges which occurred during each tone burst and plotted these counts as the length of the appropriate vertical lines in Fig. 4. Although the stimulus levels in these figures are given in relative terms of electrical input to the half-inch condenser microphone used as an earphone, they represent the approximate (+8−2) dB SPL at the tympanic membrane.

FREQUENCY SELECTIVITY IN THE NORMAL COCHLEA

Frequency selectivity of the cochlear nerve to pure tone stimuli

Pure tones of appropriate frequency and intensity evoke in single cochlear fibres an increase in their irregular spontaneous discharge rate (Fig. 1). In contrast to early reports to the contrary

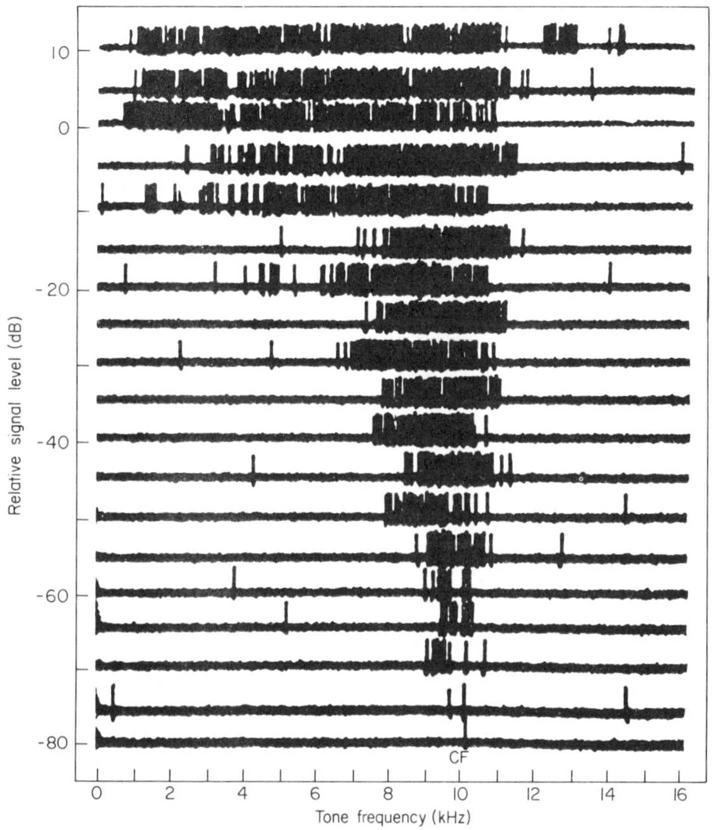

FIG. 1. Microelectrode record from single fibre in cochlear nerve of guinea-pig, in response to continuous tone swept in frequency in 5 dB steps of successively higher signal level. Alternate sweeps are in the opposite direction. Spikes are monophasically positive and 0·9 mV in amplitude. Sweep rate linear: 14 kHz/sec. The outline of the frequency response area thus defined is the frequency threshold curve (FTC). The characteristic frequency (CF) is 10 kHz (see text). After Evans (1972b).

(e.g. Tasaki, 1954; Katsuki, Suga & Kanno, 1962; Rupert, Moushegian & Galambos, 1963; Nomoto, Suga & Katsuki, 1964), this simple stereotyped excitatory response is the characteristic response obtained at this level of the auditory system to single tones, and moreover, the response properties of the population of fibres in the nerve are very homogeneous (Kiang, Watenabe, Thomas & Clark, 1965; Kiang, 1968; Evans, 1972b, 1975). The band of frequencies which evokes this response grows wider with increase in stimulus level (Fig. 1). By taking an arbitrary "threshold" response criterion (for example the boundary of the frequency-response area of Fig. 1), the frequency threshold ("tuning") curve (FTC) can be delineated. The frequency at the minimum threshold (tip) of the FTC is termed the characteristic frequency (CF) of the fibre.

A family of such FTCs for fibres originating from different points in the guinea-pig cochlea is shown in Fig. 2 (continuous and

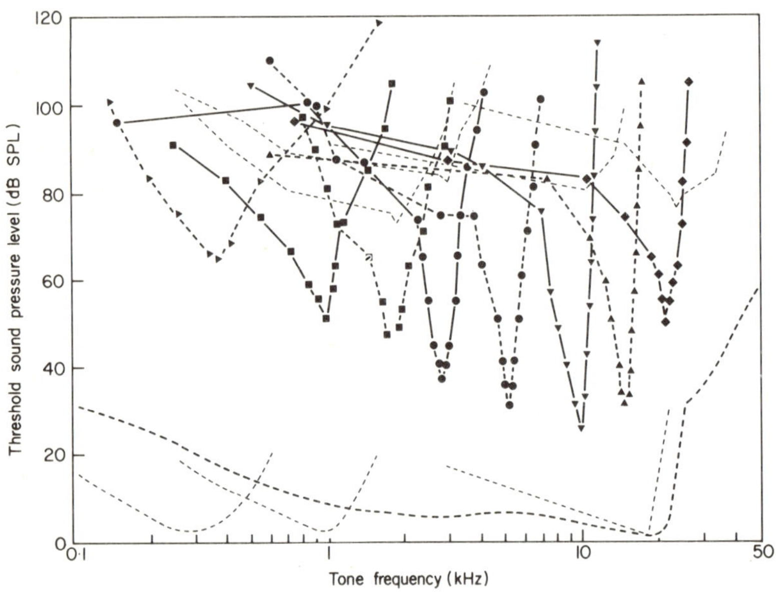

FIG. 2. Frequency threshold curves (FTCs) of single cochlear fibres recorded from normal and abnormal cochleae, compared with analogous data for the basilar membrane. Guinea-pig. Continuous and dashed lines through data points: FTCs of eight cochlear fibres from six normal cochleae. Upper thin dashed curves: FTCs of cochlear fibres from cochleae rendered abnormal by circulatory inadequacy (CFs at 1·9, 3 and 10 kHz). The thin dashed curve of highest CF (24 kHz) was obtained in an otherwise normal cochlea (see text). Lower dashed curves: analogous curves derived from the measurements of the vibration amplitude of the guinea-pig basilar membrane, by von Békésy (1944: curves with tip at 0·3 and 0·95 kHz), by Johnstone, Taylor & Boyle (1970: dashed curve at 18 kHz) and by Wilson & Johnstone (1972: dashed curve at 20 kHz). All threshold curves are corrected to sound pressure level (in dB SPL) at the tympanic membrane under closed bulla conditions. The basilar membrane curves are positioned arbitrarily on the ordinate scale. After Evans (1972b).

dashed lines drawn through symbols). The curve with the CF of 10 kHz represents the fibre of Fig. 1. Data for the cat and squirrel monkey are essentially identical (see Evans, 1975). Although small but significant variations are observed in the shapes from fibre to fibre, the threshold and tuning characteristics of the population are relatively homogeneous and are a function of the CF of the fibre. A discontinuity is, however, apparent at about 2–3 kHz: fibres with lower CFs have progressively more symmetrical FTCs; fibres with CFs higher than 2 kHz have asymmetrical FTCs, the low frequency cut-off becoming suddenly less steep at higher stimulus levels, and the high frequency cut-off becoming progressively steeper with level. Thus, fibres with high CFs have FTCs comprising two components: a sharply tuned lower threshold segment, and a broadly tuned high threshold segment extending toward the lower frequencies. The possible significance of these shapes will be referred to on pp. 157–159.

In contrast to earlier statements to the contrary from data pooled across many cats (e.g. Kiang, Watenabe *et al.*, 1965), the minimum thresholds of the FTCs in a single cochlea form one homogeneous population with a spread of less than 20 dB at any frequency, the thresholds approaching the behavioural audiogram of the animal (cat: Kiang, 1968; Evans, Rosenberg & Wilson, unpublished data; guinea-pig: Evans, 1972b) as in Fig. 3b, solid points. There is no evidence from the recent experiments, where the stimulus conditions have been well controlled, of two populations of fibres which can be distinguished on the basis of threshold, as is commonly held to be necessary to account for certain characteristics of the gross cochlear action potential in normal and abnormal ears (e.g. Davis, Deatherage, Rosenblut, Fernandez, Kimura & Smith, 1958; Portmann, Aran & Lagourgue, 1973). An alternative explanation for these characteristics will be offered on p. 155.

Frequency selectivity of the cochlear nerve to broad-band stimuli

An important question arises whether these filtering properties exhibited by the cochlear nerve to pure tones also obtain for the stimuli more often encountered in the acoustic environment, viz: broad-band noise and multicomponent stimuli, such as speech. The several studies of this problem indicate that this may well be the case, and suggest that the cochlea behaves as if it contained an overlapping bank of simple linear filters, each one represented by the characteristics of a single fibre.

The first approach, using broad-band noise stimuli, has examined the time-structure of the suprathreshold discharge pattern of cat cochlear fibres and its correlation with the waveform of the noise stimulus (e.g. de Boer & Jonkees, 1968; de Boer, 1969, 1970; de Jongh, 1972). These authors showed that under these conditions, the discharge pattern of single fibres was determined by a filtered version of the noise signal, the characteristics of the filter approximating to that of the FTC of the fibre. This result is limited to fibres with CF below 4 kHz, as beyond this frequency any "phase-locking" of the discharge pattern to the stimulus waveform becomes very difficult to detect.

Evans, Rosenberg & Wilson (1970; Evans & Wilson, 1971, 1973) used an alternative approach, namely to examine the relationship between the thresholds of single fibre responses to broad-band noise and to pure tone stimuli in the cat. The difference in thresholds observed for each fibre agreed well with those predicted on the basis that the pure tone FTC represented a linear filtering function. This therefore supports the same hypothesis as does de Boer's result, but by a completely different method, and extends it to fibres of all CFs.

Evans, Rosenberg & Wilson (1971; Wilson & Evans, 1971, Evans & Wilson, 1973) have also compared the suprathreshold responses of single cochlear fibres in the cat to multicomponent, comb-filtered noise stimuli, with the responses computed on the basis of simple linear filtering of the stimuli by the FTC filter function. Excellent agreement was obtained between observed and predicted properties. This result indicates that the fibres of the cochlear nerve respond to multicomponent stimuli in the manner of linear band-pass filters of the same relatively narrow bandwidth as the FTC; and that linear lateral inhibition is not involved in the mechanism achieving the sharp tuning. The data of Møller (1970), from the cochlear nucleus in response to double click stimulation, are consistent with this conclusion (see Evans, 1975).

The characteristics of this linear filtering process appear to be maintained over a range of signal levels of 30–60 dB above threshold (de Boer, 1969; Evans & Wilson, 1971, 1973; Wilson & Evans, 1971; Wilson, Evans & Rosenberg, 1974). Above these levels, however, there is evidence that the cochlear filter becomes increasingly nonlinear or broadly tuned or both (see Pfeiffer & Kim, 1973; Evans, 1975, for reviews).

These results taken together mean that for signals with complex spectra, and of moderate levels, the cochlear fibres act as if they were linear filters with characteristics given by the shape of their FTCs. The significance of this conclusion will be discussed on pp. 150–153.

Frequency selectivity of the cochlear nerve and cochlear partition

The lower, dashed curves in Fig. 2 are the corresponding frequency response functions for the guinea-pig basilar membrane derived from optical measurements of von Békésy (1944) and from measurements by Johnstone & Boyle (1967), Johnstone, Taylor & Boyle (1970), and Wilson & Johnstone (1972, and to be published), using the sensitive Mössbauer and the very sensitive capacitive probe techniques respectively. While these data are plotted arbitrarily on the intensity scale below the neural FTCs for convenience of comparison, they are determined at higher levels, (70 dB SPL and above) with the exception of the determinations by Wilson & Johnstone which have been made at levels as low as 40 dB SPL. The comparisons, however, are valid since all these studies found the basilar membrane vibration pattern to be essentially linear. (For more detailed discussions of this point, see Evans, 1972b; Evans & Wilson, 1973.)

The discrepancy between the relatively sharp neural FTCs and the essentially low-pass characteristics of the basilar membrane is substantial, particularly for fibres of intermediate CFs (3–15 kHz). The quantitative differences in the slopes of the low and high frequency cut-offs and in bandwidths have been described fully elsewhere (Evans, 1970a; Evans, 1972a,b; Evans & Wilson, 1973; Evans, 1975, in press). These differences amount to an order of magnitude at the intermediate frequencies.

It should be emphasized that while these cochlear nerve data from the guinea-pig confirm the observations from the cat (e.g. Katsuki, Sumi, Uchiyama & Watenabe, 1958; Kiang, Watenabe *et al.*, 1965; Kiang, Sachs & Peake, 1967) that the cochlear nerve tuning was substantially sharper than basilar membrane measurements (which exist for various animals but not the cat), they are substantially different from those of Tasaki (1954), probably for reasons which are dealt with on pp. 141–142. They thus eliminate the possibility that the discrepancy between (cat) cochlear fibre FTC and (guinea-pig) basilar membrane data arose from a species difference. Indeed, there is substantial quantitative agreement between measurements of cut-off slopes and bandwidths of cochlear fibres in cat, guinea-pig and squirrel monkey (see Evans, 1975).

Another possible explanation for the discrepancy between neural and mechanical frequency selectivity, was that the surgical opening of the cochlea necessary to measure the motion of the cochlear partition might disturb its tuning properties (Huxley, 1969). This possibility was eliminated by the demonstration (Evans, 1970b) that normally sharply tuned cochlear fibre properties could be obtained with the basilar membrane exposed, after opening the scala tympani over nearly half of its basal turn, and draining out the perilymph, thus reproducing the conditions under which Wilson & J. P. Johnstone and B. M. Johnstone and colleagues made their basilar membrane measurements.

While there is very good agreement between a variety of measures of basilar membrane frequency selectivity in the guinea-pig (Mössbauer: e.g. Johnstone *et al.*, 1970; capacitive probe: Wilson & Johnstone, 1972; laser "fuzziness": e.g. Kohlloffel, 1973; differential cochlear microphonic: e.g. Dallos, 1973), Mössbauer measurements on the squirrel monkey at a basilar membrane location corresponding to about 7 kHz are discrepant (Rhode, 1971, 1973). In particular, while the overall characteristics are very similar to the measurements in the guinea-pig (see Wilson, 1974, for detailed comparison), Rhode found a significant nonlinearity, which if exploited in constructing a curve corresponding to the neural FTC, can produce a substantially sharper curve, though still not as sharp as those of the cochlear nerve (Rhode, 1971; see Evans & Wilson, 1973).

Rhode's results have been adduced in support of a model where the sharp frequency selectivity of the cochlear nerve and certain nonlinearities evident in cochlear fibre discharge patterns have been accounted for on the basis of sharp nonlinear tuning of the cochlear partition (e.g. Pfeiffer & Kim, 1973). However, an alternative model, in which the sharp cochlear nerve tuning characteristics are determined by a "second filter" process subsequent to the low-pass filter characteristics of the basilar membrane (the first filter), is preferred for a number of reasons (Evans, 1972a,b; Wilson & Johnstone, 1972, 1973; Evans & Wilson, 1973; Evans, 1974b, 1975, in press). Firstly, there is the excellent agreement (referred to above) between the various basilar membrane measurements in the guinea-pig, none of which show the nonlinearities found by Rhode. Secondly, there is good correspondence between the capacitive probe measurements of the basilar membrane of Wilson & Johnstone (1972, 1973; see Wilson, 1974) and the differential cochlear microphonic data of Dallos (1973, 1974), in respect of amplitude, phase and the absence of a cubic intermodulation distortion product ($2f_1-f_2$), which would be expected to be present if the basilar membrane mechanics were nonlinear and sharply tuned. Thirdly, the ingenious psychophysical study by Smoorenburg (1972) of a subject with a sharply defined hearing defect appears to the present author to exclude a nonlinear sharply tuned basilar membrane as the generator of the cubic distortion product heard by this subject. Fourthly, the observed cochlear nerve nonlinearities (including the cubic distortion product) can be accounted for by a nonlinearity sandwiched between the basilar membrane and "second filter" process (see Evans & Wilson, 1973; discussion following Pfeiffer & Kim, 1973; Evans, 1974b). Fifthly, measurements of the cochlear fibre FTCs in the guinea-pig (Evans, 1972b) and cat (Evans *et al.*, 1970; Evans & Wilson, 1973; Evans, 1975) indicate that there are significant variations in the sharpness of tuning from fibre to fibre. These can amount to differences of 4:1 on the 10 dB bandwidth, in the same animal and at the same CF. Moreover, in the normal guinea-pig, high threshold abnormally broadly tuned cochlear fibres of CF above 12 kHz have been found adjacent to normally tuned fibres of the same CF (see Fig. 2). This finding may be related to the sporadic loss of outer hair cells which occurs in the basal turn of the normal guinea-pig (Wersäll, personal communication). It is difficult to reconcile such a variability of tuning with the hypothesis that sharp tuning is a function of the basilar membrane, which would on the contrary be expected to produce homogeneously distributed tuning properties in the cochlear nerve. Instead, it offers circumstantial evidence for the existence of a second sharply tuned filter "private" to each cochlear fibre. Further and more direct evidence for this hypothesis is described on pp. 144–150.

FREQUENCY SELECTIVITY IN THE ABNORMAL COCHLEA

Kiang, Moxon *et al.* (1970) were the first to observe "abnormal" FTCs of cochlear fibres in cats chronically poisoned with kanamycin, an ototoxic antibiotic. The nature of similar changes under various acute and chronic abnormal conditions of the cochlea has been investigated in detail in the present work.

Abnormal conditions having long-term effects

FTCs radically different from the normal low threshold, sharply tuned curves illustrated in Fig. 2 can be obtained in the course of recordings in the guinea-pig, under conditions where the blood supply to the cochlea is impaired, through low systemic blood pressure or partial constriction of the internal auditory artery, and where the round window-recorded whole nerve (AP) responses to click stimuli have an abnormally high threshold (Evans, 1972b). Typical abnormal curves are illustrated in Fig. 2 as the upper thin dashed curves. The thresholds of these abnormal FTCs are in excess of 70 dB SPL, and their abnormally broad tuning resembles that of the basilar membrane curves below. In fact, measurements of the cut-off slopes and bandwidths of these FTCs show close correspondence with the basilar membrane data (Evans, 1972a,b; Evans & Wilson, 1973). The one FTC shown by Tasaki (1954), being comparable to these curves, is therefore likely to be anomalous for similar reasons.

Similar curves have been obtained in the cat in the course of a series of experiments in collaboration with Dr R. Klinke, where the transient, acute effects of various antimetabolic (potassium cyanide), ototoxic (frusemide), and putative transmitter blocking agents were being investigated (see pp. 147–149). Overdosage with these agents or contamination of cochlear perfusion solutions in early experiments produced generalized or localized cochlear abnormalities which persisted for the duration of the experiment (about 24 h). In addition, similar FTCs have been obtained in cats inadvisedly exposed by commercial breeders to streptomycin during development.

One question these occurrences have helped to dispel is the suggestion that, in cochlear damage, the activity of a low threshold population of fibres is lost, leaving behind a second, high threshold population. This notion has been based on gross cochlear nerve AP data in normal subjects and patients with signs of cochlear damage (e.g. Portmann *et al.*, 1973) and on animal studies (e.g. Davis *et al.*, 1958; Keidel, 1965), and the earlier evidence referred to above (p. 137) which suggested that normal cochlear fibres exhibited a substantial range of thresholds. Figure 3 illustrates an experiment where the minimum threshold of single cochlear fibres (circles) and of the gross cochlear AP response recorded from the round window to clicks were plotted as a function of time (a) and of the CF of the cochlear fibres (b), before and after instillation of certain agents into the scala tympani via a needle inserted through the round window. The needle was inserted at the time indicated by the arrow, and the instillation, carried out inadvertently with contaminated artificial perilymph, occurred for the period indi-

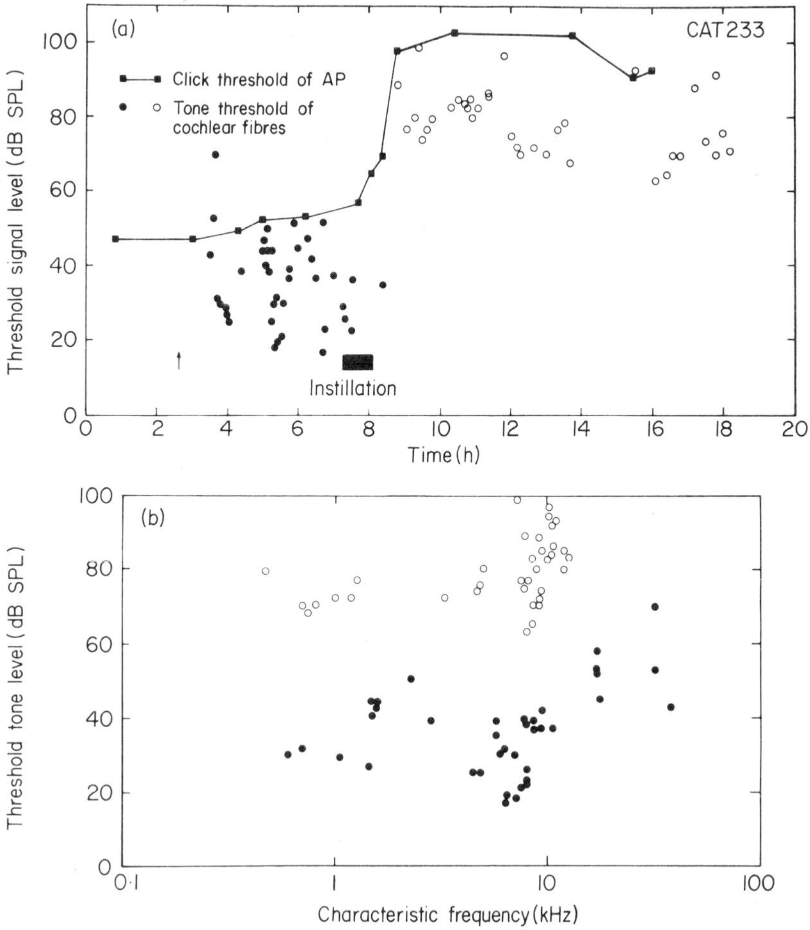

FIG. 3. Effect of acute cochlear abnormality on the threshold of single cochlear fibres and gross "action potential" (AP) responses in the cat. Circles: minimum thresholds (i.e. at the CF) of individual cochlear fibres. Squares: threshold of AP response recorded at round window to 50 μsec click (expressed in peak equivalent dB at 10 kHz). (a) Thresholds plotted as a function of time showing the uniform elevation of thresholds produced by instillation of contaminated artificial perilymph into the scala tympani of the cochlea. Instillation needle was inserted at the time indicated by the arrow and instillation occurred during the time indicated by the bar. (b) The same cochlear fibre thresholds plotted as a function of their CF. Solid circles: before cochlear contamination: open circles afterwards. From Evans & Klinke (unpublished data).

cated by the bar at 7–8 h. Hitherto, the minimum thresholds of the 38 fibres from which recordings were made, formed a single low threshold population (solid circles in Fig. 3a and b) as mentioned on p. 137 above. During the instillation, the gross cochlear AP threshold rose over 40 dB and remained at the abnormal level for the remainder of the experiment. Thereafter, all the 35 cochlear fibres recorded had high thresholds (open circles) and abnormally tuned properties.

The data of Kiang, Moxon *et al.* (1970), from cat cochleae with partial damage due to kanamycin poisoning, are consistent with these conclusions.

Measurements of the difference between the thresholds for tone and broad-band noise stimuli of such high threshold, abnormally tuned fibres, analogous to those described for normal fibres (p. 138), have been carried out (Evans, unpublished data). The normal differences are reduced by an average of 5 dB for fibres with CFs of 10 kHz, increasing to 10 dB for CFs of 1 kHz. This represents an increase in the effective bandwidth of these abnormal fibres of between three and ten times.

Abnormal conditions having short-term reversible effects

The observation that, in the guinea-pig, cochlear fibre FTCs under normal conditions of the cochlea become as broadly tuned as measurements of the basilar membrane frequency response function, provides additional—though circumstantial—evidence for a physiologically vulnerable "second filter" mechanism subsequent to the basilar membrane (see pp. 140–141). The validity of this evidence has been questioned by Rhode (1973) on the grounds that in his Mössbauer measurements of the squirrel monkey basilar membrane, reductions in the peak response were obtained after death. A series of experiments have therefore been carried out to investigate the transient effects of hypoxia on the tuning of single cochlear fibres (Evans, 1974a) and these have been accompanied by a series investigating the transient effects of antimetabolic and ototoxic agents (Evans & Klinke, 1974, and to be published).

In order to study the rapid changes which occur in the threshold and suprathreshold tuning characteristics of the fibres, the rapid, automatic analytical method illustrated in Fig. 4 was devised. Randomization of the sequence of frequency and intensity ensured that any bias introduced into the determinations by changing conditions was minimized. While the complete analysis shown in Fig. 4 took approximately 4 min to complete, sufficient information could be obtained in less than 1 min to allow an approximate FTC to be drawn. Concurrent with this analysis, the amplitudes of the cochlear microphonic and gross cochlear AP to

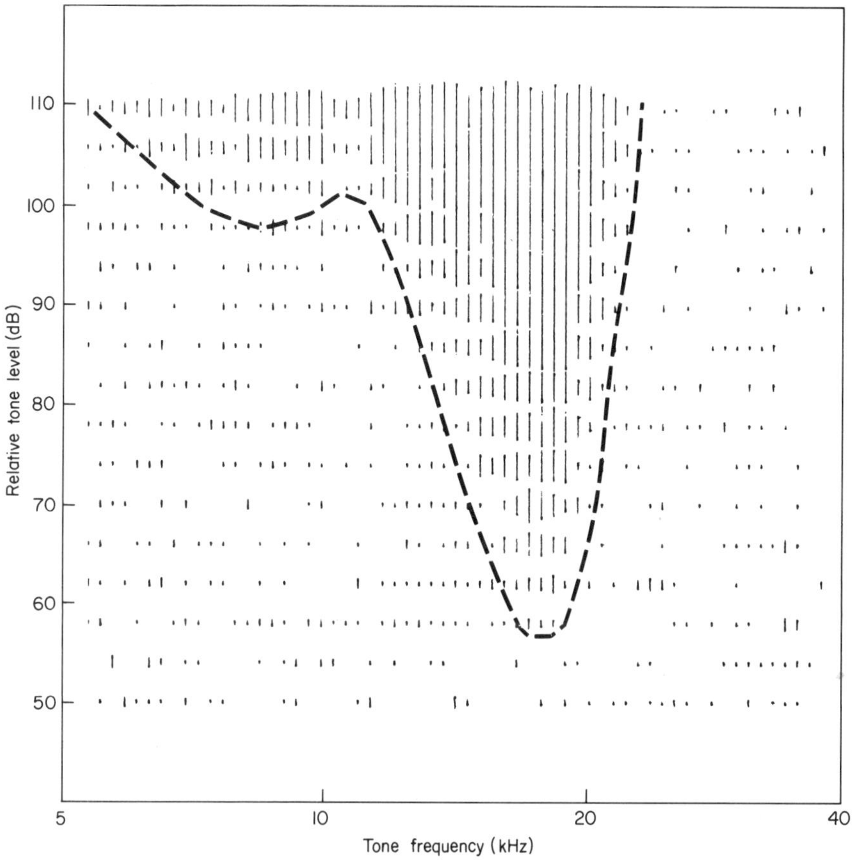

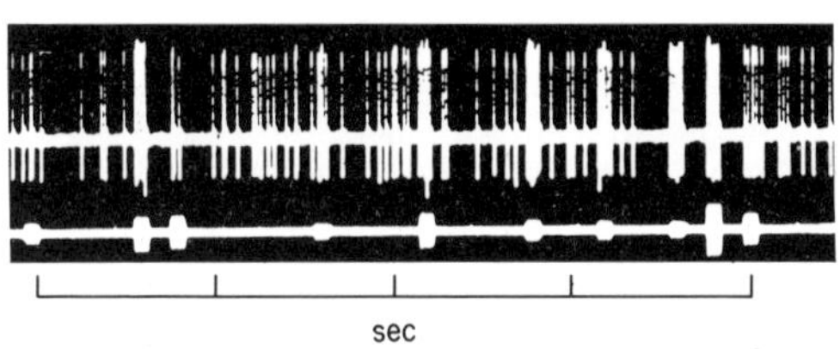

FIG. 4. Automatic method for determining the FTC of a cochlear fibre (cat). Below the graph is a sample film record of spike discharges of the fibre during analysis with round window-recorded cochlear microphonic as lower trace. Tones presented at 5/sec with random sequence of frequency and intensity controlled by on-line computer, which concurrently counts the number of spike discharges occurring during each tone presentation. The counts are displayed as length of line at the appropriate frequency-intensity locus in the main plot (see methods). FTC superimposed by hand, using arbitrary threshold criterion. (These data and those of Figs 5, 6, 7 and 9 have not been corrected for the characteristics of the sound system, but represent approximate (+8−2) dB SPL at the tympanic membrane.)

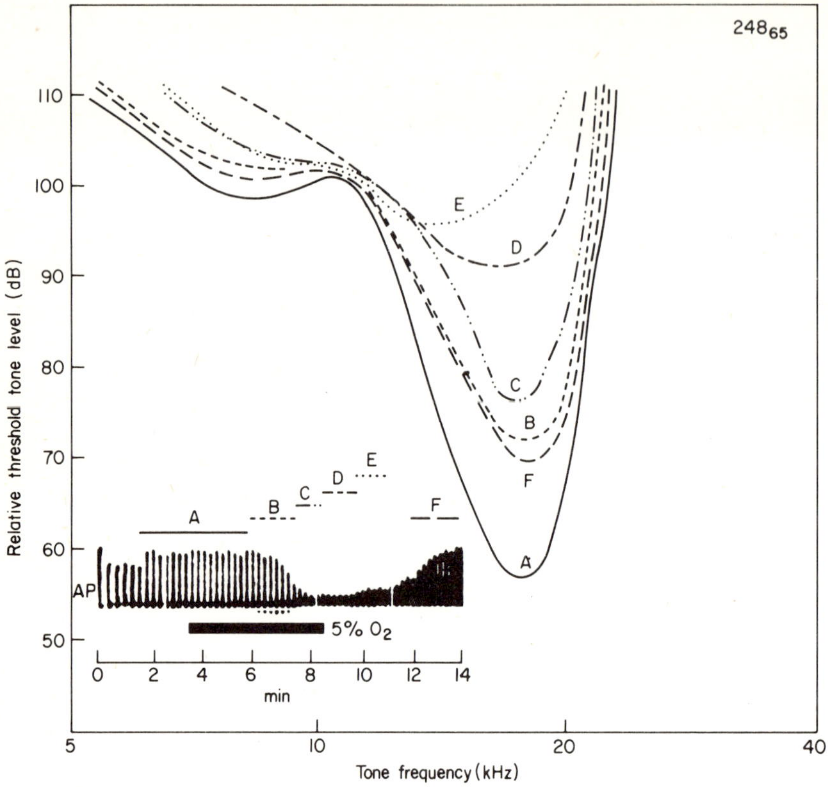

FIG. 5. Reversible effects of hypoxia on the tuning of a single cochlear fibre in the cat. Inset shows time course of gross cochlear AP amplitude in response to click stimuli of constant amplitude (approx. 40 dB above threshold) presented every 10 sec. The thick bar indicates the duration of reduction of inspired oxygen to 5%. The bars over the AP record indicate the times during which the FTCs illustrated in the main figure were determined. Curve A: control obtained before cochlear hypoxia developed (see text and Fig. 4). Curves B to E: during development of cochlear hypoxia. Curve F: during recovery from hypoxia. Note loss and partial recovery of low threshold, sharply tuned segment of FTC. After Evans (1974a,b).

click stimuli of constant amplitude (approximately 30–40 dB above threshold), were monitored from the round window (see insets: Figs 5 and 7).

Hypoxia

Figure 5 shows the reversible effects of five minutes' hypoxia on the shape of the FTC of the cat cochlear fibre illustrated in Fig. 4. The hypoxia was produced by diluting the inspired oxygen concentration to 5% with nitrous oxide. After a "latent" period of

2–3 min, cochlear hypoxia develops, as indicated by the reduction of the gross cochlear AP amplitude shown in the inset trace. The essential changes in the FTC are a progressive loss of the low threshold, sharply tuned segment and a slight shift of the CF towards lower frequencies. Substantial recovery from these effects occurred after restoration of 100% inspired oxygen as shown in analysis F, obtained before the fibre was lost. Similar effects were obtained in all fibres which could be held long enough in the presence of the considerable blood pressure changes accompanying such hypoxic manoeuvres. The cochlear microphonic (CM), however, was not substantially altered.

The time scale of these effects is much shorter and their magnitude much larger than those observed in the Mössbauer measurements of Rhode (1973), after death in the squirrel monkey, which amounted to only about 10 dB 20 min after death, when all normal cochlear potentials would have been eliminated (e.g. Butler, Honrubia, Johnstone & Fernandez, 1962). Furthermore, the fact that the cochlear microphonic is substantially unchanged during the time in which the FTCs lose 30–40 dB of their sharply tuned segment, is itself important evidence that the basilar membrane properties remain relatively unchanged (on the assumption that any distributed changes in cochlear mechanics would be reflected in the round window cochlear microphonic). Thus these data provide direct evidence for the existence of a physiologically vulnerable "second filter" in the cochlea, and this was corroborated in the following experiments.

Instillation of intracochlear cyanide

Figure 6 shows the effects on the FTC of a cochlear fibre of instilling (according to the technique of Galley, Klinke *et al.*, 1973) artificial perilymph containing 5×10^{-4} M potassium cyanide into the scala tympani of the first turn of the cat cochlea (Evans & Klinke, 1974, and to be published). Curve A was obtained prior to the instillation, which is indicated by the solid bar above the time scale. Curves B and C were obtained as the cyanide solution reached the organ of Corti (there is a delay of some two minutes). At the height of the effects (C) the FTC had almost completely lost its low threshold, sharply tuned segment. Virtually complete recovery of the normal FTC (curves D, E and F) was obtained after washing out the cyanide solution with artificial perilymph. These effects could be obtained on fibres with CFs above 7 kHz (i.e. within

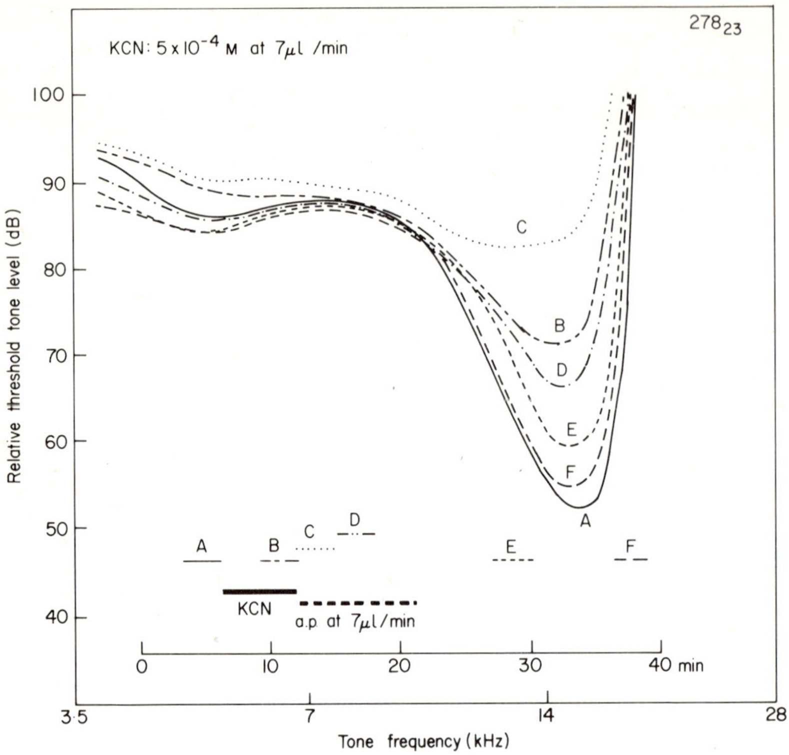

FIG. 6. Reversible effects on FTC of cat cochlear fibre of instillation of cyanide into cochlea. Inset indicates times of determinations of FTC in relation to instillation of 5×10^{-4} M potassium cyanide in artificial perilymph into the scala tympani (via a needle inserted through round window) and subsequent "washing out" with artificial perilymph (a.p.). Time for agents to reach the cochlea at instillation rate of 7 μl/min: approximately 2 min. Curve A: control FTC. Curves B and C: progressive action of cyanide eliminating lower threshold, sharply tuned segment of FTC. Curves D, E, F: almost complete recovery of FTC. After Evans & Klinke (1974).

the range of local effects of the instillation) without substantial changes in the cochlear microphonic.

Intra-arterial administration of frusemide

Frusemide is an ototoxic diuretic with a very rapid onset of action on the cochlea when administered into the vertebral circulation by injection into the subclavian artery (e.g. Brown & McElwee, 1972). The effects of such an injection on the FTC of a cat cochlear fibre are shown in Fig. 7 (Evans & Klinke, 1974, and to be published). The control FTC is the solid curve A. The second

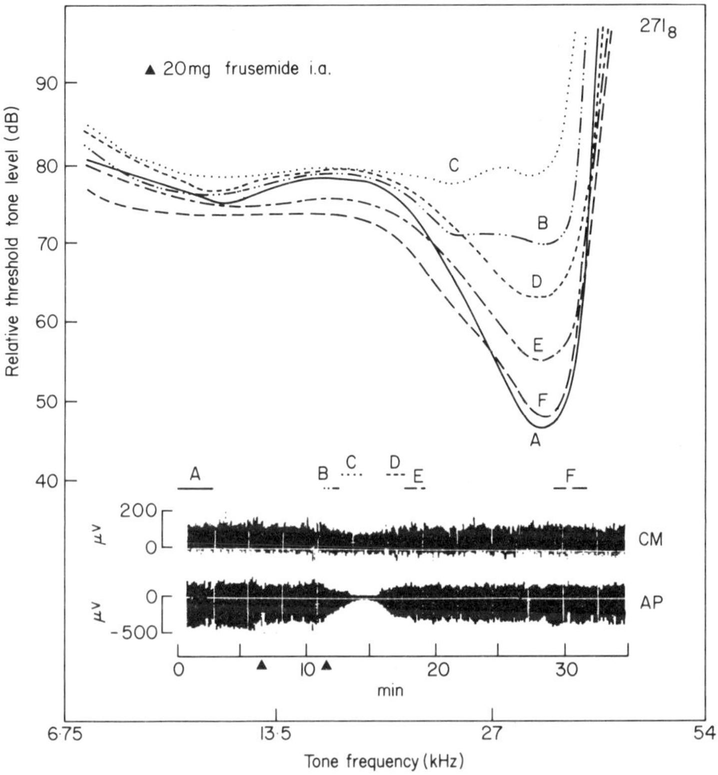

FIG. 7. Reversible effects of frusemide, a potent ototoxic diuretic, on FTC of cat cochlear fibre. Inset: cochlear microphonic (CM) and gross cochlear AP response gated from round window recording of responses to constant amplitude clicks (as Fig. 5). Arrows: injection of 20 mg frusemide into the subclavian artery of the same side as the cochlea. Note transient but substantial reduction in AP compared with slight change in the cochlear microphonic. Curve A: control FTC. Curves B, C: during action of frusemide on cochlea. Note loss of low threshold segment of FTC. Curves D, E, F: recovery. After Evans & Klinke (1974).

injection of frusemide (second arrow) was sufficiently vigorous to spill over into the vertebral circulation, and effects on the gross cochlear AP began within a few seconds. FTCs B and C were obtained during the progressive action of the agent on the cochlea, and again show complete loss of the low threshold, sharply tuned tip. Curves D, E and F were obtained during recovery. The inset records of CM amplitude show that this changed little during these effects on the FTC and AP response. With rapid re-infusion of the (sterile) urine produced during this procedure, the animal's physiological condition (e.g. blood pressure) remained stable.

IMPLICATIONS FOR NORMAL HEARING AND COCHLEAR DEAFNESS

Cochlear filtering

The above results from cat and guinea-pig indicate that the low threshold, more sharply tuned segment of the cochlear nerve FTC is very sensitive to the effects of certain antimetabolic and ototoxic procedures. These effects are reversible with sufficiently brief exposure or, if prolonged, can lead to apparently permanent loss of the sharply tuned segment of the FTCs.

The observations, (a) that the abnormal FTCs tend in shape towards the basilar membrane frequency response functions, (b) that these profound effects can be obtained reversibly within a few minutes and under conditions where it is most unlikely that the basilar membrane mechanics would be affected, and (c) that, in the case of hypoxia and frusemide action, these effects can be obtained without substantial changes in the round window cochlear microphonic (which should reflect distributed changes in cochlear mechanics), all provide direct evidence for an additional stage of filtering subsequent to the relatively broadly tuned basilar membrane. Speculation on the mechanism of this second filtering process will be deferred to pp. 155–160.

As indicated on pp. 137–139, each cochlear fibre behaves as if it were a simple linear filter, whose filtering characteristics for broad-band noise, click, and multicomponent signals of moderate levels is virtually determined only by the shape of the lowest 10 dB or less of the tip of the FTC. Measurements of the effective bandwidth of the cochlear fibre filters (Evans & Wilson, 1971, 1973) give mean values of about 0·1, 0·2, and 1 kHz for CFs at 0·2, 1 and 10 kHz respectively, i.e. they resemble 2/3, 1/4 and 1/6 octave bandpass filters respectively. (The "effective" bandwidth is the width of the equivalent rectangular filter and is derived by integrating the area under the FTC, considered as an attenuation curve in linear power and frequency coordinates; it is approximately the half-power—3 dB down—bandwidth.) It must be emphasized that although at higher levels the cochlear nerve FTCs grow wider than the above-mentioned bandwidths, their effective bandwidths remain restricted (at least for the 30–60 dB range above threshold, see pp. 137–139). That is, in terms of the influence of certain frequencies in determining the pattern of activity in a given fibre, it will be those frequencies within the effective bandwidth which are significant (see de Boer, 1970, and discussion following his paper).

Hence, it is essentially the properties of the second filtering process which will determine the filtering characteristics of the cochlear nerve to simple and complex stimuli of moderate level.

Auditory frequency selectivity in man

The working hypothesis has been developed elsewhere that the frequency selectivity of the auditory system is already developed at the level of the cochlear nerve and is limited by the tuning properties of cochlear fibres (Evans *et al.*, 1970, 1971; Evans & Wilson, 1971, 1973; Evans, 1974b, 1975, in press).

Firstly, tone-on-tone masking curves by Small (1959), Zwicker (1974), Vogten (1974) and Rodenburg, Vershuure & Brocaar (1974), using a fixed test tone level and variable masker frequency, resemble very closely the shapes of the FTCs illustrated in Fig. 2. If anything, they suggest that the discontinuity in shape from symmetrical at low CFs to asymmetrical, two segment FTCs at higher CFs, may occur at a lower frequency (about 1 kHz) in man than in cat and guinea-pig.

Secondly, there is a surprising agreement between psychophysical measures of frequency selectivity in man and analogous measures from cat cochlear fibres. The psychophysical critical band is a measure of the frequency band within which signals cannot be separated by the ear and within which signals sum and interact in various ways. The critical band can thus be understood by analogy with the bandwidth of a linear filter (Greenwood, 1961; Zwicker, 1971). The experiments mentioned on pp. 137–139 indicate that cat cochlear fibres can act as if they were linear bandpass filters. The effective bandwidth measurements we have made of these filters are in tolerable agreement with the human psychophysical critical band values (Evans & Wilson, 1971, 1973). Comparisons have also been made of more direct measurements of frequency resolving power between cat cochlear fibres and human psychophysics (Wilson & Evans, 1971; Evans & Wilson, 1973). Again the agreement is good.

Thirdly, our measurements of the cat cochlear nerve effective bandwidths represent a mean filter "Q" (i.e. CF/effective bandwidth) of about 10 for CFs from 2 to 10 kHz. This agrees well with the value derived by Duifhuis (1971) from psychophysical measurements of the temporal characteristics of the filtering ability of the human ear.

The identification of the psychophysical critical band function with the cochlear nerve filter would also account for the findings which indicate that the critical band does not require any significant amount of time to be established, apart from the "response time" of the cochlear filter (Zwicker & Fastl, 1972), because the cochlear filter characteristics hold for click stimuli also (Møller, 1970; see Evans, 1975).

Inasmuch as the cochlear nerve can be considered as a bank of overlapping cochlear fibre filters, it would be expected that speech signals would be analysed into patterns of activity within the nerve, in the manner of a speech spectrograph. Thus the individual frequency formants of the speech signal would preferentially excite fibres of the appropriate CFs. This has been directly demonstrated by Kiang & Moxon (1972), at least for speech signals of low to medium levels. At high levels, however, the limited dynamic range of individual cochlear fibres (30–60 dB) means that the rate response of most fibres will be saturated. It is therefore not obvious how fibres under these conditions can respond differentially to the components of multicomponent signals such as speech, unless some form of analysis of the temporal information carried by the discharge pattern occurs at a subsequent stage. If this were the case, however, the predominant frequency components could still be more or less limited by the effective bandwidth (see argument on p. 150).

It seems reasonable to assume that the very important second filter process will be one of the first functions to be affected in pathological conditions in man giving rise to sensorineural deafness of cochlear origin, in view of its extreme vulnerability to metabolic and ototoxic influences. This notion receives support from psychophysical measurements in patients with cochlear deafness by de Boer (1959) and Scharf & Hellmann (1966), where the critical band function was shown to be abnormally wide, indicating a breakdown of cochlear frequency selectivity. It would also be expected that the normal filtering of the formant frequency information in speech signals would be impaired in cochlear deafness, in addition to the impairment produced by selective elevation of thresholds for the high frequency components of speech. In support of this, there is evidence that a deficit in speech perception remains even after optimal amplification of speech signals (e.g. Hood & Poole, 1971).

From the experimental data reviewed above (pp. 144–149), obtained under transient abnormalities of the cochlea, it appears that the effective bandwidth (approximately 3 dB from the tip) is maintained relatively unchanged until the threshold at the CF is elevated some 30–40 dB, when it rapidly increases. This is admittedly in fibres of high CF with already high thresholds, through repeated prior exposure of the cochlea to the deleterious influences being investigated. While it remains to be seen whether similar effects obtain for fibres of lower CFs, and of lower thresholds also, it is interesting that Hood & Poole (1971) have reported that deterioration of the intelligibility of speech signals (after appropriate amplification) in patients with cochlear deafness, was not present unless the elevation in pure tone threshold exceeded an average of 31 dB; thereafter progressive deterioration occurred.

Recruitment

The hypothesis that the essential lesion in cochlear deafness is damage to the second filter mechanism, offers a simple explanation for the phenomenon of loudness recruitment and the analogous changes in the gross cochlear action potential data in man (Evans, 1972b, 1975, in press).

Most explanations for recruitment invoke two populations of cochlear fibres having high and low thresholds. In cochlear deafness it is assumed that the low threshold population is lost and that it is the high threshold population functioning alone which accounts in some way for the recruitment of loudness (e.g. Portmann *et al.*, 1973).

As has been indicated above, there is no evidence for two populations of cochlear fibres on the basis of threshold. On the contrary, the evidence reviewed here and particularly that illustrated in Fig. 3, indicates a single population of fibres, whose normally sharp tuning and low threshold properties deteriorate under conditions of cochlear pathology.

On the assumption that loudness is related to the number or proportion or spread of active fibres in the cochlear nerve, it can be schematically shown that the growth in loudness above threshold will depend upon the slopes of the cut-offs of the FTCs (Fig. 8). In the normal cochlea (left half), the size of the group of fibres activated by tone T will grow relatively slowly with intensity (lower left plot) because of the high slopes of the cut-offs, at least until the broadly tuned segment is reached in the case of fibres with higher CFs, as illustrated. If in cases of cochlear deafness, the fibres have broadly tuned FTCs (right half), then once the (elevated) thresholds are reached, the size of the active group will increase relatively more rapidly with intensity than in the normal situation.

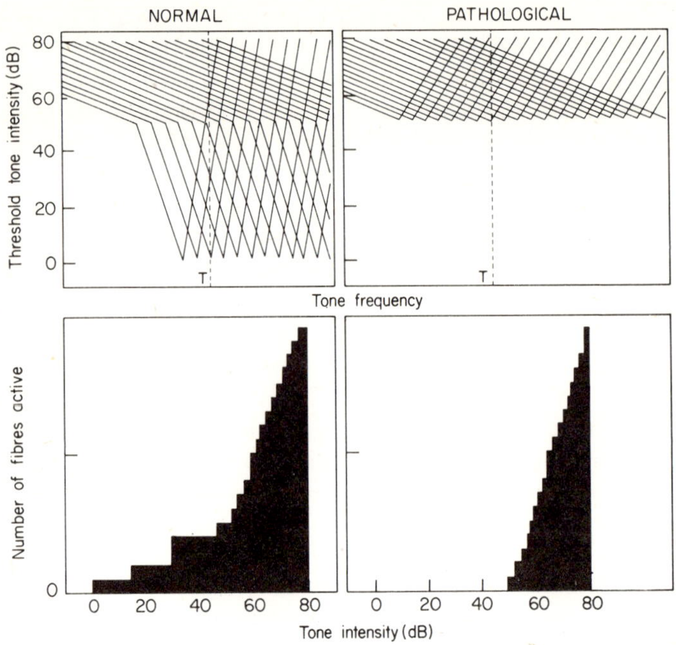

FIG. 8. Hypothetical process underlying recruitment. Schematic diagram illustrating rate of growth in number of active cochlear fibres with tone stimulus level, in normal and abnormal cochleae. Tone frequency indicated by dashed line at T. Upper half: left: sharply tuned FTCs of single cochlear fibres in normal cochlea; right: high threshold, broadly tuned FTCs of abnormal cochlea (as in Fig. 2). Lower half: growth in number of active fibres with increasing tone level, as a result of the difference in sharpness and threshold of the cochlear fibre FTCs in normal (left) and abnormal (right) cochlea. Each new fibre is added to the active group as the tone level crosses its FTC. The scales in each plot are arbitrary. In the case of the tone intensity scales in particular, the dB values are given purely for convenience in relating the upper and lower plots, and are not intended to represent absolute threshold values. After Evans (in press).

Lesions underlying sensorineural deafness of retrocochlear origin, even if they eliminated cochlear fibres, would not be expected to affect the cochlear filtering (unless the blood supply to the cochlea was impaired) because the evidence points to the second filter mechanism being entirely determined by the organ of Corti, and recruitment of loudness would not be expected, as is the case (Dix, Hallpike & Hood, 1948). On the contrary, loss of cochlear fibres would be expected to reduce the loudness of all levels relative to the normal situation.

"Pseudo-recruitment" of loudness occurs in normal subjects during and following exposure to broad-band noise. Investigations of the effects of such conditions on the shapes of FTCs indicate that while their effective bandwidth changes little with the addition of background noise, the cut-off slopes do decrease with the higher levels of noise. Similar reduction in sharpness of tuning can follow exposure of intense tones at the CF (Kiang, Moxon *et al.*, 1970).

A similar argument can be used to account for the differences observed in the growth of the gross cochlear AP response amplitude with stimulus intensity between normal and "recruiting" ears (e.g. Elberling & Salomon, 1971; Aran, 1971). Indeed there is a striking resemblance between the form of the lower plots in Fig. 8 and those obtained from determinations of the amplitude-intensity function of the gross AP generated by tone bursts in animals (e.g. Fig. 4 of Spoor & Eggermont, 1971). The complex form of the normal AP amplitude-intensity function can thus be accounted for by a single fibre population having two segment FTCs, instead of being used as evidence to the contrary (e.g. Eggermont & Spoor, 1973). It should, however, be emphasized that the gross cochlear AP represents a complex integration of cochlear and nerve potentials over cochlear space, where components will possess different time delays and thresholds and possibly polarity (e.g. Teas, Eldredge & Davis, 1962; Elberling, 1973); consequently, it is probably unwise to seek simplistic or unitary explanations of gross AP phenomena (see Evans, 1975).

Finally, studies of the growth of discharge rate with tone levels in cat cochlear fibres (see pp. 157–159) indicate that for many (but not all) fibres, the discharge rate at the CF grows more steeply after loss of the low threshold, sharply tuned tip of the FTC. This phenomenon may itself contribute to the recruitment of loudness and of the gross cochlear AP under conditions of cochlear pathology.

The origin of the second cochlear filtering process

There is as yet no satisfactory account of the mechanism(s) underlying the filtering characteristics of the cochlear nerve. Many mechanisms have been proposed to account for the substantially more sharply tuned response over that of the cochlear partition. These range from mechanical processes (e.g. "beam" hypothesis of

Huggins & Licklider, 1951; longitudinal shear model of Tonndorf, 1962; Khanna, Sears & Tonndorf, 1968), through models depending upon the pattern of innervation of the hair cells (e.g. Nieder, 1971), positive feedback models (Gold, 1948) to lateral inhibition (e.g. von Békésy, 1960, 1970; Furman & Frishkopff, 1964). Unfortunately, none of these models, in their present forms at least, appears to account quantitatively for the substantial degree of sharpening required, and at the same time to be consistent with the available physiological and anatomical data. The lateral inhibitory models appear to be specifically excluded by the comb-filtered noise experiments of Wilson & Evans (1971) and Evans & Wilson (1973).

An attractive recent model (Steele, 1973) depends upon the data which suggest that the inner hair cells, in contrast to the outer hair cells, do not make direct contact with the tectorial membrane (e.g. Lim, 1972) and therefore could be influenced by complex fluid motion through the organ of Corti. This has had some success in accounting for the FTC data.

The evidence reviewed in the previous sections suggests that any model proposed for the filtering process must include an element which is physiologically very vulnerable and which varies in characteristics from fibre to fibre (or groups of fibres). Gold's model, involving a positive feedback mechanism, has some attractions in relation to these data and to the data suggesting an inverse relationship between threshold and sharpness of tuning of cochlear fibres in the normal cochlea (Evans, 1972b; Evans & Wilson, 1973), and in relation to tone-like tinnitus. Gold's model was, however, based on estimates of cochlear "Q" which were incorrect by over an order of magnitude and were derived from incorrectly interpreted data (see Hiesey & Schubert, 1972). Furthermore, it would assume a close coupling between filter and the basilar membrane, and such a coupling is rendered unlikely by considerations concerning the $2f_1-f_2$ intermodulation products (for discussion see Wilson & Johnstone, 1973; Evans, 1975). Another class of second filter mechanism not yet explicitly treated involves some form of active "resonance" of electrochemical nature within the hair cells themselves. Such a mechanism, however, would appear to have too high a noise limitation set by Brownian motion for a system loosely coupled to the basilar membrane or other structures, if the arguments of Harris (1968) can be applied to this situation.

There are a number of lines of evidence suggesting that the cochlear nerve FTCs may be derived as a result of some kind of interaction between the inner and outer hair cells. A similar suggestion has been made by Davis (1973). The electron microscopic studies of Spoendlin (1966, 1968, 1970, 1971, 1972) in the cat cochlea indicate that no more than 5% of the cochlear nerve fibres leaving the organ of Corti originate in the outer hair cells. Thus the great majority of fibres recorded by microelectrodes in the cochlear nerve innervate the inner hair cells. Yet, while the minimum thresholds of these fibres approach the behavioural threshold, it is generally assumed (on the basis of susceptibility to acoustic damage) that it is the outer hair cells which are the more sensitive. On the other hand, this paradox would be solved if the inner hair cells or their fibres require interaction from the outer hair cells or their fibres for normal function. There are a number of lines of evidence pointing in this direction. Firstly, Kiang, Moxon *et al.* (1970), recording from cochlear fibres of cats poisoned with kanamycin, were unable to obtain normal threshold responses in regions of extensive outer hair cell loss from fibres presumed to innervate inner hair cells which were (to light microscopy at least) normal in appearance. Secondly, maximal inhibition of cochlear fibres by electrical stimulation of the crossed olivocochlear bundle occurs in fibres whose CFs correspond with the region of maximal density of efferent innervation of the outer hair cells. In contrast, the density of efferent innervation of the inner hair cell afferents appears to be more evenly distributed (Ishii & Balogh, 1968), although it must be admitted that these data are far from complete. Consequently it has been concluded (Wiederhold, 1970; Teas, Konishi & Nielsen, 1972) that the effectiveness of crossed olivocochlear bundle action corresponds more closely with the pattern of efferent innervation of the outer hair cells than with that of the inner hair cells. Thirdly, Evans (1972b) found occasional high threshold, high frequency cochlear fibres in otherwise normal guinea-pig cochleae. This could relate to the presence of sporadic outer hair cell loss in the basal turn of normal guinea-pigs (Wersäll, personal communication). Fourthly, examination of the discharge rate versus tone intensity functions at different frequencies in normal cat cochlear fibres before and during reversible elimination of the low threshold, sharply tuned segment of the FTC by hypoxia (pp. 146–147) suggests that two separate excitation mechanisms may be involved for the

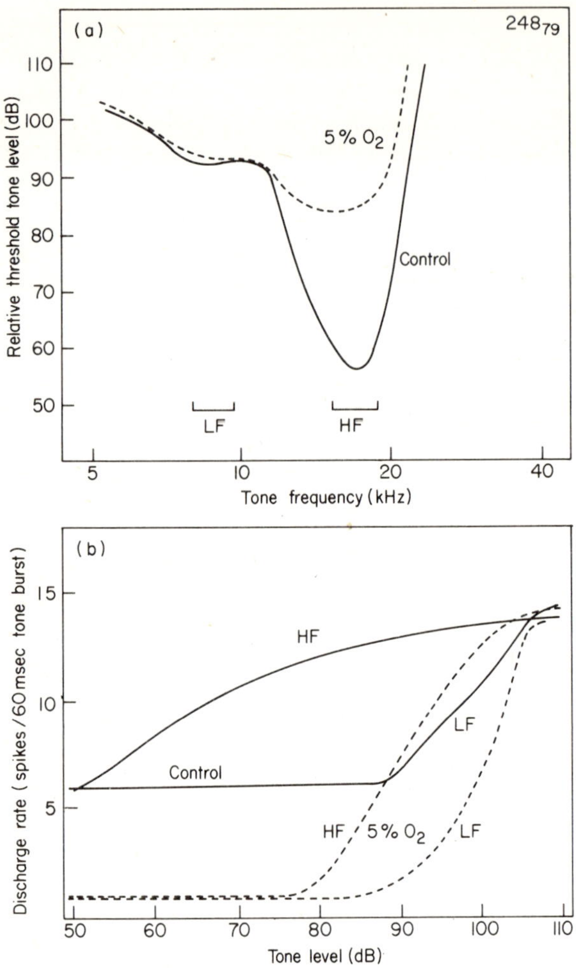

FIG. 9. Effect of hypoxia on cochlear nerve discharge rate versus tone intensity functions, in relation to changes in the FTC. (a) FTCs of single cochlear fibre in cat before (continuous line) and during (dotted line) hypoxia. Conditions similar to those of Fig. 5. (b) Plots of discharge rate of the cochlear fibre (in spikes/60 msec tone burst) as a function of the relative tone level, averaged over a band of low frequencies (LF) and a band of frequencies about the CF (HF) before (continuous lines) and during hypoxia (dotted lines), i.e. corresponding to the FTCs in the upper half of the figure. Note in the control case, the greater steepness of the low frequency (LF) function in comparison with that at the CF (HF), and the steepening of the latter under hypoxia (see text).

generation of the low threshold, sharply tuned FTC segment and the high threshold, broadly tuned segment, respectively (Fig. 9). In most (but not all) fibres, the discharge rate-intensity functions have a lower slope for frequencies corresponding to the low threshold, sharply tuned FTC segment (HF in Fig. 9), than for frequencies corresponding to the low frequency, high threshold "tail" (LF in Fig. 9). Under conditions of hypoxia, however, the rate-intensity functions for the FTC tip frequencies (HF) become steeper (dotted lines in the lower half of Fig. 9) and approximate to the slope of the function for the low frequency, high threshold tail (dotted "LF" line). Similar findings have been obtained during elimination of the sharply tuned segment by intracochlear cyanide (pp. 147–148; Evans & Klinke, unpublished data). If the low and high slope rate-intensity functions correspond to excitatory mechanisms associated with the outer and inner hair cell systems respectively, then loss of outer hair cell function would account for the change and direction of change of slope with loss of the FTC low threshold segment system in response to antimetabolic and ototoxic influences. This preferential loss of outer hair cell function is implied by their relative susceptibility to chronic ototoxic influences (e.g. Davis *et al.*, 1958; Hawkins, 1959; Ward & Fernandez, 1961; Nakai, 1971; Kaku, Farmer & Hudson, 1973), which the latter authors concluded was a result of the action of these agents on the respiratory enzymes contained within the outer cells. In this connection it is interesting that one of the effects of electrical stimulation of the crossed olivocochlear bundle is to decrease the sharpness of tuning of cochlear fibres (Wiederhold, 1970). In addition, it is evident from Wiederhold's data that the slope of the discharge rate-intensity function at the CF steepens in the same manner as shown in Fig. 9. Perhaps this indicates a selective reduction of outer cell activity by the effects of olivocochlear bundle stimulation.

For fibres of high CF, therefore, this speculative model proposes that the outer hair cells are responsible for the low threshold, sharply tuned segment and the inner hair cells for the high threshold, broadly tuned segment of the FTC.

In the absence of signs of synaptic contact or branching (*contra* Nieder & Nieder, 1970) between outer spiral and inner radial fibres, these considerations lead to the speculation that interaction between one of the former and 20 of the latter takes place in each habenular opening, where they become intimately apposed (Smith & Dempsey, 1957; Spoendlin, 1970; Arthur, Pfeiffer & Suga, 1971; Lynn & Sayers, 1970). Considerations of the length and diameter of the outer spiral fibres would appear to rule out electronic conduction (*contra* Lynn & Sayers, 1970) and lead to the suggestion that action potentials propagated along them could produce sufficient local current to generate action potentials at the inner segment region of the inner radial fibres. A feature of this interaction would be that on elimination of the outer hair cell input, and consequently the sharply tuned, low threshold segment of the FTC, the CF would be expected to shift towards lower frequencies by approximately one-fifth of an octave (corresponding to an outer-spiral fibre length of 0·6 mm, as Spoendlin, 1970). This is in fact observed in Figs 5–7.

Such a model would, of course, require the outer hair cells and/or their innervation to be sharply tuned and the inner hair cell properties to mirror those of the basilar membrane. Unfortunately, this speculation is not supported by the close correspondence between the cochlear microphonic and the filtering properties of the basilar membrane (Dallos, 1973; Wilson, 1974) if, as Dallos (1973) suggests, the outer hair cells are almost entirely responsible for the generation of cochlear microphonic. Furthermore, it is the inner cells which have the greater possibility of sharpening by hydraulic means (see above).

An alternative possibility is interaction between the inner hair cells and local electrical currents generated by the outer hair cells. These are not likely to be in phase with the inner hair cell currents (Dallos, Billone, Durrant, Wang & Raynor, 1972). If sufficient phase differences could exist between these inner and outer hair cell currents, the otherwise curious abrupt reduction at high stimulus levels in cochlear fibre discharge rate (Kiang, Baer, Marr & Demont, 1969; Kiang & Moxon, 1972), and in cochlear microphonic (Karlan, Tonndorf & Khanna, 1972) could be accounted for in terms of cancellation in the region of overlap of the outer hair cell and inner hair cell currents. Clearly, the need is for more data, before the roles of the inner and outer hair cells and their innervations can be established.

Acknowledgements

I am grateful to Drs J. P. Wilson and R. Klinke for many helpful discussions and to the former for criticisms of the manuscript. The work described has been supported in part by the Medical and Science Research Councils and the Deutsche Forschungsgemeinschaft (Kl 219).

REFERENCES

Aran, J. M. (1971). The electrocochleogram: recent results in children and in some pathological cases. *Arch. Ohr. Nas.-Kehlk.* **198**: 128–141.

Arthur, R. M., Pfeiffer, R. R. & Suga, N. (1971). Properties of "two-tone inhibition" in primary auditory neurones. *J. Physiol., Lond.* **212**: 593–609.

von Békésy, G. (1944). Über die mechanische Frequenzanalyze in der Schnecke vershiedener Tiere. *Akust. Z.* **9**: 3–11.

von Békésy, G. (1960). *Experiments in hearing.* New York: McGraw-Hill.

von Békésy; G. (1970). Travelling waves as frequency analysers in the cochlea. *Nature, Lond.* **225**: 1207–1209.

de Boer, E. (1959). Measurement of critical band-width in cases of perception deafness. *Int. Congr. Acoust.* **3 (1)**: 100–102. Amsterdam: Elsevier.

de Boer, E. (1969). Reverse correlation II. Initiation of nerve impulses in the inner ear. *Proc. K. ned. Akad. Wet.* **72**: 129–151.

de Boer, E. (1970). Synchrony between acoustic stimuli and nerve fibre discharges. In *Frequency analysis and periodicity detection in hearing*: 204–212. Plomp, R. & Smoorenburg, G. F. (eds). Leiden: Sijthoff.

de Boer, E. & Jongkees, L. B. W. (1968). On cochlear sharpening and cross-correlation methods. *Acta otolar.* **65**: 97–104.

Brown, R. D. & McElwee, T. W., Jr. (1972). Effects of intra-arterially and intravenously administered ethacrynic acid and frusemide on cochlear N_1 in cats. *Toxicol. & appl. Pharmacol.* **22**: 589–594.

Butler, R. A., Honrubia, V., Johnstone, B. M. & Fernandez, C. (1962). Cochlear function under metabolic impairment. *Ann. Otol. Rhinol. Lar.* **71**: 648–656.

Dallos, P. (1973). Cochlear potentials and cochlear mechanics. In *Basic mechanisms in hearing*: 335–372. Møller, A. R. (ed.). New York: Academic Press.

Dallos, P. (1974). Cochlear potential correlates of cubic difference tones. In *Facts and models in hearing:* 312–322. Zwicker, E. & Terhardt, E. (eds). Heidelberg: Springer-Verlag.

Dallos, P., Billone, M. C., Durrant, J. D., Wang, C.-y. & Raynor, S. (1972). Cochlear inner and outer hair cells: functional differences. *Science, Wash.* **177**: 356–358.

Davis, H. (1973). The cocktail hour before the serious banquet. In *Basic mechanisms in hearing*: 259–271. Møller, A. R. (ed.). New York: Academic Press.

Davis, H., Deatherage, B. H., Rosenblut, B., Fernandez, C., Kimura, R. & Smith, C. A. (1958). Modifications of cochlear potentials produced by streptomycin poisoning and by venous obstruction. *Laryngoscope, St Louis* **68**: 596–627.

Dix, M. R., Hallpike, C. S. & Hood, J. D. (1948). Observations upon the loudness recruitment phenomenon with especial reference to the differential diagnosis of disorders of the internal ear and VIII nerve. *Proc. R. Soc. Med.* **41**: 516–526.

Duifhuis, H. (1971). Audibility of high harmonics in a periodic pulse. II. Time effect. *J. acoust. Soc. Am.* **49**: 1155–1162.

Eggermont, J. J. & Spoor, A. (1973). Cochlear adaptation in guinea pigs. *Audiology* **12**: 193–220.

Elberling, C. (1973). Transition in cochlear action potentials recorded from the ear canal in man. *Scand. Audiol.* **2**: 151–159.

Elberling, C. & Salomon, G. (1971). Electrical potentials from the inner ear in man, in response to transient sounds generated in a closed acoustic system. *Rev. Laryng.* Suppl. [**1971**]: 691–706.

Evans, E. F. (1970a). Narrow "tuning" of cochlear nerve fibre responses in the guinea pig. *J. Physiol., Lond.* **206**: 14–15P.

Evans, E. F. (1970b). Narrow "tuning" of the responses of cochlear nerve fibres emanating from the exposed basilar membrane. *J. Physiol., Lond.* **208**: 75–76P.

Evans, E. F. (1972a). Does frequency sharpening occur in the cochlea? In *Hearing theory 1972*: 27–34. Eindhoven: IPO.

Evans, E. F. (1972b). The frequency response and other properties of single fibres in the guinea-pig cochlear nerve. *J. Physiol., Lond.* **226**: 263–287.

Evans, E. F. (1974a). The effects of hypoxia on the tuning of single cochlear nerve fibres. *J. Physiol.* **238**: 65–67P.

Evans, E. F. (1974b). Auditory frequency selectivity and the cochlear nerve. In *Facts and models in hearing*: 118–129. Zwicker, E. & Terhardt, E. (eds). Heidelberg: Springer-Verlag.

Evans, E. F. (1975). The cochlear nerve and cochlear nucleus. In *Handbook of sensory physiology* **5** (2): 1–108. Keidel, W. D. & Neff, W. D. (eds). Heidelberg: Springer-Verlag.

Evans, E. F. (in press). The sharpening of cochlear frequency selectivity in the normal and abnormal cochlea. *Audiology*.

Evans, E. F. & Klinke, R. (1974). Reversible effects of cyanide and furosemide on the tuning of single cochlear fibres. *J. Physiol., Lond.* **242**: 129–131P.

Evans, E. F., Rosenberg, J. & Wilson, J. P. (1970). The effective bandwidth of cochlear nerve fibres. *J. Physiol.* **207**: 62–63P.

Evans, E. F., Rosenberg, J. & Wilson, J. P. (1971). The frequency resolving power of the cochlea. *J. Physiol.* **216**: 58–59P.

Evans, E. F. & Wilson, J. P. (1971). Frequency sharpening of the cochlea: the effective bandwidth of cochlear nerve fibres. *Int. Congr. Acoust.* **7 (3)**: 453–456. Budapest: Akademiai Kiado.

Evans, E. F. & Wilson, J. P. (1973). Frequency selectivity of the cochlea. In *Basic mechanisms in hearing*: 519–551. Møller, A. R. (ed.). New York: Academic Press.

Furman, G. G. & Frishkopff, L. S. (1964). Model of neural inhibition in the mammalian cochlea. *J. acoust. Soc. Am.* **36**: 2194–2201.

Galley, N., Klinke, R., Oertel, W., Pause, M. & Storch, W.-H. (1973). The effects of intracochlearly administered acetylcholine-blocking agents on the efferent synapses of the cochlea. *Brain Res.* **64**: 55–63.

Gold, T. (1948). Hearing II. The physical basis of the action of the cochlea. *Proc. R. Soc.* (B) **135**: 492–498.

Greenwood, D. D. (1961). Auditory masking and the critical band. *J. acoust. Soc. Am.* **33**: 484–502.

Harris, G. G. (1968). Brownian motion and the threshold of hearing. *Int. Audiol.* **7**: 111–120.

Hawkins, J. E., Jr. (1959). The ototoxicity of kanamycin. *Ann. Otol.* **68**: 698–715.

Hiesey, R. W. & Schubert, E. D. (1972). Cochlear resonance and phase-reversal signals. *J. acoust. Soc. Am.* **51**: 518–519.

Hood, J. D. & Poole, J. P. (1971). Speech audiometry in conductive and sensorineural hearing loss. *Sound* **5**: 30–38.

Huggins, W. H. & Licklider, J. C. R. (1951). Place mechanisms of auditory frequency analysis. *J. acoust. Soc. Am.* **23**: 290–299.

Huxley, A. R. (1969). Is resonance possible in the cochlea after all? *Nature, Lond.* **221**: 935–940.

Ishii, D. & Balogh, K., Jr. (1968). Distribution of efferent nerve endings in the organ of Corti. *Acta otolar.* **66**: 282–288.

Johnstone, B. M. & Boyle, A. J. (1967). Basilar membrane vibration examined with the Mössbauer technique. *Science, N.Y.* **158**: 389–390.

Johnstone, B. M., Taylor, K. J. & Boyle, A. J. (1970). Mechanics of the guinea pig cochlea. *J. acoust. Soc. Am.* **47**: 504–509.

de Jongh, H. R. (1972). About coding in the VIIIth nerve. In *Hearing theory 1972*: 70–77. Eindhoven: IPO.

Kaku, Y., Farmer, J. C. & Hudson, W. R. (1973). Ototoxic drug effects on cochlear histochemistry. *Arch. Otolar.* **98**: 282–286.

Karlan, M. S., Tonndorf, J. & Khanna, S. M. (1972). Dual origin of the cochlear microphonics: inner and outer hair cells. *Ann. Oto-Rhino-Lar.* **81**: 696–704.

Katsuki, Y., Suga, N. & Kanno, Y. (1962). Neural mechanisms of the peripheral and central auditory system in monkeys. *J. acoust. Soc. Am.* **34**: 1396–1410.

Katsuki, Y., Sumi, T., Uchiyama, H. & Watenabe, T. (1958). Electric responses of auditory neurones in cat to sound stimulation. *J. Neurophysiol.* **21**: 569–588.

Keidel, W. D. (1965). Physiologie des Innenohres. *Hals-Nasen-Ohren-Heilk.* **3**: 235–310.

Khanna, S. M., Sears, R. E. & Tonndorf, J. (1968). Some properties of longitudinal shear waves: a study of computer simulation. *J. acoust. Soc. Am.* **43**: 1077–1084.

Kiang, N. Y.-s. (1968). A survey of recent developments in the study of auditory physiology. *Ann. Oto-Rhino-Lar.* **77**: 656–676.

Kiang, N. Y.-s., Baer, T., Marr, E. M. & Demont, D. (1969). Discharge rates of single auditory nerve fibres as functions of tone level. *J. acoust. Soc. Am.* **46**: 106.

Kiang, N. Y.-s. & Moxon, E. C. (1972). Physiological considerations in artificial stimulation of the inner ear. *Ann. Oto-Rhino-Lar.* **81**: 714–730.

Kiang, N. Y.-s., Moxon, E. C. & Levine, R. A. (1970). Auditory-nerve activity in cats with normal and abnormal cochleas. In *Sensorineural hearing loss*: 241–268. Wolstenholme, G. E. W. & Knight, J. (eds). London: Churchill.

Kiang, N. Y.-s., Sachs, M. B. & Peake, W. T. (1967). Shapes of tuning curves for single auditory nerve fibres. *J. acoust. Soc. Am.* **42**: 1341–1342.

Kiang, N. Y.-s., Watenabe, T., Thomas, E. C. & Clark, L. F. (1965). *Discharge patterns of single fibres in the cat's auditory nerve.* Cambridge, Mass: MIT Press.

Kohloffel, L. U. E. (1973). Observations of the mechanical disturbances along the basilar membrane with laser illumination. In *Basic mechanisms in hearing*: 93–113. Møller, A. R. (ed.). New York: Academic Press.

Lim, D. J. (1972). Fine morphology of the tectorial membrane. *Archs Otolaryng.* **96**: 199–215.

Lynn, P. A. & Sayers, B. McA. (1970). Cochlear innervation, signal processing, and their relation to auditory time-intensity effects. *J. acoust. Soc. Am.* **47**: 525–533.

Møller, A. R. (1970). Unit responses in the cochlear nucleus of the rat to noise and tones. *Acta physiol. scand.* **78**: 289–298.

Nakai, Y. (1971). Electronic microscopic study of the inner ear after ethacrynic acid intoxication. *Pract. Oto-Rhino-Lar.* **33**: 366–376.

Nieder, P. (1971). Addressed exponential delay line theory of cochlear organization. *Nature, Lond.* **230**: 255–257.

Nieder, P. & Nieder, I. (1970). Antimasking effect of crossed olivocochlear bundle stimulation with loud clicks in guinea pig. *Expl Neurol.* **28**: 179–188.

Nomoto, M., Suga, N. & Katsuki, Y. (1964). Discharge pattern and inhibition of primary auditory nerve fibres in the monkey. *J. Neurophysiol.* **27**: 768–787.

Pfeiffer, R. R. & Kim, D. O. (1973). Considerations of nonlinear response properties of single cochlear nerve fibres. In *Basic mechanisms in hearing*: 555–587. Møller, A. R. (ed.). New York: Academic Press.

Portmann, M., Aran, J. M. & Lagourgue, P. (1973). Testing for "recruitment" by electrocochleography: preliminary results. *Ann. Otol.-Rhinol.-Lar.* **82**: 36–43.

Rhode, W. S. (1971). Observations of the vibration of the basilar membrane in squirrel monkeys using the Mössbauer technique. *J. acoust. Soc. Am.* **49**: 1218–1231.

Rhode, W. (1973). An investigation of cochlear mechanics using the Mössbauer effect. In *Basic mechanisms in hearing*: 49–63. Møller, A. R. (ed.). New York: Academic Press.

Rodenburg, M., Vershuure, J. & Brocaar, M. P. (1974). Comparison of two masking methods. *Acustica* **31**: 99–106.

Rupert, A., Moushegian, G. & Galambos, R. (1963). Unit responses to sound from auditory nerve of the cat. *J. Neurophysiol.* **26**: 449–465.

Scharf, B. & Hellmann, R. P. (1966). Model of loudness summation applied to impaired ears. *J. acoust. Soc. Am.* **40**: 71–78.

Siebert, W. M. (1970). Frequency discrimination in the auditory system: place or periodicity mechanisms? *Proc. Inst. electr. electron. Eng.* **58**: 723–730.

Small, A. M., Jr. (1959). Pure tone masking. *J. acoust. Soc. Am.* **31**: 1619–1625.

Smith, C. A. & Dempsey, E. W. (1957). Electron microscopy of the organ of Corti. *Am. J. Anat.* **100**: 337.

Smoorenburg, C. F. (1972). Combination tones and their origin. *J. acoust. Soc. Am.* **52**: 615–632.

Spoendlin, H. (1966). The organization of the cochlear receptor. *Adv. Oto-Rhino-Lar.* **13**: 1–227.

Spoendlin, H. (1968). Ultrastructure and peripheral innervation pattern of the receptor in relation to the first coding of the acoustic message. In *Hearing mechanisms in vertebrates*: 89–119. de Reuck, A. V. S. & Knight, J. (eds). London: Churchill.

Spoendlin, H. (1970). Structural basis of peripheral frequency analysis. In *Frequency analysis and periodicity detection in hearing*: 2–36. Plomp, R. & Smoorenburg, G. F. (eds). Leiden: Sijthoff.

Spoendlin, H. (1971). Degeneration behaviour of cochlear nerve. *Arch. Ohr. Nas.-Kehlk.* **200**: 275–291.

Spoendlin, H. (1972). Innervation densities of the cochlea. *Acta otolar.* **73**: 235–248.

Spoor, A. & Eggermont, J. J. (1971). Action potentials of the cochlea. Masking, adaptation and recruitment. *Audiology* **10**: 340–352.

Steele, C. R. (1973). A possibility for sub-tectorial membrane fluid motion. In *Basic mechanisms in hearing*: 69–90. Møller, A. R. (ed.). New York: Academic Press.

Tasaki, I. (1954). Nerve impulses in individual auditory nerve fibres of guinea pig. *J. Neurophysiol.* **17**: 97–122.

Teas, D. C., Eldredge, D. H. & Davis, H. (1962). Cochlear responses to acoustic transients: an interpretation of wholenerve action potentials. *J. acoust. Soc. Am.* **34**: 1438–1459.

Teas, D. C., Konishi, T. & Nielsen, D. W. (1972). Electrophysiological studies on the spatial distribution of the crossed olivocochlear bundle along the guinea pig cochlea. *J. acoust. Soc. Am.* **51**: 1256–1264.

Tonndorf, J. (1962). Time/frequency analysis along the partition of cochlear models: a modified place concept. *J. acoust. Soc. Am.* **34**: 1337–1350.

Vogten, L. L. M. (1974). Pure tone masking; a new result from a new method. In *Facts and models in hearing*: 142–155. Zwicker, E. & Terhardt, E. (eds). Heidelberg: Springer-Verlag.

Ward, P. H. & Fernandez, C. (1961). The ototoxicity of kanamycin in guinea-pigs. *Ann. Otol.* **70**: 132–143.

Wiederhold, M. L. (1970). Variations in the effects of electric stimulation of the crossed olivocochlear bundle on cat single auditory-nerve fibre responses to tone bursts. *J. acoust. Soc. Am.* **48**: 966–977.

Wilson, J. P. (1974). Basilar membrane vibration data and their relation to theories of frequency analysis. In *Facts and models in hearing*: 56–63. Zwicker, E. & Terhardt, E. (eds). Heidelberg: Springer-Verlag.

Wilson, J. P. & Evans, E. F. (1971). Grating acuity of the ear: psychophysical and neurophysiological measures of frequency resolving power. *Int. Congr. Acoust.* **7 (3)**: 397–400. Budapest: Akademiai Kiado.

Wilson, J. P., Evans, E. F. & Rosenberg, J. (1974). Linearity of the cochlear nerve fibre filter response: a test for the influence of two-tone suppression. *Int. Congr. Acoust.* **8 (1)**: 180. Trowbridge: Goldcrest Press.

Wilson, J. P. & Johnstone, J. R. (1972). Capacitive probe measures of basilar membrane vibration. In *Hearing theory 1972*: 172–181. Eindhoven: IPO.

Wilson, J. P, & Johnstone, J. R. (1973). Basilar membrane correlates of the combination tone $2f_1–f_2$. *Nature, Lond.* **241**: 206–207.

Zwicker, E. (1970). Masking and physiological excitation as consequences of the ear's frequency analysis. In *Frequency analysis and periodicity detection in hearing*: 376–394. Plomp, R. & Smoorenburg, G. F. (eds). Leiden: Sijthoff.

Zwicker, E. (1971). Introduction to round-table-discussion on "critical bands". *Int. Congr. Acoust.* **7 (1)**: 189–192. Budapest: Akademiai Kiado.

Zwicker, E. (1974). On a psychoacoustical equivalent of tuning curves. In *Facts and models in hearing*: 132–140. Zwicker, E. & Terhardt, E. (eds). Heidelberg: Springer-Verlag.

Zwicker, E. & Fastl, H. (1972). On the development of the critical band. *J. acoust. Soc. Am.* **52**: 699–702.

Symp. zool. Soc. Lond. (1975) No. 37, 167–186.

GROSS RESPONSES FROM THE COCHLEAR NERVE IN MAN AND IN THE GUINEA-PIG

S. D. G. STEPHENS

Institute of Sound and Vibration Research, University of Southampton, Southampton, England

R. CHARLET de SAUVAGE and J.-M. ARAN

Laboratoire d'Audiologie Expérimentale, Centre Régionale de Phono-Audiologie, Université de Bordeaux II, France

SYNOPSIS

Electrical activity in the cochlear nerve was first demonstrated at the end of the last century, and its main components were separated in the late 1930s. The compound action potential response was first recorded in man in 1960, and with the application of averaging techniques considerable advances have been made over the past seven years in the understanding of this response.

Recently one of the points of interest in this response has concerned the irregular intensity-amplitude growth function, particularly in human studies. The results of various experiments investigating this problem are discussed and data on the changes in this growth function in the response to successive clicks presented in short trains are given. Analogous results in this respect are found in human studies and studies on the guinea-pig. These all show that in the human and the guinea-pig with normal responses to single clicks, there is maximal adaptation in the response to successive clicks of a train in the region of the knee-point or discontinuity in the intensity-amplitude growth function.

INTRODUCTION AND REVIEW

The first investigation of the electrical responses in the cochlear nerve was performed by Beauregard & Dupuy in 1896. They exposed the cochlear nerve of frogs and guinea-pigs and placed an electrode on this after cutting it, with a reference electrode on the tympanic membrane. On stimulating the animals with intense sounds, they obtained small deflections on a galvanometer connected between the electrodes. The deflections disappeared on the death of the animal so were assumed to be genuine action currents. These results were repeated by two subsequent workers, but the major advances in this field were dependent on the use of the valve amplifier and came with the work of Wever & Bray in 1930.

These two workers described the response known as the cochlear microphonic, which reproduces faithfully the form of the sound wave as represented at the inner ear. Subsequent work by

Tasaki, Davis & Eldredge (1954) strongly suggests that this potential is generated at the surface of hair cells. These potentials have been extensively investigated in animals by Teas, Eldredge & Davis (1962) and by Sohmer, Peake & Weiss (1971). They have been recorded during surgical intervention in man by Fromm, Nylen & Zotterman (1935), and other workers, and more recently with less traumatic techniques by Yoshie & Yamaura (1969) and by Elberling & Salomon (1973). However, perhaps because of their more limited diagnostic applications, they have been considerably less studied than the whole-nerve compound action potentials.

These action potentials were first described by Saul & Davis (1932) who separated them from the cochlear microphonics which had been originally considered by Wever & Bray (1930) to be summated action potentials. These findings were elaborated by Derbyshire & Davis (1935) and other workers, and extended in guinea-pigs by Teas *et al.* (1962) using differential electrodes placed in the scala media and the scala tympani. These workers comment on the changes in both response latency and response amplitude with increasing stimulus intensity occurring in an irregular manner. The curve for each was made up of two steeply sloping segments joined by a less steep segment. Teas *et al.* (1962) also noted that the stimulus levels at which these slopes changed were the same for amplitude and for latency.

The first recordings of the action potentials in man were reported by Ruben, Sekula, Bordley, Knickerbocker, Nager & Fisch (1960). Their recordings were made while the middle ear was exposed surgically and an active electrode could be placed on the round window for recording both the gross action potential and the cochlear microphonic.

The work of Ruben and his co-workers was limited by the poor signal-to-noise ratios obtained with such an approach, and it was the application of averaging techniques to cochlear nerve measures which led to the increased activity in this field which has come over the past seven years.

The breakthrough came in 1967 when Yoshie, Ohashi & Suzuki; Portmann, Le Bert & Aran; Sohmer & Feinmesser, and Spreng & Keidel, working independently and using different techniques, recorded human cochlear nerve compound action potentials without surgical intervention. Yoshie *et al.* (1967) inserted their active electrode beneath the skin of the external acoustic meatus close to the tympanic membrane; Portmann, Le Bert *et al.* (1967) passed a needle electrode through the tympanic membrane and on to the bony promontory; Sohmer & Feinmesser (1967) used an active electrode on the ear lobe; and Spreng & Keidel (1967) on the mastoid process or the hard palate. Since that time there has been some rationalization of the approach and most laboratories now use either the transtympanic promontory electrode or a surface-electrode approach, although Salomon & Elberling (1971) have adopted a technique in which a wire electrode is inserted beneath the skin of the external acoustic meatus from behind the pinna, thus avoiding any obstruction of the meatal cavity.

Some considerable controversy has raged since this time with regard to the relative advantages of the two main methods, and their various advantages and disadvantages have been extensively discussed elsewhere (Aran & Portmann, in press). Basically, the transtympanic approach has the advantage that large responses are obtained (up to 40 μV), so that details of their form and growth function can be more easily evaluated in terms of diagnostic information. Being large responses, they can be separated from the background noise with relatively simple computing techniques and equipment.

The advantages of the surface electrode technique are that it is less traumatic to the patient, and that responses apparently coming from higher centres in the auditory pathway may be obtained simultaneously (Sohmer & Feinmesser, 1971; Rutt, Stephens & Thornton, 1973; Terkildsen, Osterhammel & Huis in't Veld, 1973). The main disadvantage is the small size of the responses obtained, less than 0·5 μV, even at high stimulus intensities (Lieberman, Sohmer & Szabo, 1973).

The studies on the growth of response amplitude with increasing stimulus intensity in man, obtained using the transtympanic technique, showed an abrupt change in the slope of the curve in the mid-range analogous to and more pronounced than that found by Teas *et al.* (1962) and other workers in the guinea-pig. The two sections with a gradual slope at low intensities and a steeper slope at the higher intensities have been called the "L Curve" and "H Curve" respectively, by Yoshie (1968) and may be seen in Fig. 2a. This form of response has been reported by other workers in the field (Aran & Le Bert, 1968; Elberling, 1973; Eggermont & Odenthal, 1974) who have approached the problem of its origin in a number of different ways.

Aran & Portmann consider, on the basis both of the results obtained from clinical patients with well defined auditory lesions

(Portmann, Aran & Lagourgue, 1973), and from experimental guinea-pigs exposed to noise trauma (Portmann, Aran & Coste, 1973) and guinea-pigs treated with kanamycin (Aran & Darrouzet, in press), that the "L Curve" and the "H Curve" are related to the potentials arising from the outer hair cells and the inner hair cells respectively. In patients with high frequency end-organ hearing loss and relatively normal hearing at low frequencies, they find a "dissociated" electrocochleographic response with a long latency, low amplitude, broad response at low intensities and a short latency, narrow response at high intensities. The argument is that the high sensitivity outer hair cells in the basal region of the cochlea are destroyed, leaving the low sensitivity inner hair cells intact. In more apical parts of the cochlea the outer hair cells remain and thus give rise to the long latency broad response to the clicks. However, at higher intensities the inner hair cells persisting in the basal part of the cochlea give rise to the components of the short latency, high amplitude action potential. This argument is supported by the use of high frequency filtered clicks, with which only the short latency, high amplitude part of the response is obtained from these patients.

Elberling (1973), in a detailed study of the different components of the response waveform to monophasic half sinusoids using 2 dB stimulus steps, has shown that in the region of the knee-point there is a fairly abrupt transition between two peaks of different latencies. The relative amplitudes of the two peaks change at this point so that further increase in intensity results in a considerable increase in the amplitude of the earlier peak with no increase in that of the later peak. The earlier peak thus becomes the main component of the compound action potential. He argues that the two peaks arise from different parts of the cochlear partition. A subsequent study (Elberling, 1974) using half sinusoids presented together with bursts of high-pass filtered noise with various cut-off frequencies, supported the concept, arising from this latency study, that the main component of the whole-nerve action potential arises from different parts of the basilar membrane as the stimulus intensity is changed.

Eggermont & Spoor (1973) have proposed two populations of neural units differing in a number of characteristics. This was based on careful studies of the form of the latency, amplitude and action potential width stimulus-response curves in the guinea-pig, under conditions in which much adaptation had occurred. Eggermont & Odenthal (1974) point out that these differences are greater in man. They suggest two populations of "nerve fibre units"; long latency, low intensity units (0–50 dB HL in man, 0–30 dB HL in the guinea-pig) and short latency, high intensity units (40–>100 dB HL in man, 30–>100 dB HL in the guinea-pig).

In a subsequent study (Eggermont & Odenthal, in press) they used click- and tone-burst stimuli presented in a background of white noise passed through a ½-octave band-stop filter, and investigated the effects of the use of this selective filtering device on the amplitude input-output functions. (When broad-band noise is passed through a band-stop filter, the filter cuts out the frequency band centred on the frequency of the filter, but allows higher and lower frequencies to pass.) In this study they found that up to 65 dB HL the band-stop-masked responses to tone-bursts of the band-stop frequency resembled the unmasked responses to the same stimuli. Above that level, however, the amplitude of the masked responses showed little or no further increase with stimulus intensity. These results thus suggest that above a critical intensity level (65 dB HL) the contributions to further increases in action potential amplitude with increasing stimulus intensity come from activation of other parts along the cochlear partition. This would appear to be compatible with the subjective phenomenon of the threshold of overload of the ear, in which aural harmonics are produced when stimulus intensity exceeds a similar critical level (Lawrence & Yantis, 1956).

The present study was conducted to investigate further the effect of adaptation to rapidly repeated stimuli on the form of the amplitude-intensity growth function of the compound action potential responses in man, and to investigate in more detail certain aspects of this response in the guinea-pig.

MATERIAL AND METHOD

Equipment

Figure 1 shows a block diagram of the stimulus generating and recording equipment. The stimulus generator was controlled by an Intertechnique Histomat S averaging computer which gated an Ortec 4656 pulse generator. The output of this was taken to an Ortec 4654 delay and duration control unit and thence to a Racal alternate inverter. The function of this was to produce 180° phase reversal of alternate pulses. The output of this was taken through an Allison 2AB variable filter which was used as such only for the filtered click experiments described in the guinea-pig section of the

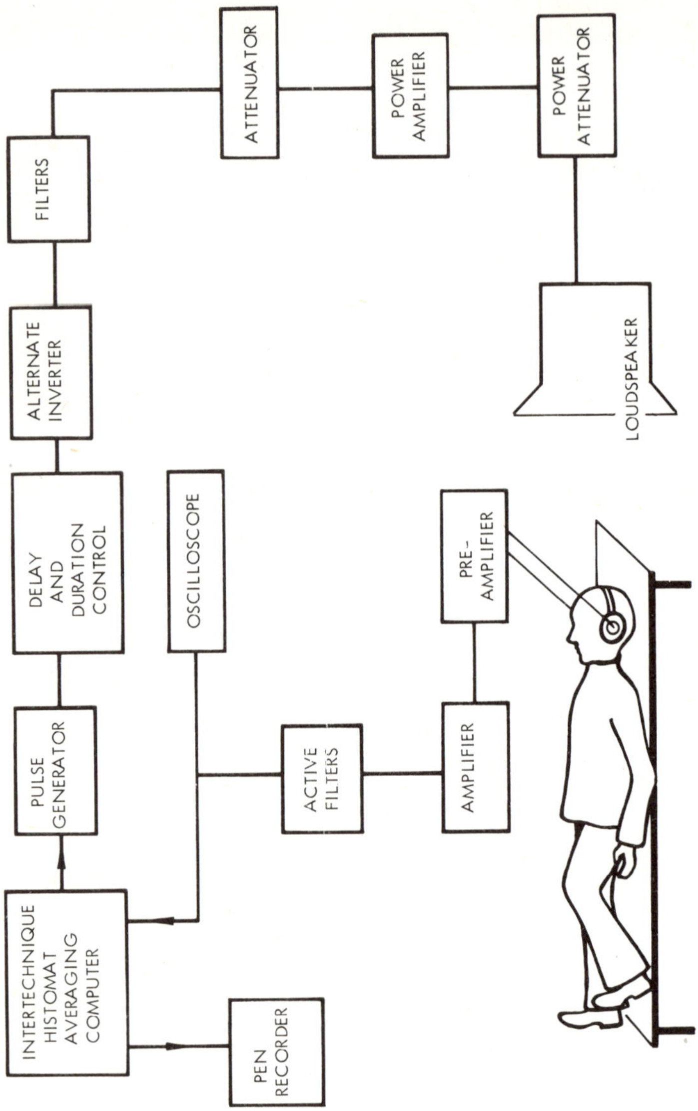

FIG. 1. Block diagram of equipment.

study. The output from this was taken to a Hewlett Packard 4437A 600Ω attenuator, which was generally used to set up stimulus levels, and then to a Bruel and Kjaer 2706 power amplifier, a 30 dB power attenuator and thence either to the loudspeaker for the human studies, or to a TDH-39 headset for the guinea-pig work.

The electrodes were connected to a Princeton PAR 113 preamplifier and then to an Intertechnique AT 210 low gain amplifier. From this the signal passed through Intertechnique ZT 210 active filters and into the Histomat S averaging computer. The input was also monitored on an oscilloscope and on a small loudspeaker to which it was taken via a monitoring amplifier. The averages from the Histomat S were recorded using an Allco A53 pen recorder.

In both human and guinea-pig studies a gain of 1000 was used on the PAR 113 preamplifier. For most guinea-pig studies a gain of 1 was used on the AT 210 amplifier although a higher gain was used when the stimulus intensity was near the response threshold. A gain of 16 was always used for the human studies. For both the human and animal studies a filter bandwidth of 3 Hz–10 kHz was used on the ZT 210 active filters.

Human study

The subjects used for this study were 11 patients tested in the audiology unit of the ENT clinic at Bordeaux, who had been found to have normal electrocochleographic responses to click stimuli. The technique used to record the cochlear responses has been described elsewhere in considerable detail (Aran & Le Bert, 1968) and involved passing a needle electrode through the posterior part of the tympanic membrane and on to the bony promontory. In adults this procedure was performed under local anaesthesia. In children under the age of eight years general anaesthesia with Ketamine was used. The reference electrode was a disc electrode placed either on the ear lobe or on the skin overlying the mastoid process.

The patient lay down in a sound-treated, acoustically insulated room with double doors, and the stimuli were produced by a screened loudspeaker.

Guinea-pig study

For this study the chronic electrode implantation technique described by Portmann, Aran & Le Bert (1966) was used. In this a wire active electrode was surgically inserted onto the round window of a guinea-pig. The reference and earth electrodes were screws holding a connector to the top of the skull. The animals were allowed to recover from the surgery and could be tested after about a week, when response amplitudes of up to 1500 μV were obtained. These animals could be subsequently tested over a period of time and a number gave good responses over a period of more than 18 months. Sixteen such guinea-pigs were used in the present study.

The stimuli were adminstered to these animals by a plastic tube inserted into the external acoustic meatus and attached to the side of the electrical connector fixed to the skull. The tube then passed to the outside of the cage where it was connected to a TDH-39 transducer in a sealed unit. The cage was within a soundproofed and electrically isolated testing box. Apart from this, the stimulus generating and response recording system was identical to that used in the human study.

Stimuli

The stimuli in both human and guinea-pig studies were alternating rarefaction and condensation clicks. The duration of the pulses fed to the loudspeaker for the human studies was 100 μsec, and the duration of the acoustical click recorded at the patient's ear was approximately 1 msec.

The duration of the electrical pulses fed to the TDH-39 transducer for the guinea-pig studies was 80 μsec for the unfiltered clicks and 64 μsec for the 8000 Hz filtered clicks. Again the acoustical waveform at the animal's ear had a duration of approximately 1 msec.

In the human study the stimulus always consisted either of a single click presented every 100 msec or a train of five identical clicks with interclick intervals of 8·5 msec, and with a silent interval of 100 msec between successive trains.

In the main part of the guinea-pig study the same stimuli were used as for the human study. In addition, trains of five clicks with interclick intervals of 16 msec and 25 msec were used. Both unfiltered and 8000 Hz $\frac{1}{3}$-octave bandwidth filtered clicks were used in this study. A number of other stimulus conditions were also investigated in subsidiary experiments and details of these stimuli are given with the experiment.

Response analysis

The averaging technique used in both the guinea-pig and the

human studies was such that the responses to the rarefaction and condensation clicks were averaged separately. These averaged responses were then added to cancel the phase-following cochlear microphonic responses and to facilitate a detailed investigation of the action potential responses. At the same time the two averages were subtracted from each other, a procedure which enhanced the cochlear microphonic and cancelled the action potential responses, thus allowing separate analysis of the cochlear microphonics.

In the present study no significant differences were found in the compound action potential and its adaptation between the responses to condensation and rarefaction clicks. All results recorded in the present study are for the summed response to clicks of alternating polarity. No adaptation of the cochlear microphonic response was found, in accord with many other studies, and, although these responses were always recorded, the results are not presented here.

In the averaging, a 1000 point average was always used. For all human studies a 40 msec window was used, thus giving 40 μsec per point. For the guinea-pig studies windows of 40, 80 or 120 msec were used according to the separation of the clicks in a train.

RESULTS

Human study

In the 11 patients with normal electrocochleographic responses to clicks, measures were obtained to the intensity-amplitude growth functions for each of the five clicks in a train. These growth functions for each click were plotted as a percentage of the maximum amplitude for that particular click in the train of clicks and an example of such a growth function is shown in Fig. 2a. Figure 2b illustrates the adaptation of the response to each click in the train, plotted in each case in terms of its amplitude as a percentage of the amplitude response to the first click of the train. This is shown as a function of the stimulus intensity.

From this figure two points emerge. The first is that the knee-point or irregularity in the intensity-amplitude growth function becomes more pronounced with the second to fifth clicks than in the response to the first click of the train. The form of the response to the first click of the train in each subject was identical to that of the response to a single click, although the absolute amplitude was slightly reduced.

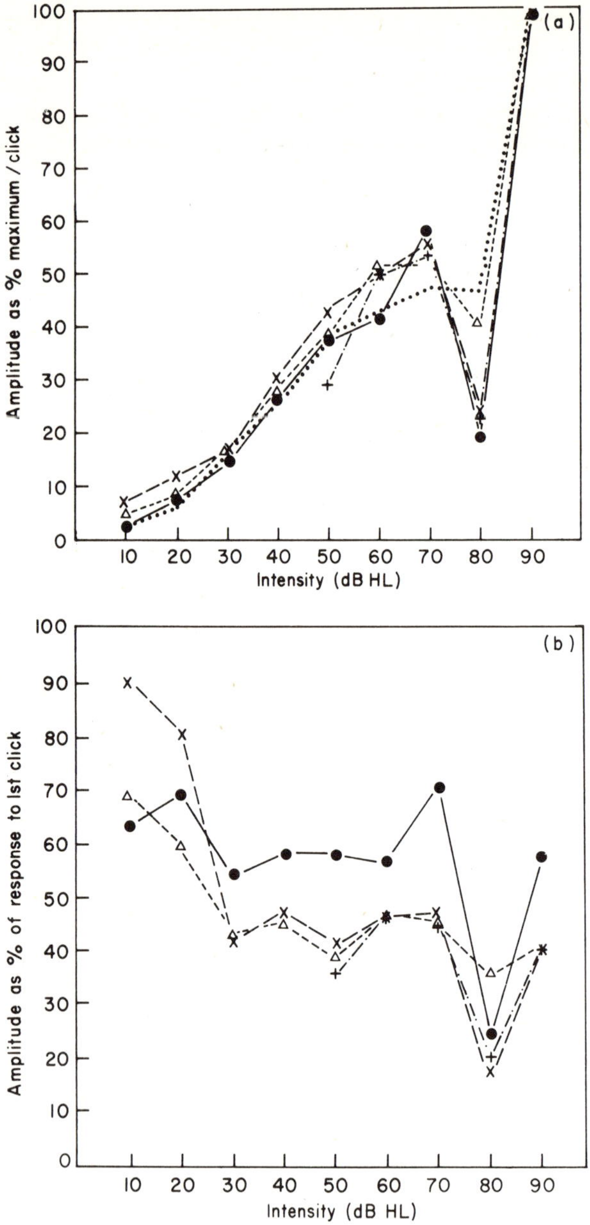

FIG. 2. (a) Subject no. 858. Form of intensity-amplitude growth function for successive clicks in train. ——— 1st click; ●———● 2nd; ×– –× 3rd; △– – –△ 4th; +–·–·+ 5th. (b) Adaptation of responses to successive clicks in train as % of amplitude of response to 1st click.

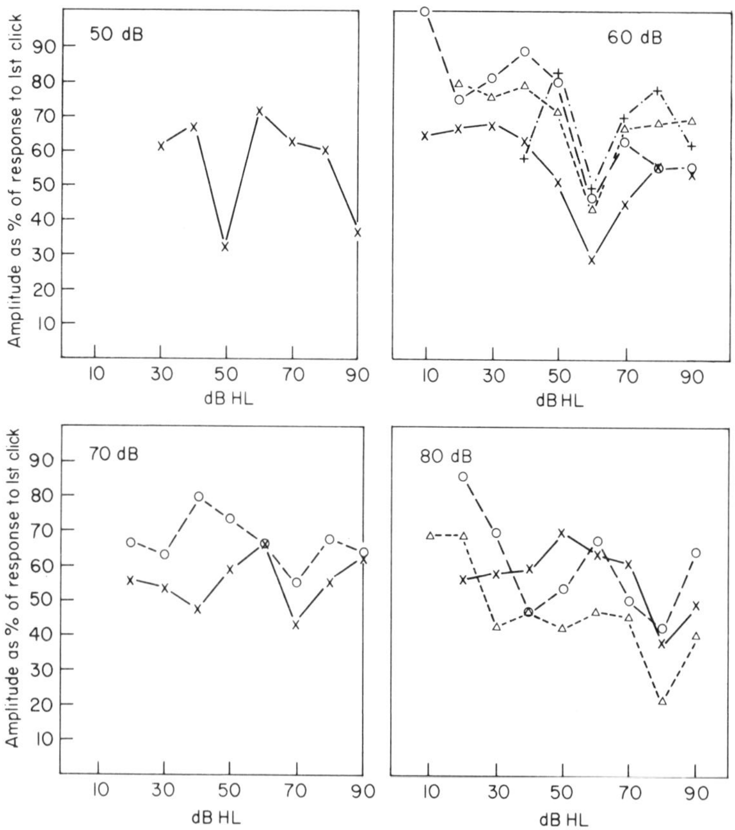

FIG. 3. Adaptation as a function of intensity in ten subjects divided according to the intensity at which maximal adaptation occurred.

The second main point in these results is that subjects show a characteristic point of maximum adaptation at one intensity, with less adaptation at higher and lower intensities. This point of maximal adaptation generally occurred at the same intensity for all the clicks in the train. In different subjects this point varied from 50–80 dB HL. Figure 3 shows the mean adaptation to clicks 3 to 5 expressed as a function of stimulus intensity for the subjects of this series, divided according to the intensity at which maximal adaptation occurred.

From Fig. 2 another point of interest is that the intensity at which maximal adaptation occurred for that particular subject coincided with the intensity at which the knee-point of the intensity-amplitude growth curve occurred. In ten of the 11 subjects

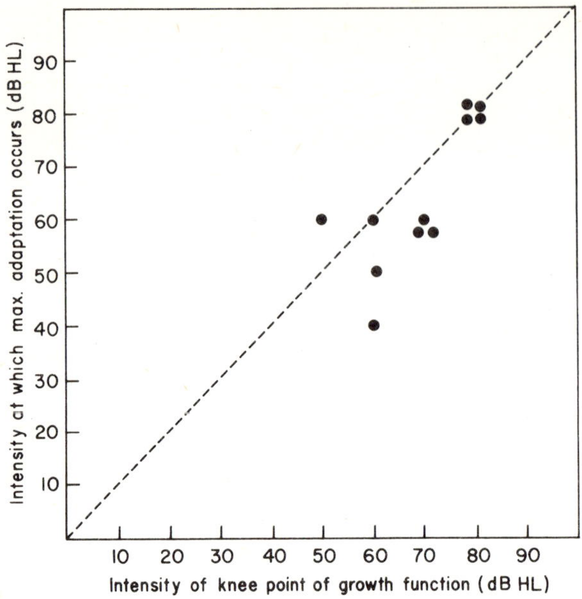

FIG. 4. Relationship between maximum adaptation and knee-point of amplitude-intensity growth function. (The dotted line represents the 45° diagonal and not the line of best fit.)

tested, the intensity at which maximal adaptation occurred was within 10 dB of the knee-point of the intensity-amplitude growth function. These results are shown in Fig. 4. Considering that these measures are somewhat approximate in that only 10 dB intensity steps were used and that in certain cases it is quite difficult to define the exact knee-point of the growth function, a good measure of accord was obtained.

If the overall mean of the results across subjects is taken, these individual differences with a sharp point of maximal adaptation disappear due to a smearing of the effect. Median results are shown in Fig. 5a. Figure 5b also shows the median adaptation across subjects plotted as a function of the click number for four different intensities and shows the same overall pattern as that reported by Eggermont & Odenthal (1974) and by Yoshie & Ohashi (1971), in that the adaptation approaches an asymptotic value by the response to the fifth click except at the lowest stimulus levels.

These results may thus be summarized as showing that the knee-point or irregularity becomes more pronounced in the adapted response (as Eggermont & Spoor, 1973, had shown in the

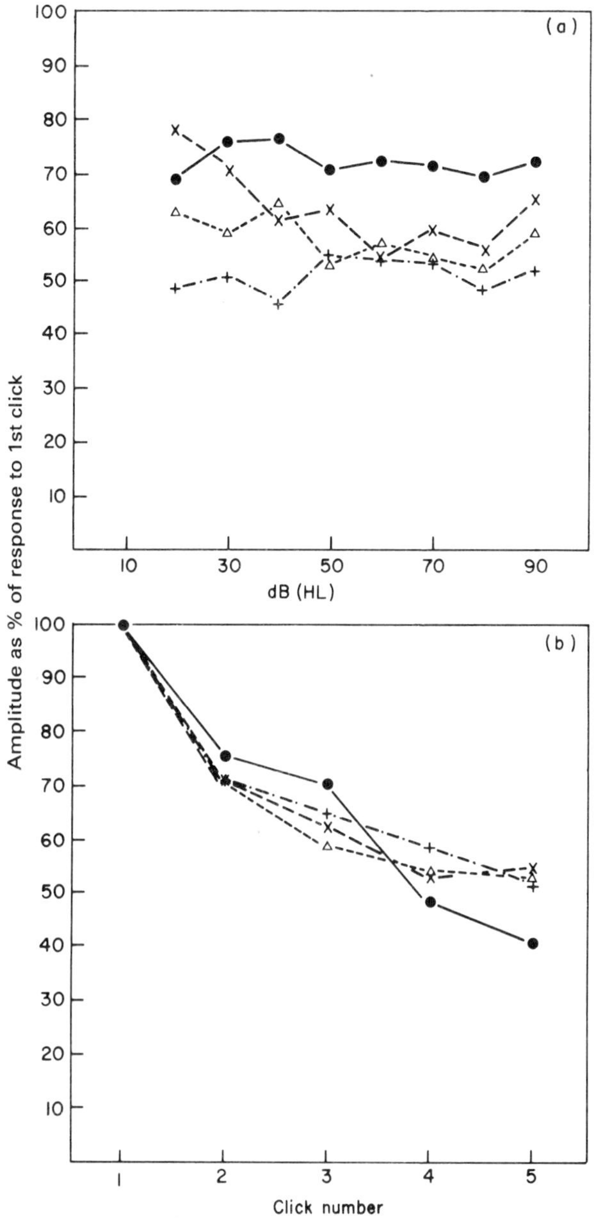

FIG. 5. Median values across subjects ($n = 11$) of (a) adaptation as a function of intensity: ●——● 2nd click; × – – × 3rd; △– – –△ 4th; + –· –· + 5th and (b) adaptation occurring with successive clicks: ●——● 30 dB HL; × – – × 50 dB HL; △– – –△ 70 dB HL; + –· –· + 90 dB HL.

guinea-pig) and that this knee-point appears to occur at approximately the same intensity as that at which the maximal adaptation occurs.

Guinea-pig studies

In order to investigate further the effect of stimulus parameters, it was necessary to search for parallels in guinea-pigs.

The 16 normally responding guinea-pigs with chronically implanted electrodes were systematically tested using clicks and also filtered clicks in which the centre frequency was 8000 Hz. The main study involved using these stimuli in trains of five clicks with interclick intervals within the train of 8, 16 and 25 msec. The silent interval between the end of one train and the onset of the next in this study was 100 msec, as in the human study. An example of the effect of the various interstimulus intervals on the shape of the intensity-amplitude growth functions is shown in Fig. 6. This shows that at the three interstimulus intervals used, an enhancement of the irregularity of the growth function occurs with the two shorter intervals, although the effect is less pronounced when an interstimulus interval of 25 msec is used.

Figure 7 shows the mean adaptation across 16 guinea-pigs as a function of intensity for both the unfiltered clicks and the 8000 Hz filtered clicks with interstimulus intervals of 8 msec. For the unfiltered clicks the results were analogous to the human results with the point of maximal adaptation varying somewhat from animal to animal and the averaging of the results producing a smearing of this effect. For the 8000 Hz filtered clicks, however, a more strikingly intensity-dependent adaptation was found. The authenticity of this as a physiological effect was confirmed by the lack of adaptation of the cochlear microphonic measures.

Smaller numbers of animals were used to examine the effects of varying a number of other stimulus parameters. Figure 8 shows the adaptation to a train of 11 clicks with 8 msec interclick intervals, and confirms that the asymptotic value of adaptation is reached by the response to the fifth click.

The adaptation was also investigated using single clicks presented at different interstimulus intervals. This study supports the finding that the irregular form of the intensity-amplitude growth function is enhanced as the cochlea shows more adaptation with shorter interstimulus intervals.

Finally, in view of the fact that the intertrain intervals (100 msec) used in this study were rather shorter than those used

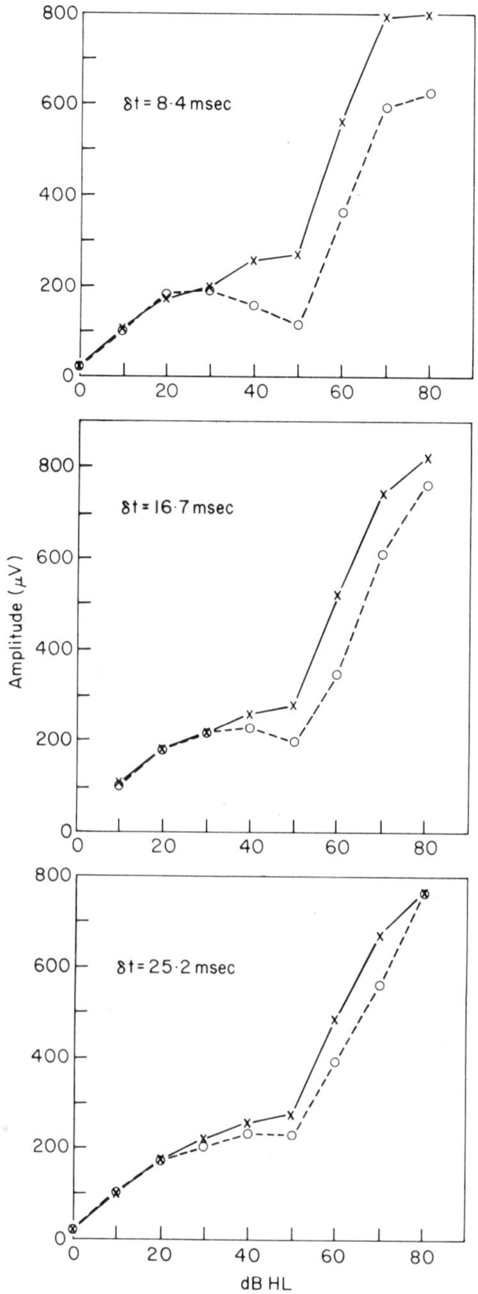

FIG. 6. Adaptation as a function of interstimulus interval. Guinea-pig no. 11 8000 Hz filtered clicks. × ——— × response to 1st click; ○ - - - ○ response to 2nd click.

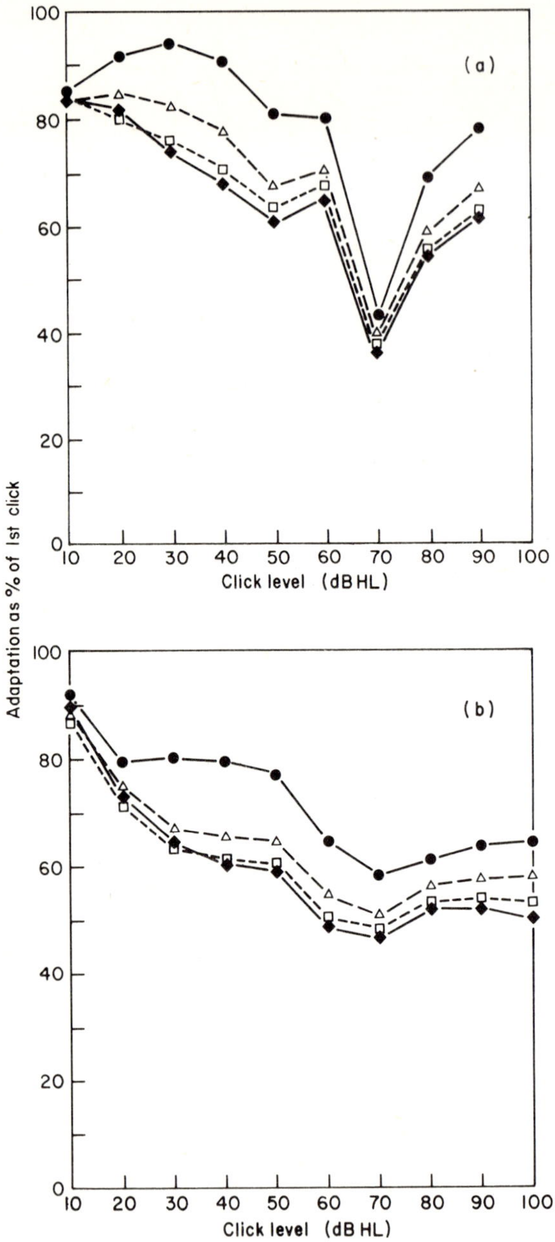

FIG. 7. Mean adaptation in 16 normal guinea-pigs. (a) 8000 Hz filtered clicks. (b) Unfiltered clicks. ●———● click 2; △– –△ click 3; □– – –□ click 4; ◆———◆ click 5.

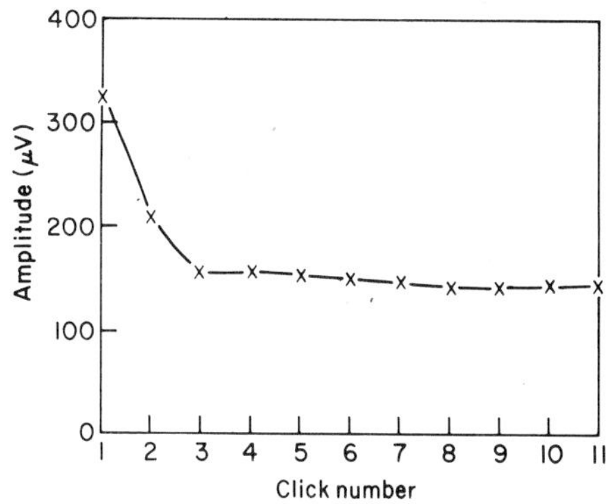

FIG. 8. Adaptation to a train of 11 clicks ($\delta t = 8{\cdot}4$ msec). Guinea-pig no. 13.

by Eggermont & Odenthal (1974a) and by Yoshie & Ohashi (1971), the effect of varying the intertrain interval from 100 to 1000 msec was studied in this investigation. No significant differences were found between the degree of adaptation or the change in the form of the intensity-amplitude growth function, with different intertrain intervals.

DISCUSSION

The results of the present study both in humans and in guinea-pigs support earlier findings of a discontinuity in the changes in the compound action potential response as a function of stimulus level. These results are in accord with the finding of Eggermont & Odenthal (in press) that there is a fairly sharp cut-off point at which this discontinuity occurs, certainly for individual subjects. The fact that there is some intersubject variability in these results is not surprising in view of the considerable intersubject variability found in the analogous subjective measure of the threshold of overload of the cochlea (Lawrence & Yantis, 1956). In view of the differential sensitivity of the cochlea to different frequencies, it is perhaps also not surprising that in the guinea-pigs a more uniform response is obtained to the filtered clicks, which might be expected to activate a more limited part of the cochlea than normal unfiltered clicks. It would seem reasonable to expect an analogous result when such filtered clicks are used in man.

Assuming that the irregularity in the intensity-amplitude growth curve is caused by two different populations of receptor units, or units from different parts of the organ of Corti, it is of relevance to ask whether the largest adaptation is due to the high-sensitivity receptors being maximally stimulated or to the low-sensitivity receptors being minimally stimulated. Apart from the analogous support, in a great variety of neurophysiological studies, of maximal adaptation occurring at high stimulus intensity levels, there are two pieces of experimental evidence to support the former concept. The first is that the high-sensitivity receptors show relatively less adaptation when stimulated at low intensity levels, so that it might be supposed that the low-sensitivity units might behave in an analogous way. The second piece of evidence comes from a parallel study which was conducted on patients with cochlear damage (Stephens, Charlet de Sauvage & Aran, 1974). In that study the mean adaptation found in patients with end-organ lesions and abnormal electrocochleographic configurations was about 75% of the amplitude of the first click, and no patient showed the pronounced adaptation found in the transitional region in the present normally-responding series. In these patients with cochlear damage there is evidence to suggest that the high-sensitivity receptors are missing.

The two population concept is also supported by the study of Charlet de Sauvage & Aran (1973) who found a more marked knee-point in adults than in children. This would suggest that in the "normal" adult population there is already some limitation of the activity of the high-sensitivity receptors when they are stimulated at the upper end of their response range.

In conclusion it may be stated that the present study would appear to support the concept that the irregular intensity-amplitude growth function in the compound action potential response is due to two populations of receptor units, on the basis of the change in the form of the response in the adapted state. It is impossible from this work to resolve the problem of the location of the two receptor units but it would appear that the maximal adaptation found at stimulus levels in the region 50–80 dB HL is occurring in the high-sensitivity receptor units which are being maximally stimulated.

Acknowledgements

The experimental work described in this study was performed at Bordeaux when Dr S. D. G. Stephens was the recipient of a Ciba Foundation/INSERM Anglo-French medical exchange bursary, and the work was further supported by Medical Research Council grant number 970/512/C, by INSERM grant number 71/1/187/10, and by DGRST grant number 72/7/0797.

References

Aran, J.-M. & Darrouzet, J. (in press). Observation of click evoked compound VIIIth nerve responses before during and over seven months after Kanamycin treatment in the guinea-pig. *Acta otolar.*

Aran, J.-M. & Le Bert, G. (1968). Les reponses nerveuses cochléaires chez l'homme. Image du fonctionnement de l'oreille et nouveau test d'audiométrie objective. *Revue Lar. Otol. Rhinol.* **89**: 361–378.

Aran, J.-M. & Portmann, M. (in press). Cochlear electrophysiology: clinical application. In *Scientific foundations of otolaryngology*: Hinchcliffe, R. & Harrison, D. F. N. (eds). London: Heinemann.

Beauregard, H. & Dupuy, E. (1896). Sur la variation électrique (courant d'action) determinée dans le nerf acoustique par le son. *Archs int. Lar. Otol. Rhinol.* **9**: 383–386.

Charlet de Sauvage, R. & Aran, J.-M. (1973). L'électrocochleógramme normal. *Revue Lar. Otol. Rhinol.* **94**: 93–107.

Derbyshire, A. J. & Davis, H. (1935). The action potentials of the auditory nerve. *Am. J. Physiol.* **113**: 476–504.

Eggermont, J. J. & Odenthal, D. W. (1974). Electrophysiological investigation of the human cochlea. *Audiology* **13**: 1–22.

Eggermont, J. J. & Odenthal, D. W. (in press). Frequency selective masking in electrocochleography. *Revue Lar. Otol. Rhinol*: suppl.

Eggermont, J. J. & Spoor, A. (1973). Cochlear adaptation in guinea-pigs: a quantitative description. *Audiology* **12**: 193–220.

Elberling, C. (1973). Transitions in cochlear action potentials recorded from the ear canal in man. *Scand. Audiol.* **2**: 151–159.

Elberling, C. (1974). Action potentials along the cochlear partition recorded from the ear canal in man. *Scand. Audiol.* **3**: 13–19.

Elberling, C. & Salomon, G. (1973). Cochlear microphonics recorded from the ear canal in man. *Acta otolar.* **75**: 489–495.

Fromm, B., Nylen, C. O. & Zotterman, Y. (1935). Studies in the mechanism of the Wever and Bray effect. *Acta otolar.* **22**: 477–486.

Lawrence, M. & Yantis, P. A. (1956). Threshold of overload in normal and pathological ears. *Archs Otolar.* **63**: 67–77.

Lieberman, A., Sohmer, H. & Szabo, G. (1973). Standard values of amplitude and latency of cochlear audiometry (electrocochleography) responses in different age groups. *Arch. klin. exp. Ohr. Nas. Kehlk.-Heilk.* **203**: 267–273.

Portmann, M., Aran, J.-M. & Coste, Ch. (1973). Vérification expérimentale sur l'animal de quelques aspects électrocochléographiques de la pathologie humaine. *Acta otolar.* **75**: 132–142.

Portmann, M., Aran, J.-M. & Lagourgue, P. (1973). Testing for "recruitment" by electrocochleography. *Ann. Otol. Rhinol. Lar.* **82**: 36–43.

Portmann, M., Aran, J.-M. & Le Bert, G. (1966). Mise en place définitive d'une électrode sur la fenêtre ronde du cobaye. *C. r. Séanc. Soc. Biol.* **160**: 551–556.

Portmann, M., Le Bert, G. & Aran, J.-M. (1967). Potentiels cochléaires obtenus chez l'homme en dehors de toute intervention chirurgicale, note préliminaire. *Revue Lar. Otol. Rhinol.* **88**: 157–164.

Ruben, R. J., Sekula, J., Bordley, J. E., Knickerbocker, G. G., Nager, T. G. & Fisch, U. (1960). Human cochlear responses to sound stimuli. *Ann. Otol. Rhinol. Lar.* **69**: 459–479.

Rutt, N. S., Stephens, S. D. G. & Thornton, A. R. D. (1973). An experiment with surface recording electrocochleography. *J. Physiol., Lond.* **232**: 109–110P.

Salomon, G. & Elberling, C. (1971). Cochlear nerve action potentials recorded from the ear canal in man. *Acta otolar.* **71**: 319–325.

Saul, L. J. & Davis, H. (1932). Action currents in the central nervous system. I. Action currents of the auditory tracts. *Archs Neurol. Psychiat., Chicago* **28**: 1104–1116.

Sohmer, H. & Feinmesser, M. (1967). Cochlear action potentials recorded from the external ear in man. *Ann. Otol. Rhinol. Lar.* **76**: 427–435.

Sohmer, H. & Feinmesser, M. (1971). Recording of averaged cochlear action potentials as a form of objective audiometry. *Int. Cong. Acoust.* **7;3**: 389–391.

Sohmer, H. S., Peake, W. T. & Weiss, T. F. (1971). Intracochlear potential recorded with micropipets. *J. acoust. Soc. Am.* **50**: 572–615.

Spreng, M. & Keidel, W. D. (1967). Separierung von cerebro-audiogramm (CAG) Neuroaudiogramm (NAG) and Otoaudiogramm (OAG) in der objektiven Audiometrie. *Arch. klin. exp. Ohr. Nas. Kehlk.-Heilk.* **189**: 225–246.

Stephens, S. D. G., Charlet de Sauvage, R. & Aran, J.-M. (1974). Adaptation de l'électrocochléogramme: note préliminaire. *Revue Lar. Otol. Rhinol.* **95**: 129–138.

Tasaki, I., Davis, H. & Eldredge, D. H. (1954). Exploration of cochlear potentials in guinea-pig with a microelectrode. *J. acoust. Soc. Am.* **26**: 765–773.

Teas, D. C., Eldredge, D. H. & Davis, H. (1962). Cochlear responses to acoustic transients: an interpretation of whole nerve action potentials. *J. acoust. Soc. Am.* **34**: 1438–1459.

Terkildsen, K., Osterhammel, P. & Huis in't Veld, F. (1973). Electrocochleography with a far field technique. *Scand. Audiol.* **2**: 141–148.

Wever, E. G. & Bray, C. W. (1930). Action currents in the auditory nerve in response to acoustical stimulation. *Proc. natn. Acad. Sci. U.S.A.* **16**: 344–350.

Yoshie, N. (1968). Auditory nerve action potential responses to clicks in man. *Laryngoscope, St. Louis* **78**: 198–215.

Yoshie, N. & Ohashi, T. (1971). Abnormal adaptation of human cochlear nerve action potential responses: clinical observations by non-surgical recording. *Revue Lar. Otol. Rhinol.* **92**: Suppl. 673–690.

Yoshie, N., Ohashi, T. & Suzuki, T. (1967). Non-surgical recording of auditory nerve action potentials in man. *Laryngoscope, St. Louis* **77**: 76–85.

Yoshie, N. & Yamaura, K. (1969). Cochlear microphonic responses to pure tones in man recorded by a non-surgical method. *Acta otolar.* Suppl. **252**: 37–69.

Symp. zool. Soc. Lond. (1975) No. 37, 187–203.

AUDITORY EVOKED RESPONSES IN SMALL MAMMALS

GILLIAN D. SALES

*Department of Zoology and Comparative Physiology, Queen Mary College, London, England**

SYNOPSIS

Studies of the auditory responses of small mammals are important in several ways. They can be used to assess the effects of noise, trauma or drugs on the mammalian auditory system, and central responses are of great interest from a comparative point of view when considering acoustic communication in animals. The present study involves assessing the auditory sensitivity of small mammals, particularly rodents, to ultrasonic signals as an indication of the communication value of these signals.

Most tests of auditory sensitivity of the central nervous system in small mammals employ either behavioural techniques, which are time-consuming, or major surgery, which may be terminal or else limits the studies to restricted periods in the animals' lives. Using terminal experiments, peaks of auditory sensitivity in the ultrasonic range have been obtained at the inferior colliculus by Brown, whose results are reviewed.

A technique for confirming Brown's results which is not terminal and which would allow auditory responses to be tested at different stages throughout the animals' lives would be useful. The averaged evoked response recorded from surface (scalp) electrodes appears to fit these criteria. There is good evidence in man that auditory evoked responses, obtained from the scalp in response to acoustic stimuli, give a good indication of the auditory sensitivity of the subject, and the technique is now regularly used in clinical practice.

The use of the scalp auditory evoked response in routinely assessing the auditory sensitivity of small mammals appears to have been first suggested by Rothenberg & Davis in 1967. So far the technique has been used by only a few groups of workers, studying responses to signals below the ultrasonic range. Their results are discussed briefly. The present attempt to study the auditory responses of rodents at ultrasonic frequencies by recording the auditory evoked response from the scalp is described, and further applications of the technique are suggested.

INTRODUCTION

The auditory system of small mammals may be studied for one of several reasons; it may be used to investigate basic mechanisms of the mammalian auditory system (e.g. Evans, this symposium pp. 135–165; Stephens, Charlet de Sauvage & Aran, this symposium pp. 167–186) or it may be studied to assess the effects of short term noise, the effects of trauma due to excessive noise or the effects of drugs on this system (e.g. Evans, this symposium pp. 133–165; Pye, A., this symposium pp. 97–105). In this paper the auditory responses of small mammals are considered from the point of view of their role in acoustic communication, and particularly in ultrasonic communication.

* *Present address*: Department of Zoology, King's College, London, England.

It is now known that both young and adults of many species of rodents from the families Muridae and Cricetidae emit ultrasounds in a variety of situations (see Sales & Pye, 1974). For example, the young of many species emit calls when removed from the nest (Sewell, 1969) and the adults of some species emit ultrasound in several different situations including aggressive behaviour (Sales, 1972a) and mating behaviour (Sales, 1972b). In rats two distinct calls have been recorded during aggressive encounters between pairs of males. Short calls 3–65 msec in duration and at frequencies of 40–70 kHz were emitted during aggressive acts—attacking, boxing or wrestling; whereas long calls up to 3400 msec and at frequencies of 22–30 kHz were produced by defeated rats. So far a communication function has been demonstrated only for the calls of the young (Sewell, 1970; Smith, 1972, this symposium pp. 317–330) but there is circumstantial evidence that the long calls of submissive rats inhibit the aggressive behaviour of the dominant animal (Sales, 1972a). At present the role of these ultrasonic calls in aggressive behaviour is being investigated by behavioural observations. However, further evidence of the importance of these, and other ultrasonic calls of rodents, in acoustic communication can be obtained by studying the auditory sensitivity of the central nervous system to ultrasonic frequencies. This can be achieved either by behavioural techniques or by physiological methods.

Observations of an animal's behavioural responses to sound indicate that the auditory information has reached certain levels of the nervous system, such as the medial geniculate body for some reflex reactions, or the cortex for voluntary reactions (see Pye, J. D., this symposium pp. 1–9). Sensitivity to ultrasounds has been demonstrated behaviourally in several groups of small mammals by using reflex responses, such as the Preyer reflex (Schleidt, 1952) galvanic skin response (Berlin, 1963) or by training the animals to respond voluntarily to sound (Gould & Morgan, 1941, 1942; Gourevitch & Hack, 1966). But behavioural techniques have many drawbacks (see Francis, this symposium pp. 237–289) and the auditory sensitivity of the central nervous system can more easily be assessed physiologically by studying either the gross response, or the response of single auditory neurones in one of the regions of the auditory system, most commonly the inferior colliculus of the mid-brain. The aim of this paper is to review the existing physiological evidence for sensitivity to ultrasonic frequencies in rodents and to evaluate a different approach that promises to be especially useful in the present study of ultrasonic communication in rodents.

EVOKED RESPONSES RECORDED INTRACRANIALLY

In rodents, sensitivity to ultrasonic frequencies within the central nervous system appears to have been investigated only from the gross response of the inferior colliculus. However, as Grinnell (1970) has pointed out "gross evoked potentials do not reflect the activity of small populations of units that might be responding with great sensitivity at other frequencies". Thus a large response at any one frequency, or a standard response that is evoked by a low intensity signal, does not necessarily mean that the auditory system of the animal is more sensitive to this frequency; it could mean that more units are responding at that frequency and so contributing to the response. This is illustrated by Grinnell's studies on the auditory responses of bats. Grinnell (1963) found that in the bat *Plecotus townsendii*, the gross response of the inferior colliculus showed two peaks of sensitivity, one at 15–35 kHz and the second at 55–65 kHz (Fig. 1), which correspond with the

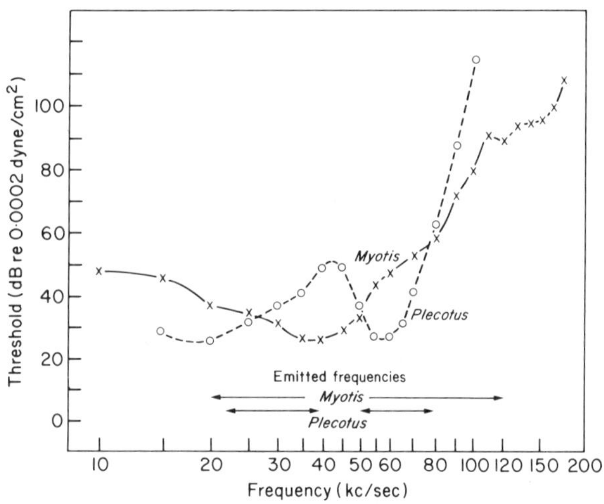

FIG. 1. Mean gross evoked response of the inferior colliculus of *Plecotus townsendii* and *Myotis lucifugus*. From Grinnell (1963).

frequencies of the fundamental and second harmonic of the echolocation cries of this species. This appears to be the first time that such bimodal peaks of sensitivity have been found in the auditory system of small mammals. However, when Grinnell looked at the responses of single units, he found that although many units were most sensitive to the frequencies corresponding to the peaks of the gross response, some units were almost as sensitive at other frequencies (Fig. 2). A peak in the gross response then may

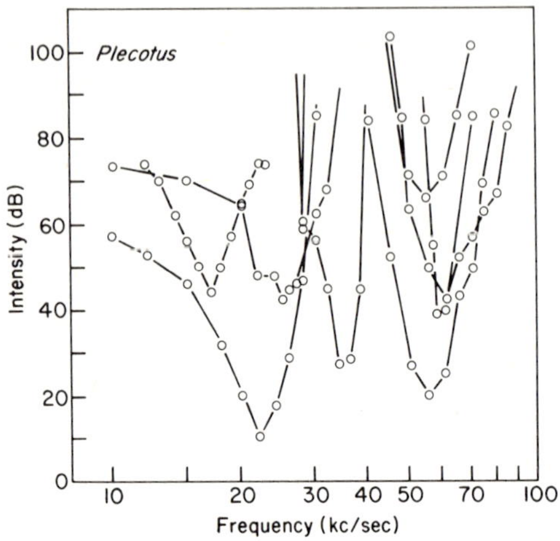

FIG. 2. Responses of single auditory units in the inferior colliculus of *Plecotus townsendii*. From Grinnell (1963).

indicate a great sensitivity to a particular frequency, or it may indicate a greater number of neurones responding. Or, of course, both factors could be operating together. Whichever is the case, the peak suggests that the frequency is important to the animal and so gives some indication of the frequencies that may be employed in communication.

The gross response of the inferior colliculus of rodents to frequencies up to 100 kHz has been studied by Ralls (1967) and by Brown (1971a,b, 1973b). Brown studied nine species of rodents and in all of them she found two peaks of sensitivity: a lower peak, generally in the audible range, and a second peak in the ultrasonic range (Fig. 3). The ultrasonic peak roughly corresponded with the frequencies emitted by the animals themselves (Table I) and the animals appeared to be more sensitive at this peak than at the lower one. As shown in Fig. 3, Brown found similar peaks of sensitivity in the cochlear microphonic response (Brown, 1970, 1973a).

Ralls (1967) studied the gross response of the inferior colliculus in two groups of rodents, *Mus musculus*, including laboratory mice and wild house mice, and in two species of *Peromyscus*, deermice.

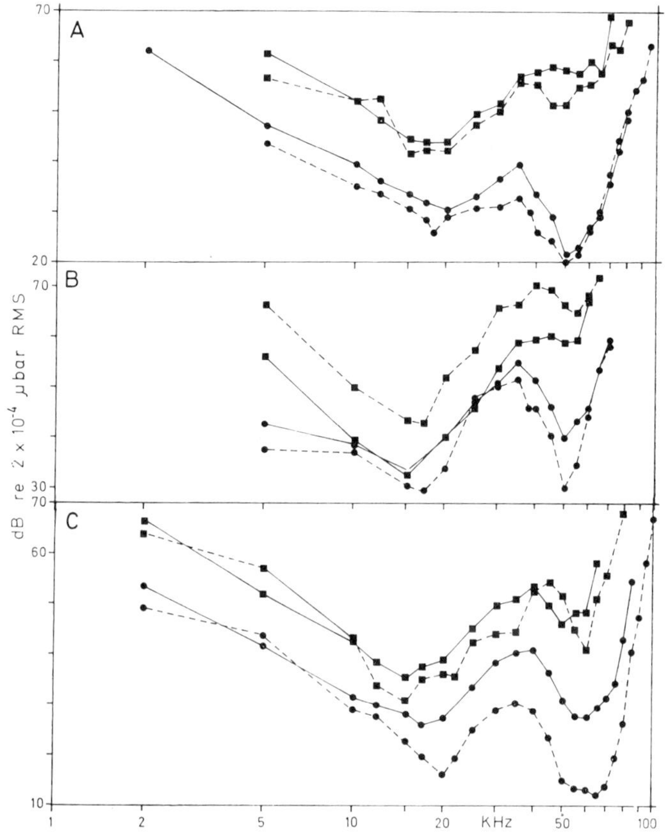

FIG. 3. Frequency response curves of A: *Mus musculus*, wild type; B: *M. musculus*, TO Swiss strain; C: *Apodemus sylvaticus*. Responses from individual animals: ● – – ● inferior colliculus; ■ – – ■ cochlear microphonic. Mean responses from 6 animals (5 for TO Swiss mice): ●——● inferior colliculus; ■——■ cochlear microphonic. From Brown (1973b).

Ralls found one peak of sensitivity in the audible range but she did not report a second peak of sensitivity in the ultrasonic range, although her graphs do indicate the presence of slight peaks at about 60–65 kHz for wild *M. musculus*. This may have been due to a failure to maintain the normal body temperature of the animals while under anaesthesia, a condition that Brown found essential for obtaining the peak responses at higher frequencies.

Brown's results are very important for studies on ultrasonic communication in rodents, for they show that in each of the species studied, the auditory system seems to be adapted to receive the signals emitted by the animals. So far there do not appear to have

TABLE I

High frequency responses at the inferior colliculus of rodents and ultrasounds produced

Species	High frequency peak of inferior colliculus response, kHz		Ultrasonic frequencies commonly produced	
	Average	Range	kHz	Animals involved
Rattus norvegicus	30	30–32	40–75	young
		40–42	40–120	adult aggression
			22–30	adult submission
			30–120	adult mating
Mus musculus (feral)	50	50–60	*c.* 70	adult
	50	45–60	(40–88	young of different strain)
			30–120	adult mating
Apodemus sylvaticus	60	50–70	40–88	young
			50–108	adult aggression
			60–105	adult mating, or exploration
Meriones shawi	30	30–40	45–130	young pure ultrasound
			35–60	adult aggression
Meriones unguiculatus	40	40	37–80	young pure ultrasound
Gerbillus sp.	40	40–45	50–85	young pure ultrasound
			40–65	adult aggression
Clethrionomys glareolus	50	50–60	20–55	young
			17–30	adult aggression
			20–55	adult mating
Microtus agrestis	50	50	25–60	young
Sigmodon hispidus	50	40–65	23–60	young

Modified from Brown (1971b) and Sales & Pye (1974)

been any studies of the responses of single auditory units to ultrasonic frequencies in rodents. So the question of whether the peaks of sensitivity that Brown found in the gross response were due to the presence of more sensitive units at these frequencies, or to more neurones, remains unanswered.

EVOKED RESPONSES RECORDED EXTRACRANIALLY

The technique used by Brown and by Ralls to record gross responses was terminal; the animals were used once only and then destroyed. For the present studies on ultrasonic communication, it would be useful to have a procedure that was not terminal and that would allow auditory responses to be tested at various stages throughout the animals' lives, particularly in relation to certain behavioural situations such as aggressive encounters in rats. The technique of averaging the evoked responses recorded from electrodes placed temporarily on the surface of the scalp appears to fit these criteria. This technique can be used to detect small auditory responses that might otherwise be lost in the background electroencephalogram (EEG).

The principle of "averaging" has several advantages and disadvantages that are relevant to its application to the study of auditory responses in small mammals. The technique uses a "signal averager", a device that adds together events presented sequentially and then gives the average value of these events. Auditory evoked responses generally occur at a certain time after the stimulus, depending on their origin; they are said to be "time locked" to the stimulus. The averager can be used to record and summate brain potentials occurring within a set period after each of a series of stimuli, and as the auditory responses always occur in the same part of the record, subsequent responses add together arithmetically. The background noise of the electroencephalogram is assumed to be random, and theoretically the amplitude of the noise increases as the square root of the number of samples taken, as does the signal-to-noise ratio (SNR). Although the suppositions on which the technique is based are not always fulfilled (see Salomon, this symposium pp. 205–236), averaging is a useful means of recording small auditory responses by electrodes on the surface of the scalp.

Averaging rapidly increases the signal-to-noise ratio over a small number of samples, but further improvements need progressively larger samples, and the process then becomes time consuming. In addition, over extended periods of time the auditory response can change so that subsequent responses no longer add arithmetically, and the response already obtained then becomes degraded. The number of samples taken is therefore generally a compromise and depends on the requirements of the experiment.

Averaged evoked responses to auditory stimuli recorded from the scalp have been studied mainly in man (e.g. Davis, 1965; Milner, 1970; Regan, 1972). Two classes of auditory responses have been obtained from the human scalp using this technique (Davis, Engebretson, Lowell, Mast, Satterfield & Yoshie 1964). The first is a "fast" response with a latency of less than 50 msec to its highest peak and is generally evoked by click stimuli. This response

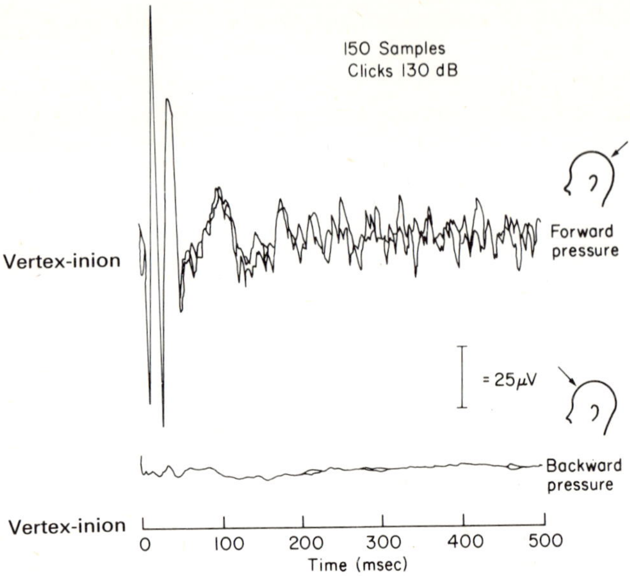

FIG. 4. Averaged myogenic responses to click stimuli recorded from the human scalp. The myogenic origin of the response is suggested by the fact that the amplitude is increased when pressure is applied to the back of the head, increasing the tension in the neck muscles (above), but the response is abolished when pressure is applied to the forehead, relaxing the neck muscles. Two traces in each line. From Bickford *et al.* (1964) with permission.

may have one or more peaks (Fig. 4) and it is thought to arise in the muscles of the head and neck (Davis, Engebretson *et al.*, 1964; Bickford, Jacobson & Cody, 1964). The second class of response has a longer latency, approximately 160 msec to the largest peak and 50–60 msec to the earliest peak (Davis, Engebretson *et al.*, 1964; Regan, 1972), although the exact latencies and relative amplitudes of the peaks appear to depend on the age and

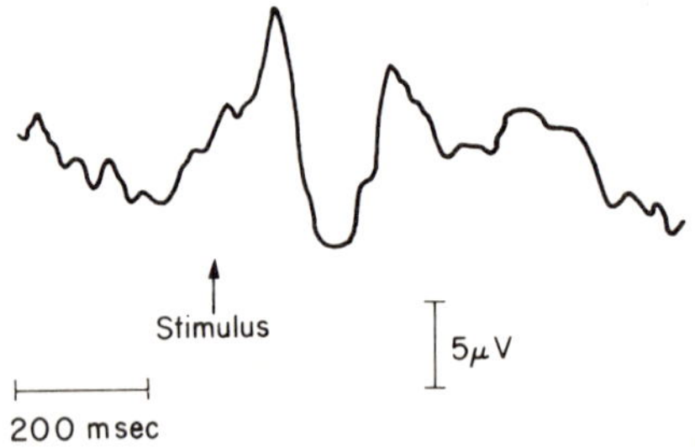

FIG. 5. Averaged responses to tone stimuli of 1 kHz recorded from the vertex of the author. (Recorded by H. Beagley of the Institute of Laryngology and Otology, London.)

behavioural state of the subject as well as on certain parameters of the stimulus (Regan, 1972). This "slow" response is thought to arise from the auditory cortex. It is best recorded from the vertex and so is often called the "vertex" or "V" potential (Fig. 5).

There is now a large body of evidence that in man the vertex potential can give a good indication of the auditory sensitivity of the subject (e.g. Davis, 1966; Davis, Hirsh, Shelnutt & Bowers, 1967; Rapin & Graziani, 1967), and this technique of "evoked response audiometry" is now regularly used in clinical practice (e.g. Davis, 1965, 1973; Beagley, 1973), although some authors suggest caution in the clinical use of data obtained in this way (Rose, Keating, Hedgecock, Miller & Schreurs, 1972; Salomon, this symposium pp. 205–236).

In small mammals, averaged evoked responses to acoustic stimuli have been recorded from surface electrodes by several

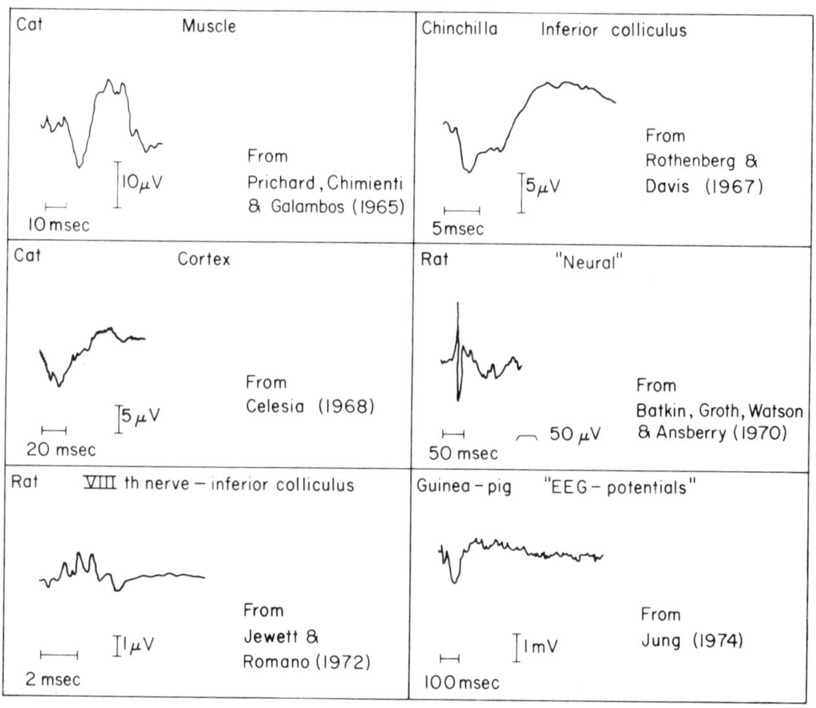

FIG. 6. Some examples of averaged auditory evoked responses recorded from the scalps of small mammals, with suggested origins of the responses. From various authors as shown.

groups of workers, but the techniques employed and the responses obtained appear to differ somewhat, as do the suggested origins of the responses (Fig. 6). Prichard, Chimienti & Galambos (1965) used pairs of bipolar stainless steel electrodes hooked into various muscles on the head of cats and they also implanted electrodes in different parts of the cortex. They found that short latency responses to click stimuli could be recorded extracranially when the animals were awake and alert, but not when they were sleeping naturally, when under light chloralose anaesthesia or when partially paralysed, although large cortical responses could be obtained under all these conditions. The authors therefore suggested that the extracranial responses arose in the muscles, and the latency indicated that they may possibly be associated with the pinna reflex which in the guinea-pig has a similar latency of 5–7 msec (Totsuka, Nakamura & Kirikae, 1954).

However, Celesia (1968), who used electrodes in somewhat similar positions, but with the reference electrode screwed to a skull bone, obtained rather different results. Celesia recorded auditory evoked potentials to click stimuli from both intra- and extracranial sites in alert cats, as well as in cats that were asleep or under barbiturate or deep chloralose anaesthesia. He found that in alert animals the extracranial response did not correlate with cortical responses and he agreed with Prichard and his colleagues that these responses probably arise in the cranial muscles. But Celesia obtained good responses during sleep and anaesthesia and he believed that these reflected potentials arising in the cortex. He showed that muscle potentials were not contributing to the response, for when he paralysed the muscles of anaesthetized animals, the evoked response remained unchanged. Celesia suggested that the differences between his own results and those of Prichard *et al.* (1965) could have been due to the depth of chloralose anaesthesia used. He pointed out that the light level of anaesthesia used by Prichard and his colleagues is often associated with generalized myoclonic reflex reactions to various stimuli, including auditory stimuli, but that the higher dose which he used produces anaesthetic depression of the central nervous system similar to the effect of barbiturates, and without myoclonic reflexes.

Rothenberg & Davis (1967) used a platinum needle electrode inserted subcutaneously over the centre of the skull and a silver cup in the mouth of animals under barbiturate anaesthesia. From the latency and shape of the response they obtained, they concluded that it arose mainly in the inferior colliculus, while Jewett & Romano (1972) suggested that the potentials which they recorded between the scalp and tongue of young and adult rats and cats under halothane or barbiturate anaesthesia, arose within the auditory structures of the brain stem from the VIIIth nerve to the inferior colliculus.

Batkin, Groth, Watson & Ansberry (1970) and Jung (1974) did not suggest specific sites of origin for the responses they obtained. Batkin and his co-workers used platinum needle electrodes subcutaneously over the area of the auditory cortex and in the pinna of rats anaesthetized with barbiturates and Jung used small silver-silver chloride disc electrodes placed on the pinna, on the centre of the scalp and over the temporalis muscle of guinea-pigs. Jung was able to obtain electroencephalogram potentials from the scalp of his animals without sedation, although previous workers (Cody, Gutrecht & Bickford, 1965; Kern, Cody & Bickford, 1969) had found that cortical responses could be recorded from scalp electrodes in guinea-pigs only during curarization; under other conditions myogenic potentials were obtained.

Obviously the origins of these auditory evoked potentials are by no means clear and the position of the electrodes may determine which structures contribute to the recorded response. Latencies alone do not help a great deal here as potentials from various structures from the cochlear nucleus to the inferior colliculus, as well as some muscle reflexes, occur within a similar time period, 3–25 msec after the onset of the acoustic stimulus (Davis, 1973). However, the comparatively longer latency of the potentials recorded by Jung may indicate that these are comparable with the human vertex potential.

Recently Kevanishvili & Kajaia (1973) have investigated systematically the origin of the auditory evoked response recorded from the scalp of anaesthetized cats. The authors recorded potentials from the scalp using silver disc electrodes and compared these with potentials recorded simultaneously from various levels of the auditory pathway such as the cochlea, the cochlear nuclei, the inferior colliculus and the auditory cortex. In addition, the scalp potentials were studied after various regions of the auditory pathway had been sectioned or removed. Kevanishvili & Kajaia found that in intact, anaesthetized cats the scalp response to click stimuli consisted of five to eight "fast" deflections and two "slow" deflections. The first two fast deflections occurred before the slow waves, with a latency of 0·8–1·0 msec to the first deflection. The other fast deflections were superimposed on the first slow wave which had a latency of 2–3 msec. The scalp response to tone stimuli consisted of slow waves only. These had latencies of 40–50 msec when the stimuli were near the threshold intensity, and of 25–30 msec for louder stimuli of 90–100 dB SPL. From their experimental results the authors concluded that scalp responses to auditory stimuli do not express the evoked activity of a certain level of the auditory pathway, but rather have a complex origin and in fact represent the algebraic sum of evoked responses that spread by volume conduction from various levels of the system, from the intracranial part of the cochlear nerve up to the auditory cortex. They suggested that the principal sources of these potentials were the auditory fibres and nuclei of the bulbar, pontal and mesencephalic levels.

Several groups of workers (Rothenberg & Davis, 1967; Batkin *et al.*, 1970; Jewett & Romano, 1972; Kevanishvili & Kajaia, 1973) have pointed out the potential value of the technique of averaging scalp responses in animal audiometry. So far it has been used to obtain audiograms in three species: the chinchilla by Rothenberg & Davis (1967), the rat by Batkin and his colleagues (1970) and the cat by Kevanishvili & Kajaia (1973); while Jewett & Romano (1972) have used the technique to study the development of the auditory responses to click stimuli in cats and rats, and Jung (1974) used it to determine the effect of certain drugs on the auditory evoked responses of guinea-pigs. It is clear, however, that averaged responses to auditory stimuli that are recorded from the scalps of small mammals need to be studied in more detail before the results of different workers can be compared directly.

All of the above studies used signals in the audible range; in this laboratory the use of the scalp recorded averaged evoked response is being investigated as a means of assessing the auditory sensitivity of rodents to audible and to ultrasonic frequencies. It has been found that while similar responses can be obtained from different species, the response may also vary between individuals of the same species. Figure 7a shows the response to a 7 kHz signal obtained from a woodmouse *Apodemus sylvaticus*, under barbiturate anaesthesia. Platinum electrodes were used subcutaneously in the centre of the scalp between the ears, and behind the right ear. The response of a different animal of this species to a 10 kHz signal is illustrated in Fig. 7b. Again platinum electrodes were used, this time in the centre of the scalp and in the temporalis muscle. The difference in form of the responses may be due to the difference in

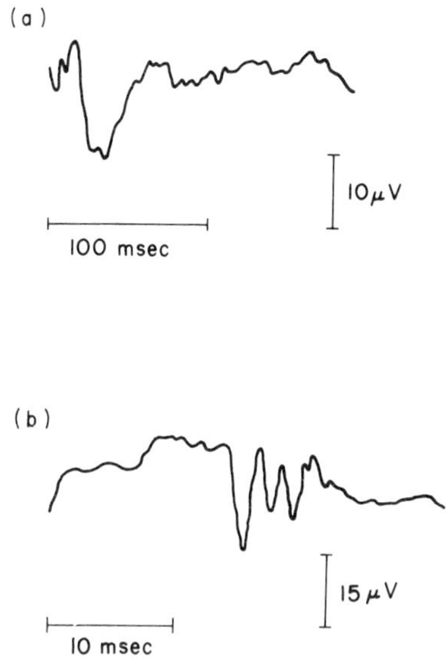

FIG. 7. Averaged evoked responses recorded from the scalp of *Apodemus sylvaticus*. (a) Response to a 7 kHz signal recorded between the centre of the scalp and behind the right ear; (b) response to a 10 kHz signal recorded between the centre of the scalp and the temporalis muscle.

the position of the electrodes. The depth of anaesthesia has also been found to affect the response. Too deep a level abolishes it, but with too light a level the response is often contaminated, apparently with potentials from random muscle twitches, and a greater number of samples is needed to give a clear response.

Responses to signals in the audible range have been obtained from four species of rodents: rats *Rattus norvegicus*; mice *M. musculus* (albino strain); woodmice *A. sylvaticus*, and bank voles *Clethrionomys glareolus*. Responses to signals up to 40 kHz have been obtained from bank voles and up to 50 kHz from mice. It is obviously possible to record averaged responses to ultrasonic signals from the scalp of rodents and although complete audiograms have not yet been obtained, it is anticipated that this will be achieved very soon. Comparison with Brown's results described earlier, particularly in the presence or absence of bimodal peaks of sensitivity, should then be very interesting.

DISCUSSION

Although there have been comparatively few detailed studies of scalp recorded, averaged evoked potentials in small mammals, the technique appears to be potentially very exciting. As Rothenberg & Davis (1967) first suggested, it can be used for regularly assessing the auditory sensitivity of small mammals and it can also be used for studying the effects of acoustic trauma (Eldredge, 1973). In man, certain hearing impairments have been associated with the loss of sensory cells at different locations in the cochlea (Bredberg, 1968), but Ward & Duval (1971) found that in the chinchilla, complete destruction of the outer hair cells in the first $1\frac{1}{2}$–2 turns did not affect the absolute behavioural threshold at all. In this laboratory, part of the investigation into the importance of ultrasonic signals in rodent behaviour involves the use of acoustic trauma in an attempt to reduce or eliminate a rat's sensitivity to a particular frequency band (22–30 kHz, or 40–70 kHz). The effect of this sensory loss on aggressive or submissive behaviour is then studied, but clearly it is necessary to check whether exposure to loud tones does affect the animal's auditory sensitivity and if so, at the correct frequency. This could be done most easily using the averaging technique.

The technique could also be used to study the acoustic responses of animals in different hormonal conditions. Smith has found that rodents of differing hormonal states, for example lactating females or virgin females, show different levels of behavioural responsiveness to certain acoustic stimuli (Smith, 1974; this symposium pp. 317–330). In man the ability to determine absolute pitch varies with hormonal cycles (Wynn, 1972) and it is possible that changes in auditory sensitivity occur with hormonal changes in rodents. This and other problems, such as the development of auditory sensitivity to ultrasonic frequencies, could be investigated using evoked responses recorded from the surface of the scalp. Thus this technique, although in its early stage as far as small mammals are concerned, appears to have great promise as a research tool in many aspects of research on auditory responses.

SUMMARY

Bimodal peaks of sensitivity, including peaks in the ultrasonic range, have previously been reported in the gross response of the inferior colliculus of nine species of rodents by Brown. The significance of these peaks in ultrasonic communication is discussed. The averaged auditory evoked response recorded from the scalp appears to be a useful technique for assessing auditory sensitivity repeatedly in small mammals. This technique is regularly used in man and it has been applied to small mammals by a few groups of workers whose results are described. Although the origin of these scalp recorded responses is not entirely clear, the technique has several advantages over intracranial recording methods and promises to be a valuable research tool for future research on acoustic responses in small mammals. It is currently being used to assess the auditory sensitivity of rodents to ultrasonic frequencies as an indication of the importance of these signals in acoustic communication.

Acknowledgements

The author wishes to thank Professor J. D. Pye, Dr A. Pye and Dr J. C. Smith for their help and discussions in the preparation of this manuscript. Permission to reproduce figures is gratefully acknowledged from the following: Figs 1 and 2, A. D. Grinnell and Cambridge University Press, London; Fig. 3, A. M. Brown and Springer-Verlag, Berlin; Fig. 4, R. G. Bickford and the New York Academy of Science; Fig. 5, H. Beagley; Fig. 6, S. Batkin, D. L. Jewett, J. W. Prichard and Elsevier Publishing Co., Amsterdam, G. G. Celesia (*Arch. Neurol.* 1968, **19**: 430–437) copyright 1968, American Medical Association, W. K. Jung and Springer-Verlag, Berlin, S. Rothenberg and the Psychonomic Press.

References

Batkin, S., Groth, H., Watson, J. R. & Ansberry, M. (1970). Effects of auditory deprivation on the development of auditory sensitivity in albino rats. *Electroenceph. clin. Neurophysiol.* **28**: 351–359.

Beagley, H. A. (1973). Neuro-physiological tests in audiometry. In *Disorders of auditory function*: 47–55. Taylor, W. (ed.). London: Academic Press.

Berlin, C. I. (1963). Hearing in mice via GSR audiometry. *J. Speech Hearing Res.* **6**: 359–368.

Bickford, R. G., Jacobson, J. L. & Cody, D. T. R. (1964). Nature of average evoked potentials to sound and other stimuli in man. *Ann. N.Y. Acad. Sci.* **112**: 204–218.

Bredberg, G. (1968). Cellular patterns and nerve supply of the human organ of Corti. *Acta otolar.* suppl. **236**: 1–135.

Brown, A. M. (1970). Bimodal cochlear response curves in rodents. *Nature, Lond.* **228**: 576–577.

Brown, A. M. (1971a). High frequency responsiveness in rodents at the level of the inferior colliculus. *Nature, Lond. (New Biol.)* **232**: 223–224.

Brown, A. M. (1971b). *Auditory responses to high frequencies in small mammals.* Ph.D. Thesis: University of London.

Brown, A. M. (1973a). High frequency peaks in the cochlear microphonic response of rodents. *J. comp. Physiol.* **83**: 377–392.

Brown, A. M. (1973b). High levels of responsiveness from the inferior colliculus of rodents at ultrasonic frequencies. *J. comp. Physiol.* **83**: 393–406.

Celesia, G. G. (1968). Auditory evoked responses. *Arch. Neurol.* **19**: 430–437.

Cody, D. T., Gutrecht, J. A. & Bickford, R. G. (1965). Averaged evoked potentials to sound stimulation in guinea pigs. Abstract *Fedn Proc. Fedn Am. Socs. exp. Biol.* **24**: 206.

Davis, H. (ed.) (1965). The young deaf child: identification and management. *Acta otolar*, suppl. **206**: 5–258.

Davis, H. (1966). Validation of evoked response audiometry (ERA) in deaf children. *Int. Audiol.* **5**: 77–81.

Davis, H. (1973). Classes of auditory evoked responses. *Audiology* **12**: 464–469.

Davis, H., Engebretson, M., Lowell, E. L., Mast, T., Satterfield, J. & Yoshie, N. (1964). Evoked responses to clicks recorded from the human scalp. *Ann. N.Y. Acad. Sci.* **112**: 224–225.

Davis, H., Hirsh, S. K., Shelnutt, J. & Bowers, C. (1967). Further validation of evoked response audiometry. *J. Speech Hearing Res.* **10**: 717–732.

Eldredge, D. H. (1973). Anatomical and physiological correlates of threshold shifts in the chinchilla after exposure to noise. In *Disorders of auditory function*: 103–114. Taylor, W. (ed.). London: Academic Press.

Gould, J. & Morgan, C. T. (1941). Hearing in the rat at high frequencies. *Science, N.Y.* **94**: 168.

Gould, J. & Morgan, C. T. (1942). Auditory sensitivity in the rat. *J. comp. Psychol.* **34**: 321–329.

Gourevitch, G. & Hack, M. H. (1966). Audibility in the rat. *J. comp. physiol. Psychol.* **62**: 289–291.

Grinnell, A. D. (1963). The neurophysiology of audition in bats: intensity and frequency parameters. *J. Physiol., Lond.* **167**: 38–66.

Grinnell, A. D. (1970). Comparative auditory neurophysiology of neotropical bats employing different echolocation signals. *Z. vergl. Physiol.* **68**: 117–153.

Jewett, D. L. & Romano, M. N. (1972). Neonatal development of auditory system potentials averaged from the scalp of rat and cat. *Brain Res.* **36**: 101–115.

Jung, W. K. (1974). Eine einfache Methode zur EEG-Audiometrie am Kleintier. *Arch. Oto-Rhino-Lar.* **206**: 181–190.

Kern, E. B., Cody, D. T. R. & Bickford, R. G. (1969). Vertex-response thresholds to pure tones in guinea-pigs. *Arch. Otolar.* **90**: 315–325.

Kevanishvili, Z. Sh. & Kajaia, O. A. (1973). On the origin of the auditory averaged evoked responses recorded from the scalp in the anaesthetized cat. *Acta otolar.* **76**: 98–108.

Milner, B. A. (1970). *The physical study of bioelectric potentials generated in response to auditory stimuli.* Ph.D. Thesis: University of London.

Prichard, J. W., Chimienti, J. & Galambos, R. (1965). Evoked responses from extracranial sites in the cat. *Electroenceph. clin. Neurophysiol.* **18**: 493–499.

Ralls, K. (1967). Auditory sensitivity in mice: *Peromyscus* and *Mus musculus*. *Anim. Behav.* **15**: 123–128.

Rapin, I. & Graziani, L. J. (1967). Auditory-evoked responses in normal, brain damaged, and deaf infants. *Neurology* **17**: 881–894.

Regan, D. (1972). *Evoked potentials in psychology, sensory physiology & clinical medicine.* London: Chapman & Hall.

Rose, D. E., Keating, L. W., Hedgecock, L. D., Miller, K. E. & Schreurs, K. K. (1972). A comparison of evoked response audiometry and routine clinical audiometry. *Audiology* **11**: 238–243.

Rothenberg, S. & Davis, H. (1967). Auditory evoked response in chinchilla: application to animal audiometry. ***Perception & Psychophysics*** **2**: 443–447.

Sales (née Sewell), G. D. (1972a). Ultrasound and aggressive behaviour in rats and other small mammals. *Anim. Behav.* **20**: 88–100.

Sales (née Sewell), G. D. (1972b). Ultrasound and mating behaviour in rodents with some observations on other behavioural situations. *J. Zool., Lond.* **168**: 149–164.

Sales, G. D. & Pye, J. D. (1974). *Ultrasonic communication by animals.* London: Chapman & Hall.

Schleidt, W. M. (1952). Reaktionen auf Töne hoher Frequenz bei Nagern. *Naturwissenschaften* **39**: 69–70.

Sewell, G. D. (1969). *Ultrasound in small mammals.* Ph.D. Thesis: University of London.

Sewell, G. D. (1970). Ultrasonic communication in rodents. *Nature, Lond.* **227**: 410.

Smith, J. C. (1972). Sound production by infant *Peromyscus maniculatus* (Rodentia: Myomorpha). *J. Zool., Lond.* **168**: 369–379.

Smith, J. C. (1974). *Sound communication in some myomorph rodents with special reference to adult–infant relationships.* Ph.D. Thesis: University of London.

Totsuka, G., Nakamura, K. & Kirikae, I. (1954). Studies of the acoustic reflex. Part 1. Electromyographic studies of the acoustic-auricular reflex. *Ann. Otol. Rhinol. Lar.* **63**: 939–949.

Ward, W. D. & Duval, A. J. (1971). Behavioural and ultrastructural correlates of acoustic trauma. *Ann. Otol. Rhinol. Lar.* **80**: 881–896.

Wynn, V. T. (1972). Measurements of small variations in "absolute" pitch. *J. Physiol., Lond.* **220**: 627–637.

Symp. zool. Soc. Lond. (1975) No. 37, 205–236.

AUTOMATIC EVALUATION OF HUMAN AUDITORY EVOKED POTENTIALS FROM THE VERTEX

G. SALOMON

ENT Department, Gentofte University Hospital, Denmark

SYNOPSIS

A computerized automatic procedure, scoring presence of sound evoked vertex potentials, is described. The test, based on preliminary studies, depends on an estimate of stable reproduction of responses and on a template evaluation. Quantitative automatic scoring in normal children and adults proved that automatic and first rate subjective scorings were of equal quality. Acceptable thresholds were obtained in 95% of deaf children, when no medication was used. In lightly sedated children the rate of errors was unacceptably high and in anaesthetized normal children the evoked or electric response audiometry (ERA) threshold does not reflect the psycho-acoustic threshold.

Only a fraction of repetitive sound stimulations at moderate intensities produce a recognizable ERA response in normal children. In contrast to subjective scoring, the automatic ERA could partially manage this problem using repetitive tests. The presence of a response in either combined computation or single testing in automatic scoring is indicative of a response. The rate of positive scores increases with intensity but does not increase with increasing interstimulus intervals beyond two seconds, or with an increasing number of sweeps.

DEVELOPMENT OF A RELIABLE METHOD OF AUTOMATIC EVALUATION OF VERTEX POTENTIALS FOR CLINICAL AND EXPERIMENTAL USE

Introduction

It is well known that evaluation of bioelectric recordings is a subjective procedure. Davies (1958) found only 90% individual consistency at the second evaluation of electrocardiograms (ECGs) by cardiologists and 80–90% consistency between observers. A much greater variance must be expected when turning from well-defined pulse shapes in ECG to the highly variable potential patterns of the vertex ERA responses (Lentz & McCandless, 1971; Cohen, Rapin, Lyttle & Schimmel, 1971; Rose, Keating, Hedgecock, Schreurs & Miller, 1971). This variance calls for an objective technique of response evaluation.

Experiments with parametric statistics

In 1968 we made an attempt to evaluate ERA responses objectively using the Student's t test (Salomon, 1969). Figure 1 shows how this test was adapted for ERA. A series of 50 sampling sweeps, each consisting of $n = 150$ sampling points (see also Salomon, 1970), is outlined. All values at a certain sampling point, e.g. no. 3, throughout all the sweeps are regarded as a population. This forms a total of 150 populations consisting of 50 values each. The t test will estimate whether the mean value in each population shows a statistically significant deviation from zero. This is crucial because, if no such deviation is present and if the values of the populations are normally distributed, no response is present.

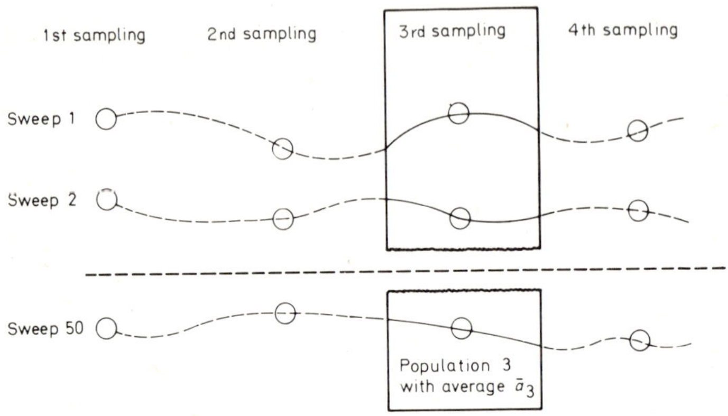

$$\bar{a}_3 = \frac{\sum_{n=1}^{50} a_{3,n}}{50} \quad ; \quad \varepsilon_3 = \frac{\text{Std}_3}{\sqrt{50}}, \text{ where}$$

$$\text{Std}_3 = \sqrt{\frac{\sum_{n=1}^{50} \left(\bar{a}_{3,n}\right)^2}{50} - \left(\frac{\sum_{n=1}^{50} \bar{a}_{3,n}}{50}\right)^2}$$

$$t_3 = \frac{\bar{a}_3}{\varepsilon_3}$$

If population 3 is normally distributed, and if

$t_3 > 2{\cdot}7$: $\bar{a}_3$ is deviating from 0 indicating response

FIG. 1. Calculation procedures adapting Student's t test for ERA.

In Fig. 1 an example of the calculation procedure is outlined for population 3. The mean value for 3, $\bar{a}_3$, is calculated as well as the error of the mean ε_3 which is an expression of the variance (uncertainty) of the value $\bar{a}_3$. The Student's t value can now be calculated on the premis that the population called 3 is normally distributed. If we have used 50 sweeps or more and if t is greater than 2·7, then $\bar{a}_3$ is not zero, indicating a response.

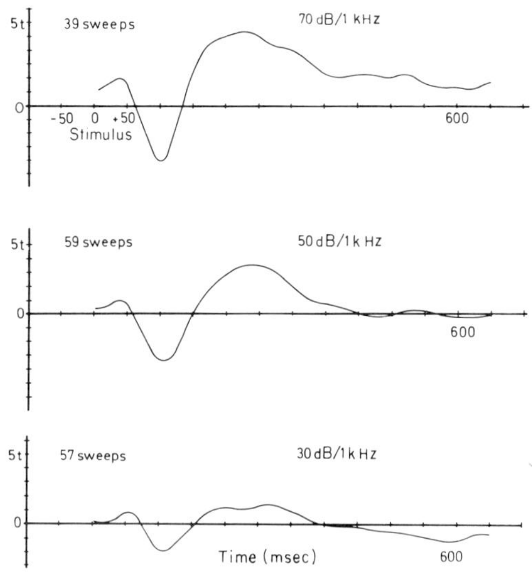

FIG. 2. T values of ERA based on Student's t test. The curves represent the t values calculated for each measure point. At 70 dB HL and 50 dB HL stimulation, the t values exceed 2·7, indicating a response.

Figure 2 shows an example with the t values plotted for each sampling point at three different intensities. In areas where responses could be expected, these responses are reflected in large t values ($t > 2·7$).

The Student's t test was unfit for clinical use, especially in children (Salomon, 1971) where the assumption that each population is normally distributed was not valid (Skogstad Nielsen, Salomon & Elberling, 1971; Spreng, 1971). The fact that the $\bar{a}_n$ can also deviate from zero as a result of a combination of long-lasting bioelectric d.c. shifts on the surface of the brain (Caspers, 1963) and electrical offset drift in the recording system is an additional explanation of the failure of the t test in ERA evaluation.

Experiments with non-parametric tests

In 1971 we introduced the multi-channel write-out system described by Burian & Gestring (1971) and by Skogstad Nielsen *et al.* (1971). Figure 3 shows an example. Besides an average curve of

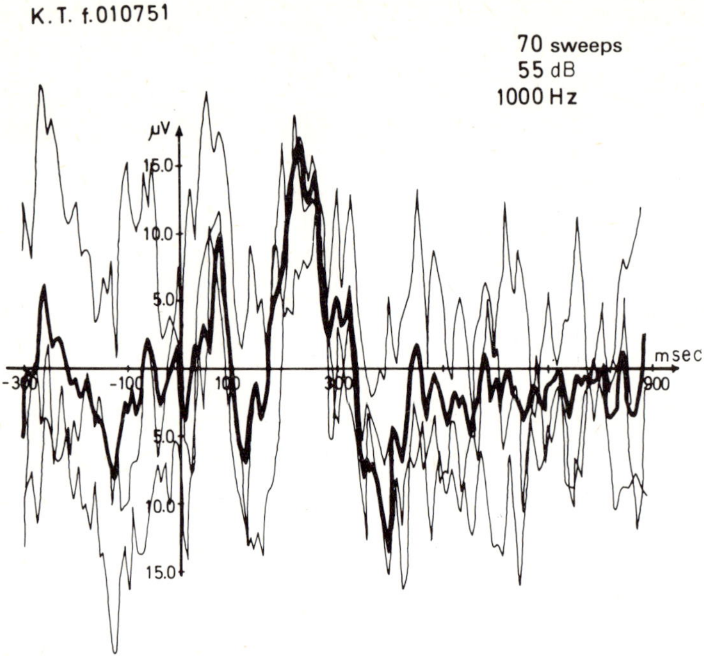

FIG. 3. Average curves in response to sound stimulation. The thick line indicates the average of all 70 sweeps. The three thin lines indicate the three light averages of sweeps 1–23, 24–46 and 47–70 respectively.

all sweeps, indicated by a thick line, average curves for the first, second and last third of the total amount of sweeps are also indicated by thin lines (light averages). This permits control of reproducibility of a pattern as well as recognition of an expected response pattern before judging presence or absence of response. To evaluate the relative importance attached to reproducibility in response recognition or template resemblance, two scoring systems were designed (Salomon, Beck & Elberling, 1973). Following a three-channel write-out of light averages (Fig. 4, upper left) one scoring point was assigned for each maximum or minimum that was common to the three light averages. This score, called reproductiveness score, indicated the consistency of a response. Correspondingly, a template score of one point was assigned for each maximum or minimum that was common to an individual template (Salomon *et al.*, 1973) and the average curve (Fig. 4) was calculated. Table I shows the results of this scoring in 13 patients. As expected, the scores increase with intensity, and although approximately the

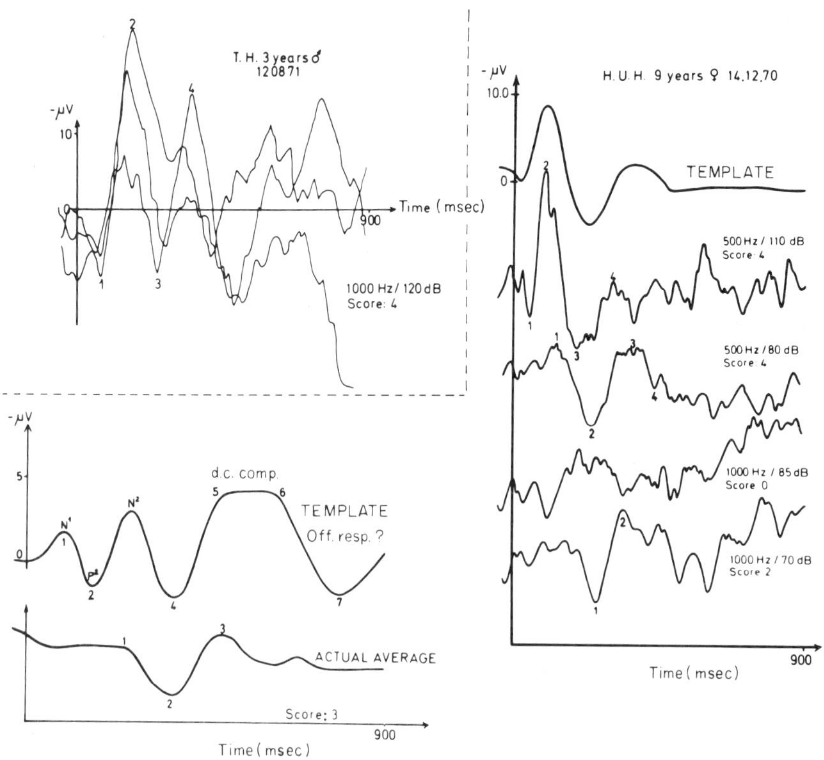

FIG. 4. Definition of 3/3 reproductiveness score and template score. The upper left corner shows the 3/3 reproductiveness scoring. Four angular bends are reproduced in all three part averages. The template score indicated by the number of identical angular bends between the template and the average curve is outlined lower left in a schematic drawing and on the right from a clinical case. After Salomon, Beck & Elberling (1973).

same amount of information is given by the two types of scores, the reproductiveness score shows the best separation between suprathreshold and subthreshold values, and the reproductiveness score alone can be used as an indication of response.

The reproductiveness scoring is a procedure where the same weight is attached to a concordant maximum or minimum in the light averages independent of amplitude. The deciding factor is the temporal location of the relatively highest and lowest value in the time domain. Consequently, we could predict that a rank correlation (coefficient of concordance) between the light averages should be able to detect the presence of responses. Using Friedman's test (Friedman, 1937), the results shown in Fig. 5 were obtained.

TABLE I

Template score and 3/3 reproductiveness score in deaf children

Age (years)	Subthreshold stimulation		Suprathreshold stimulation not exceeding 30 dB		Stimulation exceeding threshold more than 30 dB	
	Mean	s.d.	Mean	s.d.	Mean	s.d.
TEMPLATE SCORE						
4–10	1·33	1·20	2·38	1·60	4·00	1·10
0·5–4	1·58	1·70	3·25	1·40	3·70	1·00
3/3 REPRODUCTIVENESS SCORE						
4–10	1·08	1·20	2·07	1·10	4·83	1·20
0–4	0·70	1·20	2·55	1·50	3·29	1·40

The scores increase with increasing stimulation intensity in both age groups and in both types of scoring. The best separation between subthreshold stimulation and intense stimulation is obtained with 3/3 reproductiveness score in the young children. After Salomon, Beck & Elberling (1973).

To the left are shown averages in a test where the response is clearly visible; note the concordance between the thin lines. The P_F value (obtained from Friedman's test) accordingly is 0·992, indicating a response. The most characteristic features of the response are the slopes. Therefore the Friedman test was also performed on the first time-derivative of the curve and a slightly higher P value was obtained. The basis for our reproductiveness scoring, the maxima and minima of the curves, represents the second time-derivative of the curves (thin lines) and the Friedman test on the second time-derivative was also calculated as indicated on Fig. 5. Although the P_F values increase in the derivatives, no significant change in response evaluation takes place. To the right are shown recordings where a response is just visible on the thick curve but subjective reproducibility is barely seen. In this case only P_F values for the first and second time-derivatives indicate the response.

The difference in P_F values calculated from the original data also corresponds to the visual impression that only on the left drawing is an electronegative d.c. deflection of the electroencephalogram (EEG) present in response to the stimulus. A systematic investigation, judging whether rank correlation is most reliable when based on original values (d.c. response), the first time-

70 sweeps
500 msec

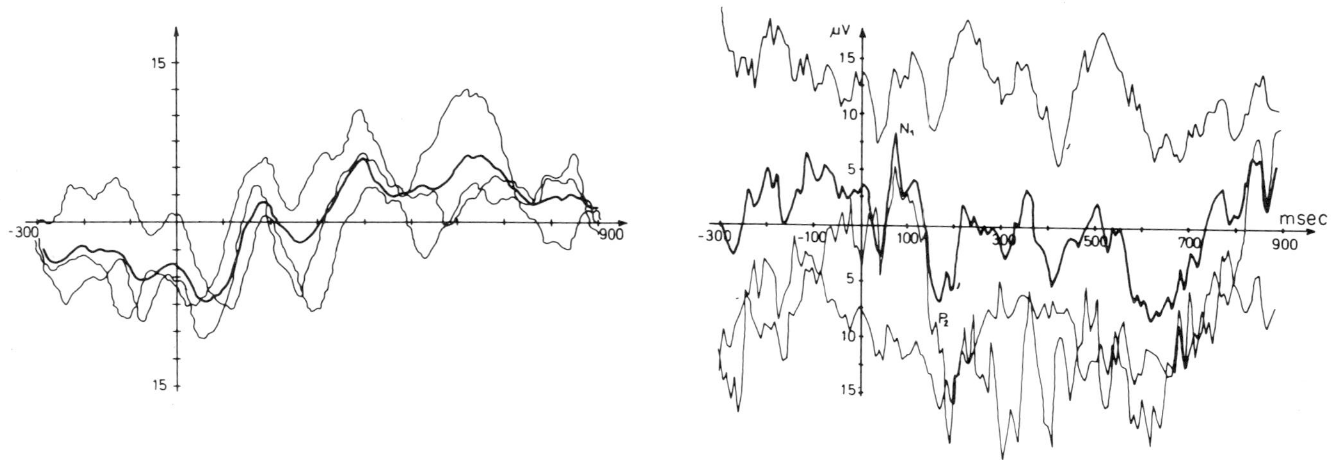

Friedman test for 3 light averages

Fractiles
0 Deriv. 0·992
1 " " 0·994
2 " " 0·999

Friedman test for 17 light averages

Fractiles
0 Deriv. 0·597
1 " " 0·994
2 " " 0·999

FIG. 5. Average curves in response to sound stimulation and the corresponding P_F values. To the left an average response where concordance between the three light averages is prominent. In this child the response consists of "early" components (P_{75}, N_{175}, P_{250}, N_{400} and a vertex electronegative deflection with a latency of 100 msec (d.c. response). Lower left the corresponding P values (see text). To the right an average evoked response where concordance is barely seen. In this case the P_F values are calculated on 17 light averages consisting of four sweeps each (not outlined on the figure). Upwards indicates vertex negative against mastoid.

derivative (response slopes) or second time-derivative (peaks of the response) showed that P_F values based on the first derivative were the most reliable (Salomon, 1974). The fact that rank correlation between three light averages could apparently establish the presence of a response raised the question: what is the optimum number of light averages to use? It was found that rank correlation on 17 light averages, each consisting of four sweeps, resulted in higher P_F values than eight light averages consisting of eight sweeps. This in turn was better than three light averages consisting of 23 sweeps. However, the use of 70 sweeps one by one was the

TABLE II

Pilot studies for strategy using rank correlation

Comparison when using different values		
A		B
I Selected sampling values (for instance every second or third sampling)	⟷	All sampling values participate individually
II Average values of 2 or 3 successive samplings	⟷	All sampling values participate individually
III Light averages (average of 3 or more successive sweeps)	⟷	All sweeps participate individually

Higher P values obtained in column B = best choice

best choice (see Table II). Although a reduction of the number of measured values used in the rank correlation was expected only to exclude information of minor value in exchange for a considerable reduction of the computing time, a pilot study showed that such a reduction impeded the efficiency of the rank correlation (see Table II).

The final elaboration of the automatic test

The method, which is described by Salomon (1974), is based on the principle that it is possible to normalize the values in a sweep by substituting measured values with ranked values. Figure 6 shows how ranking substitutes the highest digitized value with the rank value 1, the second highest with the value 2 and so on. In this way all

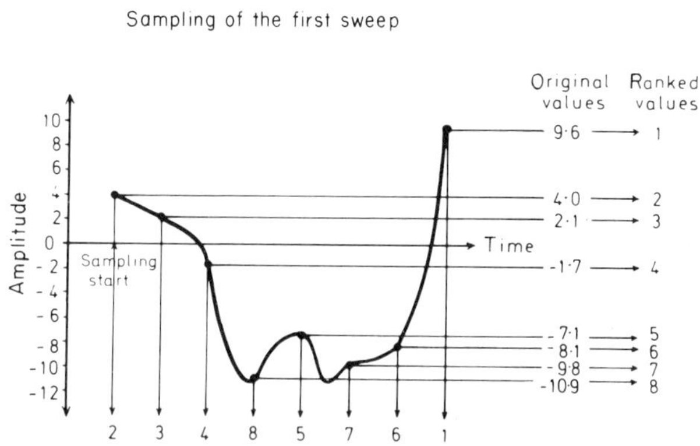

FIG. 6. The ranking procedure. Description of 1st sweep in ranked values. The ranking procedure is performed by substituting the original values obtained from A-D conversion with ranked values. The sweep is thus described in ranked values. From Salomon (1974).

values in a certain sampling point from all sweeps are converted into a well-defined population of numbers, and the mean value of this population is calculated. Because of the ranking, these mean values will be normally distributed if no response is present, as shown in Fig. 7a. If a response is present a deviation from a normal distribution is found, e.g. as shown in Fig. 7b. The probability of a vertex response in a series of sweeps is expressed by a calculated P value for deviation from normal distribution called P_F (Friedman, 1937).

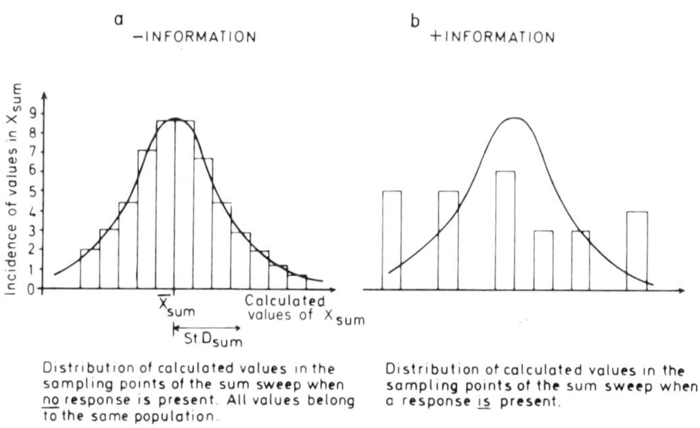

FIG. 7. Examples of distribution of the sum values in each sampling point after ranking. Distribution (a) where no response (− information) and (b) where a response (+ information) is present. From Salomon (1974).

To test the signal detection efficiency of rank correlation based on the first time-derivative (Salomon, 1974), P_F values were calculated in 159 tests, 80 tests subthreshold (stimulus minus 20 dB HL) and 79 tests suprathreshold, in ten normal children and ten normal adults. Calculating each measured value separately, using each sweep one by one and using $P_F \geqslant 0{\cdot}970$ as criterion, in all 17% errors were produced (in 159 tests), 18% false positive and 16% false negative responses. An exact way to express the efficiency of an evaluation method for ERA responses is the signal-to-noise ratio (SNR) (Schimmel, Rapin & Cohen, 1974). Using Schimmel's nomogram, the SNR was found to be 1·9 for the first time-derivative. An improvement was obtained by introducing a template in the calculation of the P_F values: from experience we know that the potentials of the average response can be expected in the time interval from 50 to 400 msec after the stimulus. A heavier statistical significance was assigned during this time using more sampling points during this interval. The number of errors was thus reduced to 10% (Salomon, 1974), 9% false positive and 11% false negative scorings. The corresponding SNR (after Schimmel *et al.*, 1974) was 2·6, and was the highest obtained in this mixed group of children and adults. Using subjective scoring the corresponding SNR was 2·3 (using Table III in Salomon, 1974).

Before we settled for the final version of the clinical test we raised two questions. (a) Is the same template suitable for adults and children? (b) How do subjective and objective scorings compare in children and adults?

With regard to the first question, in my 1974 publication the statistical influence of values after 400 msec was reduced by using only every fourth sampling point in the calculations. In children the latency of the ERA components is nearly 100 msec longer than in adults (Benning, Karnahl & Strange, 1972; Richter, 1972) and the off-response (stimulus duration 500 msec) is often more pronounced (personal experience; I. Rapin, personal communication). In spite of that, the use of a higher statistical influence during the time period 400–900 msec poststimulus did not enhance the automatic evaluation in children. Calculations of the SNR using every fourth and the SNR using every second measure point during this interval poststimulus resulted in exactly the same SNR of 1·9 with identical confidence limits.

With regard to the second question, Table III shows the SNR obtained subjectively and objectively in 17 normal children and ten normal adults at different intensities. There is a good agreement

TABLE III

Signal-to-noise ratios (SNR) (in arbitrary units) of ERA responses at different intensities in ten normal adults and 17 normal children

Intensity dB HL	SNR adults		SNR children	
	Automatic	Subjective	Automatic	Subjective
80	2·4	2·1	2·2	1·6
45	2·5	2·4	1·5	1·8
25	1·9	2·2	0·6	1·0

The ratios are based on Schimmel, Rapin & Cohen (1974), where SNR is found on a nomogram based on % of type I errors and % of type II errors. Percentage errors are calculated in the following way: in tables for "exact confidence limits" (Geigy Scient. Tabl.), the 95% confidence limits of the percentage type I and type II errors at 0·990 and at 0·997 criteria respectively are found. Using the mean value between these limits, a SNR is found for each criterion value (Schimmel 1974) and the mean SNR is presented. The SNR thus obtained are smaller than direct nomogram recordings (e.g. adults at 80 dB up to SNR = 5·0) but highly confident. The 25 dB ratios are based on 50 tests, 45 dB on 30 and 80 dB on ten tests in each group.

between the results in these normal adults and my clinical experience that the subjective scoring of ERA is of nearly equal ease at all intensities in adults (McCandless & Lentz, 1968). Table III shows that response recognition is more difficult but possible in children at high and moderate intensities, but at lower intensities in children, where the SNR is very low, response scoring becomes troublesome (Schimmel *et al.*, 1974). Before jumping to conclusions from the concordance between the subjective and objective SNR, it must be emphasized that the subjective scoring is done by a highly trained observer (six years experience) using d.c. recordings (Salomon, 1970), after on-line rejection of artifacts (Skogstad Nielsen *et al.*, 1971), after classification and only using sweeps with similar frequency and amplitude parameters (Salomon *et al.*, 1973), and using an evaluation after the multi-channel write-out system (Burian & Gestring, 1971; Salomon, 1971). The SNR (Table III) are evaluated slightly higher after automatic scoring both in adults and in children at high intensities and slightly lower at medium and low intensities. The general conclusion, however, is that automatic and subjective scorings in normal adults and children are surprisingly equal. The only notable difference is that the gradual increase in SNR (0·6–1·5–2·2) with intensity in children scored automatically, could not be produced by subjective scoring although it probably reflects a biological fact (see later Fig.

11). The fact that the observer knew the stimulation intensities may have added a positive bias towards a correct subjective scoring.

The SNR conveys the ability of the tests to indicate presence of signals, i.e. vertex responses. This presence is scored with P_F values, e.g. values above 0·970 were used as indicator in my 1974 paper. A certain percentage of false positive errors (type II errors) will be included in our plus scores, and this percentage can be reduced when using a higher P_F criterion such as 0·990 and 0·997 (see Table IV). This reduction in type II errors corresponds to an increasing

TABLE IV

Number of errors and signal-to-noise ratio (SNR) using criterion values $P \geqslant 0{\cdot}970$; $P \geqslant 0{\cdot}990$ and $P \geqslant 0{\cdot}997$

Stimulus intensity	n	Criterion, P value		
		0·970	0·990	0·997
− 20 dB	80	9	5	3
Effective + stimulus	79	11	16	19
	SNR	2·6	2·6	2·8

Note that the numbers of false positive scores (type II errors) decrease from nine to three while the numbers of "misses" (type I errors) increase with increasing criterion value. Note that the SNR is independent of the criterion value.

number of "misses" (type I errors). Note that the SNR is constant. In clinical practice type II errors are the most dangerous because they will deprive a deaf child of therapy. Therefore we normally use values of 0·990 or 0·997 as indicators in the clinic. In subjective scoring it is impossible to govern this type of bias.

Summary

In the search for an automatic evaluation of sound-evoked vertex potentials, parametric statistical evaluations appeared to be unreliable, especially in children.

Analysing the foundations of subjective response scorings showed that the recognition of stable reproduction of responses was a sufficient basis for ERA, although a template evaluation would enhance the evaluation. On this basis Friedman's coefficient of concordance was tested as a tool with satisfactory preliminary results. The Friedman test was adjusted to ERA use by trial and error. Quantitive automatic scorings and optimal subjective scorings proved equal.

EVALUATION OF CLINICAL RELIABILITY OF THE RANK CORRELATION TEST

Introduction

In 1971 and 1972, before the automatic evaluation was developed, the ERA test, besides a clinical score, also produced punch cards, where all measured values obtained during the test were stored in digital form. The present investigation covers all consecutive, ERA patients, aged one to four years, who in 1971 and 1972 could not produce a psycho-acoustic threshold, but since then and before the first of March 1974 were able to produce such a threshold. Using the punch cards from 1971 and 1972 we have made a retrospective automatic scoring. In this way a subjective scoring without knowledge of the correct threshold can be compared with an automatic scoring, and under the assumption that no significant change in hearing threshold has taken place, both scorings can be compared with a psycho-acoustic threshold determination (Price, 1969).

Material and method

Figure 8 shows an example of how the values from a patient are obtained: the subjective binaural ERA scoring in 1971 at 500 Hz was minus at 50 dB, and plus at 70 dB. The automatic scoring using $P_F \geqslant 0{\cdot}997$ as the criterion produces the same results. The threshold in both cases is 60 dB, which is in accordance with the psycho-acoustic threshold produced later on. At 1000 Hz the subjective ERA threshold is 65 dB, the automatic ERA threshold is 80 dB, whereas the psycho-acoustic threshold is 75 dB. At 2000 Hz both ERA thresholds are 65 dB versus a psycho-acoustic threshold of 75 dB.

The patients in this investigation are divided into two groups: in 19 patients no medication was used during the test and in only five patients was a light sedation with vallergan used, because of the uncertainty of the results after medication (Salomon *et al.*, 1973). For comparison we also calculated P values in five normal children

Frequency Hz	Intensity dB HL	Subjective scoring	Automatic scoring
500	50	−	0·715 −
	70	+	0·997 +
1000	60	−	0·733 −
	70	+	0·018 −
	70	(+)	0·686 −
	90	+	1·000 +
2000	55	−	0·091 −
	75	+	0·999 +

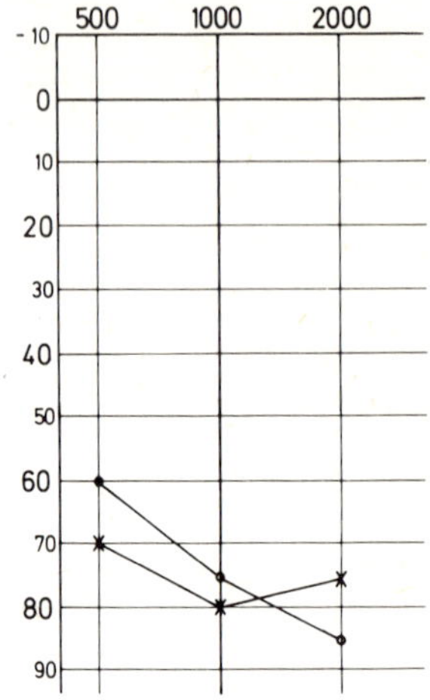

FIG. 8. Example of subjective and automatic ERA threshold determination and psycho-acoustic threshold two years later.

tested during light anaesthesia (Salomon, 1974). In these normal children we used Suzuki's mean psycho-acoustic threshold value (Suzuki & Origughi, 1969) as an indicator of hearing threshold.

Results

Table V shows the results in the 57 tests from non-medicated patients. As indicated by the psycho-acoustic thresholds from 1971 and 1972, the children had medium to severe hearing losses. The next two columns show the deviation of ERA thresholds from the psycho-acoustic thresholds: to the left the subjective ERA threshold and to the right the automatic ERA threshold. The mean deviation from psycho-acoustic threshold is only 1 dB after subjective scoring and 8 dB after automatic scoring. Excluding "clinical errors" these mean values are 1·3 dB and 6·5 dB respectively. The choice of stimulus intensities used during the test procedure was governed by the subjective evaluation of ERA responses of on-line averaging. This, of course, resulted in a choice of more adequate stimulus intensities for the subjective scoring, giving a positive bias towards this procedure. In spite of that, Table VI shows that subjective and automatic scorings are of equal value in the clinic. If mistakes greater than 30 dB are regarded as clinically unacceptable (Salomon *et al.*, 1973), more than 95% of the

TABLE V
ERA thresholds and psycho-acoustic thresholds in non-medicated children

Test frequency Hz	Psycho-acoustic threshold (dB HL)	Subjective scoring of ERA in dB re psycho-acoustic threshold	Automatic scoring of ERA in dB re psycho-acoustic threshold
250	80	0	0
	65	0	0
	100	−10	0*
	>100	0	0
	90	10	10
500	110	−23	5
	60	0	0
	95	−23	0*
	97	10	18
	55	0*	0*
	75	−5	−5
	105	0	15
	30	20	30*
	75	2	2
	65	15	15
	50	0	0
	20	0†	80*
	70	30	30
	>120	−10	0
	95	10	10
	110	−5	0
1000	120	−23	−8
	75	−10	5
	105	−15	0
	97	−7	−7
	50	0†	0†
	60	5	10
	115	−5	5*
	50	15	15
	85	−35	−5
	65	17	17
	105	5	15*
	75	5	5
	75	−30	0*
1000 (control)	50	2	15‡
	65	20	35
	>120	0	0
	100	2	2
	100	10	10
	35	10	25

continued

TABLE V (*continued*)

2000	105	−8	7
	75	−10	−10
	100	−18	0*
	85	10	25*
	55	5	5
	75	−10	10
	110	0*	0*
	65	27	27
	80	0†	−5
	85	−13	−5
	70	5	5
	70	0	0
	65	27	27
	60	15	15
	>120	0	0
	85	5	5
	105	15	15
Mean deviation from psycho-acoustic threshold		1 dB (1·3 ‖)	8 dB (6·5 ‖)

* Stimulus intensity resulting in a positive score was not used.
† Stimulus intensity below threshold was not used.
‡ At a former less successful test 110 dB was negative.
‖ Excluding values >30 dB.

tests are clinically valuable, and 85% of threshold determinations are within 20 dB.

The SNR reported in Table III were calculated from normal children. The SNR for medium and high intensities were equal using subjective and automatic scorings. This equality between subjective and automatic scoring in normal children is reproduced in this group of 19 deaf children (Table VI). The perfect agreement between subjective and automatic scorings in the normal and the deaf indicates that the deaf children have some degree of recruitment: ERA thresholds are established by plus and minus scoring at intensities differing by ten to 20 dB. The lowest intensity resulting in a positive score in normal children would therefore be a low sensation level where automatic scorings are significantly inferior in SNR. The equal status of automatic and subjective scoring in Table VI shows that, with ten to 20 dB SL (sensation level) stimuli, medium sensation levels are reached in

TABLE VI

Difference (in dB) between subjective scoring and psycho-acoustic threshold and between automatic scorings and psycho-acoustic threshold in deaf children

	Subjective		Automatic	
Difference (in dB)	*n*	%	*n*	%
≤10	39	68	40	70
10< ≤20	10	18	9	16
20< ≤30	7	12	6	11
30<	1	2	2	4
Total	57	100%	57	101%

the deaf children because not until these sensation levels are reached do automatic and subjective scorings show equal SNR.

The difference between the mean psycho-acoustic threshold and the mean subjective and automatic thresholds is of the same order as reported for four- to 13-year-old children (Davis 1954, 1966). The ERA thresholds seem to be more accurate with increasing hearing loss.

The time lapse between ERA threshold determination and subsequent psycho-acoustic threshold was one to two years. During this interval an increase of hearing loss can be expected. If this increase is in the order of 5 dB, the mean deviation in subjective and automatic scorings are both within 5 dB. Table VII shows the corresponding values from the sedated children. Both in subjective and in automatic scorings, only three tests out of 11 have determinations of ERA within 10 dB of the psycho-acoustic threshold. It is obvious that the size as well as the number of errors is nearly unacceptably high, and the mean deviation from the psycho-acoustic threshold increases from the range of 5 dB to the range of 15 dB. Taking into account both the fact that the rate of errors in both subjective and automatic scoring increases with sedation, and our knowledge of the success in scoring in non-medicated children, we may state that the errors in sedated children are due to the changes in the biological substrate, namely sedation, and not to methodical short-comings in ERA.

Table VIII shows results from five anaesthetized children. Like the sedated patients, these anaesthetized children showed an unacceptably high percentage of errors with a mean deviation

TABLE VII

ERA thresholds and psycho-acoustic thresholds in sedated children

Test frequency Hz	Psycho-acoustic threshold (1972–1973) dB HL	Subjective scoring of ERA (1971) in dB re psycho-acoustic threshold	Automatic scoring of ERA (1971) in dB re psycho-acoustic threshold
250	75	15	25*
	85	15†	15*
500	90	17	30*
	30	10*	10*
	95	25*	25*
1000	95	0	0‡
	50	60*	40‖
	105	15*	15‖
	20	25	25
2000	>120	0	0
	65	7¶	7¶
Mean deviation from psycho-acoustic threshold		17·2	14·7

* Stimulus intensity resulting in a positive scoring was not used.
† Stimulus intensity below threshold was not used.
‡ At a former less successful test 120 dB was negative.
‖ At a former less successful test 110 dB was negative.
¶ At a former less successful test 120 dB and 90 dB were negative.

TABLE VIII

ERA thresholds and psycho-acoustic thresholds in anaesthetized normal children

Age (years)	Psycho-acoustic threshold in normal children (COR; Suzuki & Origughi, 1969) dB HL	Subjective scoring of ERA (1971) in dB re psycho-acoustic threshold	Automatic scoring of ERA (1971) in dB re psycho-acoustic threshold
2·5	18	4	24
1	25	30*	30*
1	25	30*	30*
2·5	18	24	44*
1	25	−10	10
Mean deviation from psycho-acoustic threshold		16	28

* Stimulus intensity resulting in a positive scoring was not used.

between psycho-acoustic and ERA threshold from 16 to 28 dB (Rapin, Schimmel & Cohen, 1972). The patients in the non-sedated and sedated groups were of equal pathology and represent a mixed group of mono-symptomatic deafness and multi-handicapped children. Although no systematic proof can be given, it is our impression that ERA results (Table VI) were confirmed after cochleography (Salomon, 1972; Salomon & Elberling, 1971; Elberling & Salomon, 1973) in patients where no reliable psycho-acoustic threshold has so far been established.

Summary

In non-medicated one- to four-year-old children, a comparison between psycho-acoustic threshold, ERA threshold based on subjective evaluation of recordings and ERA threshold based on automatic evaluation was performed. In non-medicated children, both types of scorings were of equal efficiency and gave good results in 85% of the cases and acceptable thresholds in 95% of the tests. In sedated children, the same comparison showed that the mean error increased and although automatic scoring was superior to subjective scoring, the rate of errors was unacceptably high. In anaesthetized normal children, ERA threshold does not reflect the psycho-acoustic threshold. The comparison between subjective and automatic scoring in deaf children is in perfect agreement with results from normal children.

STUDIES OF PHYSIOLOGICAL PARAMETERS IN SOUND-EVOKED VERTEX POTENTIALS

Introduction

In the previous chapter it has been demonstrated that automatic scoring of vertex potentials permits a clinical evaluation of hearing thresholds in adults and children. Besides being an improvement in clinical audiometry, the automatic scoring also offers an objective method of testing physiological data in EEG audiometry. Examples of such data are the consistency of the response pattern from stimulus to stimulus over a period of time, the optimum number of stimuli in a test, the clinical reliability of ERA with increasing stimulus intensity and the optimum length of interstimulus interval. These parameters have been studied extensively in adults (Onishi & Davis, 1968; Davis & Zerlin, 1966; Beagley & Knight, 1967; Keidel & Spreng, 1965) but in young children, where there is a great need for such information, the variability of the vertex potentials (Blair, 1968; McCandless & Lentz, 1968) has impeded such studies.

Method

Seventeen normal children, aged one to four years, were tested during a 60–90 min session according to the procedure shown in Fig. 9. Each test consisted of 70 stimuli, and only a few test

Test procedure

Test order	1	2	3	4	5	6	7	8	9	10	11
Stimulus intensity dB HL	-20	25	80	-20	45	25	-20	25	45	-20	25
Interstimulus interval, sec		2	2		2	3		3	3		5

Examples (patients 2 and 15) of scores at 25 dB HL

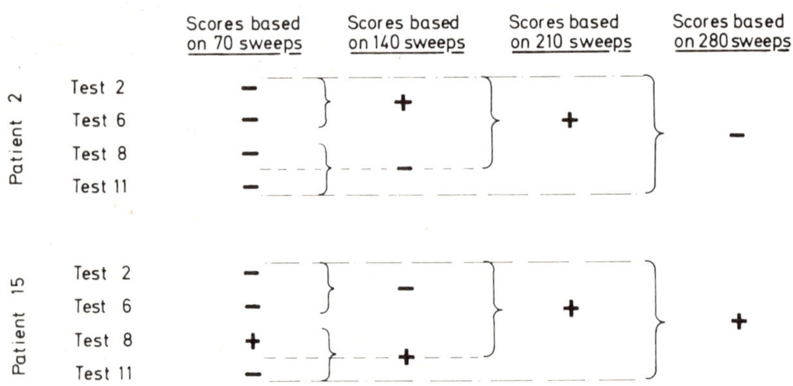

FIG. 9. Stimulation procedures during the test session and examples of results of combined scoring in normal children. All children were stimulated with the intensities and interstimulus intervals in the order outlined. Plus and minus on the lower part are scored according to a P_F criterion of 0·990.

intensities were used, to allow a scoring based on combined processing of many sweeps from different tests with the same intensity. That is, minus 20 dB HL, which means no stimulus, could be scored using 70 sweeps, and also by combined processing of 140 sweeps, 210 sweeps and 280 sweeps. Twenty-five dB HL could be tested in the same way. The tests were also performed with

TABLE IX

Type I errors (false negative scores) and type II errors ("misses") in normal children

	Stimulus intensity											
	−20 dB HL			25 dB HL			45 dB HL			80 dB HL		
No. of sweeps	No. of tests	Positive scoring ($P_F \geqslant$ 0·997)	Positive scoring ($P_F \geqslant$ 0·990)	No. of tests	Positive scoring ($P_F \geqslant$ 0·997)	Positive scoring ($P_F \geqslant$ 0·990)	No. of tests	Positive scoring ($P_F \geqslant$ 0·997)	Positive scoring ($P_F \geqslant$ 0·990)	No. of tests	Positive scoring ($P_F \geqslant$ 0·997)	Positive scoring ($P_F \geqslant$ 0·990)
70	57	1	2	57	5	6	30	9	14	15	9	11
140	30	0	0	30	3	4	14	7	7			
210	15	0	0	15	2	2						
280	17	0	0	15	2	3						

Number of sweeps and positive scoring after clinical criterion ($P_F \geqslant 0{\cdot}997$) and statistical criterion ($P_F \geqslant 0{\cdot}990$)

a varying interstimulus interval permitting evaluation of the influence of this parameter. The lower part of Fig. 9 shows examples of scores from two patients in response to 25 dB/1000 Hz stimuli. In the left column are shown scores after 70 sweeps and the other columns show scores after the combined processing of more sweeps. *P* values exceeding 0·990 indicate positive scoring. In patient 2, all four tests consisting of 70 sweeps show negative scores. Testing the sweeps from test 2 and test 6 combined, a total of 140 sweeps, resulted in a plus score, but testing tests 8 and 11 together resulted in a minus score. A plus score was produced by 210 sweeps and a minus score by 280 sweeps. In patient 15 we see the same type of fluctuation between plus and minus scores.

Results

Table IX shows the scores from all 15 patients.* At minus 20 dB only two positive scores ($P_F \geqslant 0{\cdot}990$) are present in 57 trials using 70 sweeps. No positive scores occur with combined processing. The Table demonstrates that the higher the stimulus intensity the higher the rate of positive scoring.

Table X shows the rate of positive scores $P \geqslant 0{\cdot}990$ with increasing stimulus intensity and increasing number of sweeps in

TABLE X

Rate of positive scores with increasing stimulus intensity and number of sweeps

Number of sweeps	Stimulus intensity			
	−20 dB HL	25 dB HL	45 dB HL	80 dB HL
70	0·04 $n = 57$	0·11 $n = 57$	0·47 $n = 30$	0·73 $n = 5$
140	0·00 $n = 30$	0·13 $n = 30$	0·50 $n = 14$	
210	0·00 $n = 15$	0·13 $n = 15$		
280	0·00 $n = 17$	0·25 $n = 15$		

combined processing. In the 25 dB and 45 dB columns no significant change in scoring with increasing number of sweeps beyond 70 sweeps can be demonstrated.

Only two different hypotheses can possibly explain the constant rate of positive scores.

* Two unco-operative patients were abandoned.

The first hypothesis postulates that some children do show an ERA response after only 70 stimuli and sweeps. These same children continue to show the same response during a longer stimulation period composed of more sweeps. Other children do not show a response and this absence of response does not change with an increasing number of stimuli over a period of time.

The second hypothesis postulates that some children change from minus response to plus response with an increasing number of stimuli in the time domain, and a corresponding number of children change from plus response to minus response with increasing number of sweeps. Figure 10 suggests the second hypothesis to be valid.

Results of combined processing of sweeps.

25 dBHL						45 dBHL		
Scores after 2 tests with 70 sweeps	Scores after combined processing of 140 sweeps	Number of tests	Scores after 2 tests with 140 sweeps	Scores after combined processing of 210 sweeps	Number of tests	Scores after 2 tests with 70 sweeps	Scores after combined processing of 140 sweeps	Number of tests
−, −	−	23	−, −	−	11	−, −	−	4
+, −	−	3	+, −	−	2	+, −	−	3
+, −	+	1	+, −	+	2	+, −	+	5
−, −	+	2	−, −	+	0	−, −	+	0
+, +	+	1	+, +	+	0	+, +	+	3

FIG. 10. Combined processing of sweeps in normal children. + indicates P_F values exceeding 0·990. Braces indicate tests used for combined scoring. In this way the same sweeps can be evaluated using 70 or 140 stimuli.

Figure 10 shows the result of combined processing of sweeps. The most common occurrence is that two negative scores, each based on 70 sweeps, produce a minus score after combined processing of 140 sweeps. But we also observe that many different combinations occur, e.g. that two individual minus scorings after combined processing produce a plus score; a plus and a minus scoring combined can produce either plus or minus scores.

When two negative scores, based on 70 sweeps each, result in a positive score after combined processing, it is probably caused by a low SNR not sufficient to produce a positive score with 70 sweeps; but the information in the signal is constant enough over a sufficient time to be detectable after 140 sweeps with combined scoring. An example of such low-grade signals could be the off-response. On the other hand, clear information e.g. resulting in a positive score after 70 sweeps, may disappear during the next 70 sweeps, resulting in a minus score. The information is not only on a lower level but is sometimes even changed during the next 70 sweeps, resulting in a negative combined score of plus and minus. The intermittent presence of response indicates that averaging is not only functioning as a SNR improvement technique, under the premise that a time-locked constant signal is concealed in a stochastic noise; averaging is also producing a mean signal in the time domain. This corresponds to the clinical impression that the on-line signals are fluctuating, increasing and decreasing several times during a test with low intensity stimulus.

Using Schimmel's nomogram, as described in Table III, SNRs were calculated in 17 children after automatic scorings had been performed on 9, 16, 25, 36, 70, 140, 210 and 280 sweeps, always using the first sweeps in a test during a session. The results are shown in Fig. 11. As expected, the SNR at 80 dB HL shows a steep

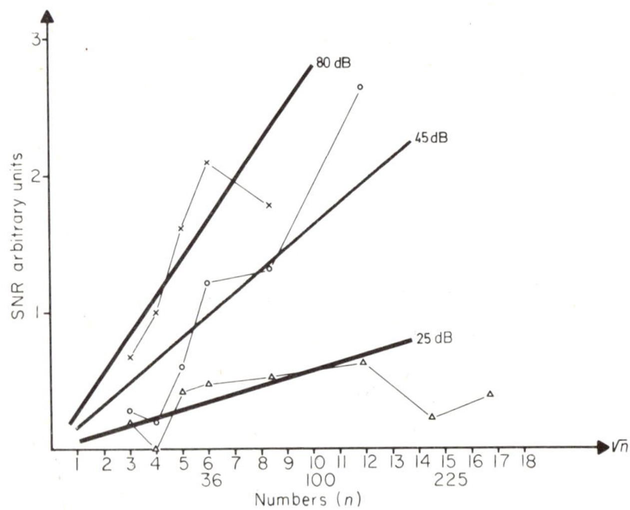

FIG. 11. Signal-to-noise ratios (SNRs) at 80 dB HL, 45 dB HL and 25 dB HL with increasing number of stimuli. × : 80 dB HL; ○: 45 dB HL and △: 25 dB HL indicate SNRs calculated as indicated on Table III. The straight lines at 80 and 45 dB are drawn as the best fit through zero. The straight line at 25 dB is the best fit for values up to 140 sweeps.

proportionality with $\sqrt{n}$, reaching satisfactory clinical values of 2·0 after approximately 60 sweeps. The same type of proportionality is found at 45 dB HL stimulus intensity, but only after approximately 140 sweeps is a satisfactory SNR reached. The increase of the SNR at 45 dB is produced in spite of the constant rate of positive scores in Table X because a reduction in type II errors is brought about with an increasing number of sweeps. At 25 dB HL it is uncertain whether the SNR is proportionate to $\sqrt{n}$, leaving two hypotheses open.

(a) Proportionality is not present. This hypothesis implies that the noise must be dependent on the signal. This does not satisfy the conditions for averaging, so that averaging must be avoided.

(b) Proportionality is present at 25 dB. If this is the case, the steepness of the curve must be even lower than outlined on Fig. 11, e.g. with the straight line only reaching a SNR of 0·5 after 225 sweeps. Clinically satisfactory SNR therefore can first be expected after 3600 sweeps. This makes averaging an impossible proposition (see also Table III).

In conclusion it has been demonstrated that averaging cannot be used to demonstrate responses to low intensity stimulus in normal children, possibly because the statistical conditions for averaging are not satisfied. In contrast, rank correlation technique clearly demonstrates an ability to extract useful information for physiological studies after only a few sweeps, where subjective evaluation of the average is purely guesswork. The information containing the deciding factors in the scoring is composed of several components and these are fluctuating in time and appearing both alternately and synchronously. To improve response recognition in the future we must therefore concentrate our further efforts on finding demarcations in time-separating areas with constant information.

At 25 dB HL, the SNR is significantly higher than zero but still only one in ten tests is positive, which is a low rate for clinical purposes. This calls for an improved technique.

The positive scores were randomly obtained in each patient at 25 dB, giving an impression that the threshold is fluctuating when scored in this way. Clinical threshold using combined scoring changes the picture. Table XI shows all the different combinations of scorings obtained. Only the seven children above the dotted line always produced negative scores. In the other eight children, we obtained some positive scores after combined processing. Recalling that only two false positive scores appeared in 57 subthreshold tests (95% confidence limit 0–14%) of 70 sweeps and none in combined processing with subthreshold stimulation, all children below the dotted line can be scored with a positive reaction to 25 dB HL stimulation. The method raises the ratio of positive scores at 25 dB HL intensity from one in ten to eight in 15. The method has some

TABLE XI

Classification of patients based on scores processed with an increasing number of sweeps (25 dB HL)

Classification according to combination of scores	Number of positive scores: 70 sweeps (positive of 4 trials)	140 sweeps (positive of 2 trials)	210 sweeps (positive of 1 trial)	280 sweeps (positive of 1 trial)	Number of patients with combination in question
1	0	0	0	0	7
2	1	0	0	0	3
3	0	1	0	0	1
4	0	0	0	1	1
5	1	1	1	1	1
6	2	1	0	1	1
7	0	1	1	0	1
				Total:	15

The seven children above the dotted line always produced negative scores (see text). Eight children produced six different combinations of positive scoring indicating response.

resemblance to part averages (Salomon *et al.*, 1973) produced in the multi-channel write-out system, but the control of the rate of type II errors reverses the use: any positive score indicates hearing regardless of the result of the other part P values. Theoretically the multi-channel write-out system is only effective when used over a limited time and number of sweeps, and this limited number of sweeps makes subjective scoring doubtful.

Table XI shows that in response to 25 dB HL stimulation, either sufficient information or alternatively no information is present when measured with our method. This is probably caused by periodical reduction of response signal and by periodical production of a type of information which our Friedman test cannot utilize. The reduction of information at low intensities can be explained by fluctuating attention (Davis & Zerlin, 1966). The same changes in attention can also account for changes in the information as shown on Fig. 12. An adult normal person is tested at different intensities while reading a book. The average evoked responses to tone pips of different intensities are shown in the left column. The on-responses with components between 50 and 400 msec are fairly stable in the reading mode. The electronegative

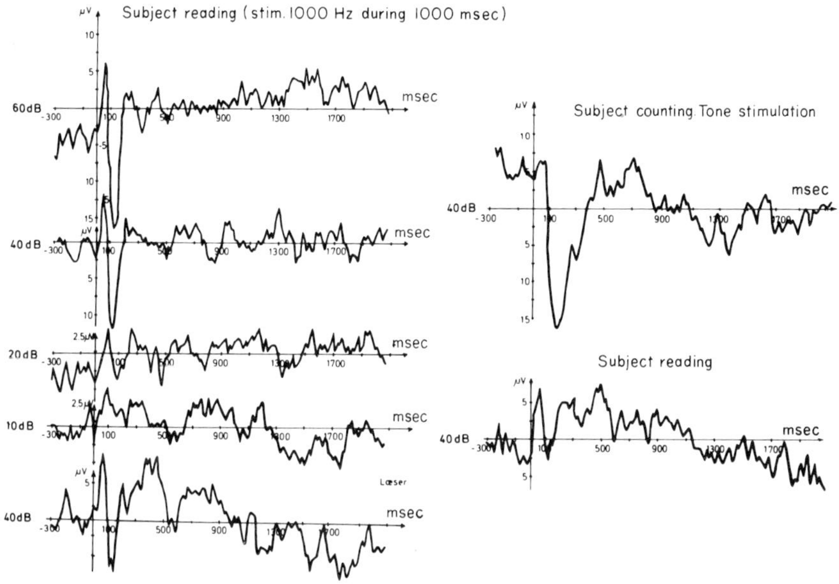

FIG. 12. "Spontaneous" fluctuations of the electronegative sound-evoked vertex potentials and fluctuations of the a.c. component of the vertex potential with attention. The left column shows successive testing of a normal adult person during reading. Note that the N_1P_2 configuration is fairly constant, showing decreasing amplitude with decreasing intensity. Raising the intensity to 40 dB HL (lower left) shows fair reproducibility. Immediately after the five tests to the left, the subject is submitted to a new test. During this test he counts the tone stimulations silently, avoiding any synchronous movements. The diverting shape after this procedure disappears during the last test, where the subject resumes reading. Note the fluctuating appearance of the d.c. components; d.c. components are most prominent in the time interval from 100 msec to 1000 msec on the lower two tracings.

d.c. responses (vertex negative upwards) fluctuate between absence (upper 40 dB curve) and different degrees of presence. Attention to the stimulus, by silently counting the tone stimulation, completely changes the on-response and this change is reversed immediately when the subject goes back to reading. If this type of change appears frequently during the test, no positive information can be obtained by averaging and possibly only a very small amount of information is detected by Friedman tests, which are sensitive to time-locked change of the variance although the signal changes in shape. In children below four years of age, the attention cannot be directed but part P values will catch periods of attention giving positive scores. Possibly other methods of attention-raising procedures, like scrambling of frequencies, can enhance response recognition.

Table XII shows the rate of positive scores with increasing interstimulus interval. The stimuli were given at random time intervals. The numbers indicate the mean interval. The rates of positive scores do not improve when the interstimulus interval is increased beyond two seconds. This is in good agreement with our clinical experience and is probably caused by the fluctuation of information.

TABLE XII

Rate of positive scores with increasing interstimulus interval

Stimulus intensity (dB HL)	Interstimulus interval (sec)	Number of tests	Number of positive scores	Rate of positive scores
25	2	15	3	0·2
	3	30	3	0·1
	5	12	0	0·0
45	2	15	7	0·5
	3	15	7	0·5

Summary

The recognition of ERA responses in normal children after repetitive stimulation at moderate intensities fluctuates. This indicates that the brain periodically does not produce sufficient information to make a positive score with automatic evaluation. The reason for this fluctuation is discussed.

The rate of positive scores increases with intensity but for clinical use it is important that the rate does not improve with an increasing interstimulus interval beyond two seconds or with an increasing number of sweeps.

The free choice of criterion in automatic evaluation permits a minimum risk for false positive evaluations. Therefore ERA could be improved by repetitive tests where presence of response in either combined computation or single testing is indicative for a response.

The SNR was calculated. At moderate and high stimulation intensities the SNR was proportional to $\sqrt{n}$, where n is the number of sweeps. The SNR showed a steep increase with $\sqrt{n}$ at high intensities. At low intensities the SNR did not produce a definite trend, raising the serious question whether the average technique is a reasonable method of choice.

Acknowledgement

This work was supported by the Danish Medical Research Council.

References

Beagley, H. A. & Knight, J. J. (1967). Changes in auditory evoked response with intensity. *J. Lar. Otol.* **8**: 861–873.

Benning, C. D., Karnahl, Th. & Strange, G. (1972). ERA: response characteristics and latencies in fifteen normal-hearing children. *Arch. klin. exp. Ohr. Nas.-Kehlk. Heilk.* **201**: 222–230.

Blair, A. W. (1968). Some experiences with electrically evoked somatosensory potentials in newborn infants. *Nord. Med.* **82**: 1259–1260.

Burian, K. & Gestring, G. F. (1971). Discrepancies between subjective and objective acoustic thresholds. *Arch. Ohr. Nas.-Kehlk. Heilk.* **198**: 73–82.

Caspers, H. (1963). Relations of steady potential shifts in the cortex to the wakefulness sleep spectrum. *UCLA Forum med. Sci.* **1**: 177.

Cohen, M. M., Rapin, I., Lyttle, M. & Schimmel, H. (1971). Auditory evoked response (AER). *Arch. Otolar.* **94**: 214–220.

Davies, L. G. (1958). Observer variation in reports on electrocardiograms. *Br. Heart J.* **20**: 153–160.

Davis, H. (1954). Slow cortical responses evoked by acoustic stimuli. *Acta otolar.* **59**: 179–185.

Davis, H. (1966). Validation of evoked-response audiometry (ERA) in deaf children. *Int. Audiol.* **5**: 77–81.

Davis, H. & Zerlin, S. (1966). Acoustic relations of the human vertex potential. *J. acoust. Soc. Am.* **39**: 109–116.

Elberling, C. & Salomon, G. (1973). Cochlear microphonics recorded from the ear canal in man. *Acta otolar.* **75:** 489–495.

Friedman, M. (1937). The use of ranks to avoid the assumption of normality implicit in the analysis of variance. *J. Am. stat. Ass.* **32**: 675–701.

Keidel, W. D. & Spreng, M. (1965). Audiometric aspects and multisensory power functions of electronically averaged evoked cortical responses in man. *Acta otolar.* **59**: 201–208.

Lentz, W. E. & McCandless, G. A. (1971). Averaged electroencephalic audiometry in infants. *J. Speech Hearing Disord.* **36**: 19–28.

McCandless, G. A. & Lentz, W. E. (1968). Evoked response (EEG) audiometry in non-organic hearing loss. *Arch. Otolar.* **87**: 123–128.

Onishi, S. & Davis, H. (1968). Effects of duration and rise time of tone bursts on evoked V potentials. *J. acoust. Soc. Am.* **44**: 582–591.

Price, L. L. (1969). Evoked response audiometry: some considerations. *J. Speech Hearing Disord.* **34**: 137–141.

Rapin, I., Schimmel, H. & Cohen, M. M. (1972). Reliability in detecting the auditory evoked response (AER) for audiometry in sleeping subjects. *Electroenceph. clin. Neurophysiol.* **32**: 521–528.

Richter, E. (1972). *Untersuchung der Latenzen akustisch evozierter Potentiale bei Kleinkindern mittels der "Electric-Response-Audiometry"*. Inaug. Dis. der Medizinischen Fakultäten der Albert-Ludwigs-Universität, Freiburg.

Rose, D. E., Keating, L. W., Hedgecock, L. D., Schreurs, K. & Miller, K. E. (1971). Aspects of acoustically evoked responses. *Arch. Otolar.* **94**: 347–350.

Salomon, G. (1969). Automatic EEG-audiometry on a "general purpose" process data machine (IBM-system 1800). *Nord. Med.* **81**: 440–441.

Salomon, G. (1970). Clinical "electrical response audiometry" on a general-purpose computer (IBM 1800). *Acta otolar.* **263**: 238–241.

Salomon, G. (1971). Discussion. First internat. Symp. of the ERA club. *Arch. klin. exp. Ohr. Nas.-Kehlk. Heilk.* **198**: 70–71.

Salomon, G. (1972). L'Elettrocochleogramma, Simposio XII Congresso Nazionale della Societa Italiana Audiologia e Fonetria. *Boll. Audiol. Fonetr.* **21**: 9–34.

Salomon, G. (1974). Electric response audiometry (ERA) based on rank correlation. *Audiology* **13**: 181–195.

Salomon, G., Beck, O. & Elberling, C. (1973). The role of sedation in ERA from the vertex. *Audiology* **12**: 150–166.

Salomon, G. & Elberling, C. (1971). Cochlear nerve potentials recorded from the ear canal in man. *Acta otolar.* **71**: 319–325.

Schimmel, H., Rapin, I. & Cohen, M. M. (1974). Improving evoked response audiometry with special reference to the use of machine scoring. *Audiology* **13**: 33–65.

Skogstad Nielsen, S., Salomon, G. & Elberling, C. (1971). Evoked response audiometry from the vertex using an IBM 1800 timesharing system in clinical use. *Computer Progr. Biomed.* **2**: 24–25.

Spreng, M. (1971). Small computers in evoked response audiometry (ERA). *Arch. klin. exp. Ohr. Nas.-Kehlk. Heilk.* **198**: 50–70.

Suzuki, T. & Origughi, K. (1969). Averaged evoked response audiometry (ERA) in young children during sleep. *Acta otolar.* Suppl. **252**: 19–28.

Invited Discussion

B. McA. Sayers: I am impressed by Dr Salomon's work and I congratulate him on making the clearly successful attempt to impose a real objectivity on the detection problem in evoked response audiometry. Without doubt, the best way to relieve *a posteriori* anxiety about the results of an experiment is to draw heavily on information available *a priori*. Dr Salomon has shown one way in which this can be done, but other ways exist, drawing on initial information about the signal component or about the noise component.

Considering the noise first, it is well known that successive lengths of noise (for example, acquired from successive poststimulus sweeps) may be correlated or uncorrelated; in the former case the signal-to-noise ratio does not improve in direct proportion to the number of sweeps (at least this is generally true), whereas in the latter case this proportionality does apply. It is certainly true in many ERA measurements that the noise is certainly correlated from one sweep to the next.

There is no bar to averaging, but it is vital to know something about the noise in order to take full advantage of the noise properties, for example in varying stimulus spacing to achieve the best possible noise reduction. Without a knowledge of the noise characteristics, it is neither satisfactory nor justifiable to question the likely efficacy of averaging.

When it comes to the signal, *a priori* information may well be more difficult to obtain. Dr Salomon has demonstrated one way in which any available information could be used, but this procedure, as I understand it, like that of many workers who do not employ such sophisticated methods, does presume that the response obtained from averaging is necessarily identical whenever successfully evoked—that is, available in that precise form within each individual sweep for which a response could be demonstrated had the noise been sufficiently small. I question this assumption. Indeed, the observations of Mr Beagley and myself do show, convincingly to us, that, at least in some cases, positive responses resulting from an averaging procedure grow up not so much as the steady enhancement of an unvarying response component through background noise, but as the summation of, sometimes, several different kinds of component. In this situation it clearly makes no sense at all to talk of signal-to-noise ratio improvement, since the signal only results in its final form as the summation of a series of perhaps independent components. I am very much of the opinion that further work on the nature of the response within individual sweeps is certainly necessary and I believe that it will be found that the effect I mention above is quite common. If this kind of information about the signal were then available to be built into an objective procedure like that described by Dr Salomon, much enhanced results might well be anticipated. At present, the success of Dr Salomon's method is the more impressive because, as I see it, the technique is artificially limited at present because it utilizes an assumption which is common but which may not be generally valid.

SALOMON: I agree with you that *a priori* an optimum of information on the nature of the signal is desirable. But, as in many biological mechanisms, the signal generator is very complex. Therefore the gain obtained by further investigations may not be proportionate to the invested efforts.

We have obtained *a priori* information from: (a) the literature (e.g. Beagley & Knight, 1967: "The slopes are a characteristic feature of the response"); (b) special experiments (e.g. Salomon, Beck & Elberling, 1973: testing the relative importance of the reproductiveness and concordance with a template); (c) the rank correlation itself (Salomon, 1974). After this type of *a priori* information had been obtained, I had no anxiety about choosing the method presented. On the other hand, I agree that further knowledge on the signal will undoubtedly provide us in the future with more powerful signal detectors than here presented.

Concerning the noise, of course we tested whether the noise characteristics fulfilled the conditions for the Friedman test: a population of sweeps without stimuli were produced and the test observation Z from the Friedman test (Salomon, 1974: 185) was calculated. This enabled us to establish that the number of false positive responses did not differ from the theoretically expected number, and that in spite of the fact that a low-pass 15 Hz filter was used. This experiment showed that no correlation was present between the noise in the different sweeps, when the Friedman test was used as a tool.

In this presentation I have not decided on an attitude towards the evaluation of "the likely efficacy" of averaging at 25 dB. I only state that, contrary to 45 dB and 80 dB stimulus intensity (where proportionality between SNR and $\sqrt{n}$ is present), averaging is not a reasonable suggestion for clinical use.

I am well aware of your information that Friedman test and averaging are dependent on time-locked, constant responses. This is clearly stated in the presentation! In my paper I also have clearly stated that this condition is not always fulfilled and examples of this are shown. However, some fluctuation of latency tends to produce blurred average responses that might be missed; whereas the Friedman test to some degree will be able to recognize a change in the distribution of the sum sweep from the normal distribution, although the change is very small in all the individual time-addresses over which the response has been blurred.

In my paper I have stated: "The information containing the deciding factors in the scoring is composed of several components and these are fluctuating in time and appearing both alternately and synchronously". I am pleased that your experiments also support this observation. I notice that you found that the use of the Friedman test was a success. As mentioned in the paper, the Friedman test was used under certain limiting conditions. However, reasonably good results were obtained with a limited effort, and the method is available for practical clinical work until more efficient methods have been developed.

Symp. zool. Soc. Lond. (1975) No. 37, 237–289.

BEHAVIOURAL AUDIOMETRY IN MAMMALS: REVIEW AND EVALUATION OF TECHNIQUES

R. L. FRANCIS

Department of Zoology and Comparative Physiology, Queen Mary College, London, England

SYNOPSIS

Four types of behavioural techniques have been used to measure intensity thresholds in infrahuman mammals: unconditioned, respondent, and both avoidance and reward operant procedures. The characteristics and different forms of each of these types are described. A comparative evaluation of the advantages and disadvantages of these different types is made, with special attention to precision, reliability, speed, and other practical considerations. Material is drawn from a review of approximately 170 audiometric studies and other relevant literature.

A new study of rodent species, which are suspected of being very sensitive to ultrasonic frequencies, is described briefly.

INTRODUCTION

The principle of behavioural audiometry techniques is that a change in the auditory stimulus produces a change in the subject's behaviour. This response may be either an "innate" reflex, presumably moulded by evolutionary survival pressures, or a conditioned response, largely designed by an experimenter for efficient training and testing. Consequently, audiometric research has used an assortment of widely varied behaviour patterns, unequally suited to the purpose.

This paper reviews and evaluates the techniques of one branch of behavioural audiometry: intensity threshold measurement in infrahuman mammals. The material is distilled from a review of approximately 170 of the most commonly cited papers appearing since the introduction of electrical oscillators and loudspeakers to animal audiometry. Prior to this there was no generally available method for the quantification of auditory sensitivity. This paper has two basic aims: first, to survey the full range of species and techniques in the literature; second, to present some guidelines for selecting a technique appropriate for a given purpose.

"Behavioural" technique has been defined in a variety of ways, just as "behaviour" itself is open to a number of definitions. Sometimes "behavioural" is used synonymously with "conditioned", thus excluding techniques which employ unconditioned responses. This review includes studies in which the subject responded to auditory stimulation with an effector external to the auditory system. These responses are referred to as indicator responses. Studies using the response of middle ear muscles or the electrical potentials of the cortex are excluded from this review.

Thresholds are usually measured in one of two types of investigation. Either the absolute sensitivity is used to understand further the biology of a species itself, or a threshold level is used to investigate auditory physiology. A typical physiological experiment consists of three steps: threshold is determined; some specific component of the auditory system is treated (by a means such as nerve section, presentation of very loud sounds of specific frequencies, or administration of ototoxic drugs); and finally threshold is re-determined to analyse the effects of this treatment on hearing. Many studies have followed this scheme. Their stereotyped requirements have produced a few standard methods for threshold determination. However, in different situations, for example when the hearing of a particular species is of interest for its own sake (e.g. Elder, 1934; Sidman, Ray, Sidman & Klinger, 1966), the behavioural peculiarities of intended subjects and the inventiveness of individual experimenters have produced an assortment of unique and unlikely behavioural techniques.

The aim of all these techniques is to establish "stimulus control" of the animal's behaviour: stimulus change elicits behaviour change. A procedure can be classified as "unconditioned", "respondent", "avoidance operant" or "reward operant", depending on the type of stimulus control. Each of these types has inherent properties which make it more efficient or otherwise better suited to particular requirements.

Procedures which employ conditioned responses, which are the most common methods for threshold assessment, are mainly means of training the subject to respond reliably during threshold tests. A test itself may require a few hours, whereas the training may require months or even years. With human subjects, a few minutes of verbal instruction usually accomplishes the same task.

Likewise, the non-verbal subject cannot describe an auditory sensation; this must be inferred from the animal's behaviour. Under specific conditions, a behavioural change indicates that a sensation occurred; the absence of any change implies that no sensation occurred. However, because behaviour is a function of many internal and external stimuli, all behavioural techniques are subject to influences which weaken the validity of these inferences.

In some cases, the experimenter attempts to control undesirable stimuli. In other cases, such influences are ignored because they are assumed to be insignificant or because they are unidentified.

Behavioural procedures themselves incorporate various types of non-auditory stimulation which may also invalidate inferences about auditory function. For example, it is common practice to punish wrong responses, but the effects of punishment remain largely unanalysed. Two studies of the lemur (Mitchell, Gillette, Vernon & Herman, 1970; Mitchell, Vernon & Herman, 1971) unintentionally provided one of the rare experimental assessments of this practice. With essentially the same subjects, the two studies reported significantly different threshold values for each frequency tested. The experimenters attributed this to the different punishments employed in the two studies. The report of the second study was appropriately entitled: "What does the lemur really hear?" (Mitchell, Vernon *et al.*, 1971).

The physiological experimenter is not interested in absolute threshold values; he is concerned only with the relative values before and after treatment. Therefore, his results are completely valid if he can maintain the same behavioural influences throughout the experiment. Unfortunately, as is illustrated on pp. 270–271, this may not always be possible, because of interactions between motivation and the experimental treatment.

Theoretically, the intensity threshold, or absolute limen, may be defined as the minimum sound intensity capable of producing an auditory sensation. In fact, owing to variations in the sensory system, the sound-field and the subject's attention and body position, there is a small range of stimulus-intensity values which sometimes produce a sensation, and sometimes do not. An identical stimulus repeatedly presented to the same subject is thought to produce a Gaussian distribution of sensations (Guilford, 1954; D'Amato, 1970). This threshold range may be widened by variation in the probability that the subject will respond, which is a process quite separate from sensation, as illustrated by the two lemur studies above. "Threshold" is a value, selected by the experimenter, within this range. Operationally, to "determine a threshold" is to measure an unknown quantity, but the experimenter's definition of "threshold" and the experimental procedures used, both "determine" the results obtained.

To locate the threshold range on the sound intensity continuum and to determine the distribution of responsiveness within this threshold range, the experimenter systematically varies the stimulus intensity between test trials. The schedule used is one of two basic "psychometric methods", the method of constant stimuli or the method of limits. Various modifications of these are described in many experimental psychology textbooks. Guilford (1954) presents a complete description and evaluation of these, based on data from human psychophysics. There are very few such data for animal subjects. Miller, Watson & Covell (1963) and Stebbins (1970b) provide the most comprehensive experimental comparisons of these methods encountered in this review.

The remainder of this paper is divided into three sections. The first (pp. 240–258) is a description of each of the four types of behavioural procedures; Tables I–IV are chronological, cross-referenced listings of studies according to the indicator response used in each study; an Appendix reclassifies the studies according to species. The next section (pp. 258–273) is a discussion of the relative advantages and disadvantages of these procedures. The final section (pp. 273–274) is a description of the application of one of these procedures to a new study of small mammals which are thought to be very sensitive to ultrasounds.

DESCRIPTION OF TECHNIQUES

Procedures with unconditioned responses

As suggested by Fig. 1a, the indicator response is a reflex elicited directly by the auditory stimulus. This response is a disturbance of some baseline activity, the normal rate or pattern of which is established before acoustic stimulation. In order to stabilize the baseline, the subject may be physically restrained, chemically sedated or made to perform a simple behavioural act such as licking or chewing steadily. A number of unconditioned responses have been used as indicator responses. These are described below. Table I chronologically lists the studies which have used each response.

The galvanic skin response (GSR), otherwise known as the electrodermal response (EDR) or psychogalvanic reflex (PGR), is a decrease in the normal electrical resistance of the skin measured between two electrodes attached to a hairless portion of the subject's body.

Respiration is monitored either through the mechanical expansion and contraction of the animal's body (Corbielle & Baldes, 1929a; Bradford, McKinley, Rousey & Klein, 1973), or through

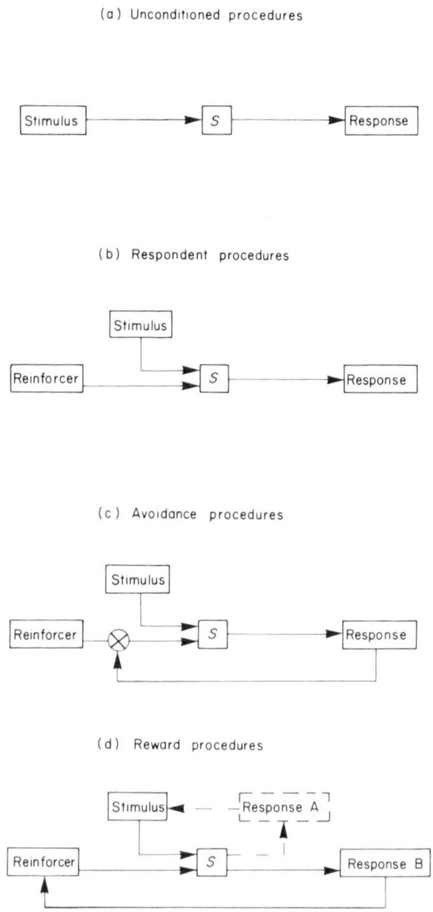

FIG. 1. The four behavioural procedures for threshold measurement. The diagram of each indicates how the auditory stimulus, Stimulus, controls the behaviour, Response, of the subject, S. In (d) response A is the optional observing response, and response B is the indicator response.

the resultant pressure changes in an enclosed environment (Criborn, 1970). The responses to a low intensity sound are usually a decrease in amplitude or duration of the breathing cycle, or a running together of two cycles (Bradford *et al.*, 1973). However, the response may depend on the subject's baseline respiration rate, according to the "law of initial values" (Gautier, 1972).

Heart rate, recorded with two electrocardiograph electrodes attached to different limbs (Corbielle & Baldès, 1929b) or to skin

TABLE I

Unconditioned procedures. Each study and its subjects are listed chronologically under the appropriate indicator response, which is described in the text

Study	Subject
GSR	
Berlin (1963) R	mouse
Anderson & Wedenberg (1965) R,R,A,Rw	guinea-pig
Berlin, Gill & Leffler (1968)	mouse
Gill & Berlin (1968)	mouse
Soliman (1969)	mouse
Respiration	
Corbielle & Baldes (1929a)	rabbit, dog, cat
Koyama (1964)	rabbit
Criborn (1970)	mouse
Bradford, McKinley, Rousey & Klein (1973)*	dog
Heart rate	
Corbielle & Baldes (1929b)	rabbit
Blood pressure	
Corbielle & Baldes (1929c)	dog
Operant suppression	
Ray (1970) R	mouse
File (1973)	rat
Preyer reflex	
Schleidt (1948, 1951) U	red-backed vole
Dice & Barto (1952) U,A	deermice
Schleidt (1952)* U	rodents
Deol & Kocher (1958)	mouse
Denes & Kocher (1961)*	mouse
Murphy & Harris (1961)*	rat
Alford & Ruben (1963)*	mouse
Herman & Clack (1963)*	rat
Mikaelian & Ruben (1964, 1965)*	mouse
Clack (1966b)* Rw	rat
Powers, Warfield & Ruben (1966)*	mouse
Hack (1968)*	mouse
Henry, K. R. (1972)* U	mouse
Henry, K. R., Wallick & Davis (1972)* U	mouse
Crifo (1973, 1974)* R	guinea-pig
Seizure	
Morgan & Gould (1941)	rat
Dice & Barto (1952) U,A	deermice
Henry, K. R. (1972) U	mouse
Henry, K. R. *et al.* (1972) U	mouse

continued

Study	Subject
Movement	
Kellogg & Kohler (1952)	bottlenosed porpoise
Kellogg (1953)	bottlenosed porpoise
Ames & Arehart (1972)*	sheep
Startle	
Fleschler (1965)*	rat
Immobility	
Miller & Murray (1966)*	guinea-pig
Vibrissae twitch	
Schleidt (1952)* U	rodents
Dorsal skin twitch	
Schleidt (1948, 1951) U	red-backed vole
Head jerk	
Chang (1936)*	hedgehog
Poduschka (1968)	hedgehog

The letters following the dates of a study indicate which, if any, other procedures were used in the study: U = unconditioned. R = respondent, A = avoidance, Rw = reward. This designation does not necessarily mean that thresholds were determined with more than one procedure.
In some cases a single study used two forms of the same class of procedure (e.g. GSR and Preyer reflex unconditioned procedures); this is indicated with the appropriate number of letters (e.g. "U,U"). Some studies using unconditioned procedures did not determine thresholds. * Indicates studies in which thresholds of the unconditioned response were determined; studies of hearing development are also included. The species' common name is that used in the Appendix. A common name in the plural indicates that more than one species in the group were used.

near the heart (Jamison, 1951), generally decreases with weak stimuli and increases with intense stimuli (Graham & Clifton, 1966).

Blood pressure is measured with a pressure gauge via an intra-arterial cannula; individual acoustic stimuli may produce either an increase or a decrease in pressure (Corbielle & Baldes, 1929c).

In the operant suppression procedure the experimenter initially trains the subject to lick a spout steadily. This is done by requiring many licks between each liquid reward. A sound stimulus briefly interrupts this electrically-recorded licking (Ray, 1970).

The Preyer reflex, or pinna reflex, is a twitch of the pinna, and is detected visually by the experimenter. Seizure, or audiogenic seizure, is a graded progression of sound-produced disturbances, from "wild running" through epileptic-like convulsions to death (K. R. Henry, 1972). The experimenter may visually or automatically monitor individual occurrences, rate their severity, and time their latency.

Movement, again detected visually, is either a disruption of a regular movement pattern (Kellogg & Kohler, 1952; Kellogg, 1953), or a movement, such as pricking the ears or facing the sound source, by a previously motionless animal (Ames & Arehart, 1972).

The startle response (of rats) is actually a wave of muscle contraction spreading from the animal's mouth through its entire body (Fleschler, 1965). The response is detected visually or electromechanically through a movement-sensitive platform.

The sound-induced dorsal head-jerk of hedgehogs (Chang, 1936) and twitches of the vibrissae or dorsal skin of rodents (Schleidt, 1951, 1952) have been observed visually.

The immobility or "freeze" response of the guinea-pig begins with interruption of activity and may quickly become complete catatonia if the stimulus is sufficiently intense and novel. In hearing tests, sound-produced interruption of lettuce-chewing by a water-deprived subject has been observed by the experimenter, who then rated his confidence in his observation on a four-point scale (Miller & Murray, 1966).

Most of these responses can be associated with either the orientation reflex (also called the orientating reflex), or the defence reflex (Sokolov, 1960; Graham & Clifton, 1966). Each "reflex" is a group of simultaneous, autonomic and somatic reactions which adjusts the animal's senses to new environmental situations. Weak stimuli elicit the orientating reflex, which is thought to enhance sensory responsiveness; intense or painful stimuli produce the defence reflex, which is thought to reduce sensitivity.

Procedures with conditioned responses

Conditioning is the formation or strengthening of a response through a temporal association of events. Respondent conditioning associates two stimulus events, a sound stimulus and a reinforcement that produces an "innate", characteristic response. Operant conditioning associates a response with the reinforcement it produces. In both cases, the name suggests the relationship between the response and the reinforcement (Catania, 1968).

Respondent procedures

In respondent conditioning (Fig. 1b), otherwise known as classical or Pavlovian conditioning, the indicator response is initially elicited by the reinforcer, not by the auditory stimulus. In all studies reviewed, the reinforcer has been electric shock except in one study of the rat which used ammonia fumes (Jamison, 1951). The conditioning association, which ultimately enables the auditory stimulus to elicit the indicator response, results from repeated stimulus-reinforcer pairings: the auditory stimulus is presented for a few seconds, and is concluded with a brief delivery of the reinforcer. Eventually, the stimulus alone elicits a conditioned response, which resembles the unconditioned response to the reinforcer for all responses except respiration. The conditioned respiration response consists of very fast, short breaths in a pattern which, oddly enough, may be more consistent than normal breathing or breathing following electric shock (Upton, 1929a). Coincidentally, respiration was the first and only conditioned response used in threshold determination for a number of years. Table II lists respondent indicator responses and studies.

The methods for treating subjects and recording responses are similar to the unconditioned procedures for the GSR, respiration, heart rate and operant suppression responses. For conditioning, electric shocks are delivered through electrodes attached to the subject, or through a floor grid.

In shiver suppression, the animal is placed on a vibration-sensitive platform in front of a stream of cold air or in a cold room. After several minutes a baseline shivering pattern emerges. To date, this procedure has been used only with guinea-pigs. The subject is held rigidly by a collar. This induces catatonic-like immobility which, paradoxically, facilitates accurate testing. Anderson & Wedenberg (1965) re-induced this state with a few electric shocks if a subject revived during tests.

Vocalization, a squeak elicited by electric shock, was used in one study with the rat (Cowles & Pennington, 1943). A recent study with the mouse (Markl & Ehret, 1973) used a conditioned eye-blink response.

Operant procedures

Operant, or instrumental, procedures involve either the avoidance of a noxious reinforcer, or the acquisition of a rewarding reinforcer. In both cases, "reinforcement" is contingent on the

TABLE II

Respondent procedures

Study	Subject
GSR	
Berlin (1963) U	mouse
Semenoff & Young (1964)	pigtailed monkey
Anderson & Wedenberg (1965) U,R,A,Rw	guinea-pig
Dalton (1968) R	primates
Bragg & Dreher (1969) R,A	capuchin
Respiration	
Upton (1929a) A	guinea-pig
Upton (1929b)	guinea-pig
Wever (1930)	cat
Horton (1933, 1934)	guinea-pig
Alexander, Githler, Diess & Rodwin (1955)	guinea-pig
Sherrick & Bilger (1959)	guinea-pig
Heart rate	
Jamison (1951)	rat
CONDITIONED SUPPRESSION:	
Shiver suppression	
Anderson & Wedenberg (1965) U,R,A,Rw	guinea-pig
Crifo (1973, 1974) U	guinea-pig
Operant suppression	
Sidman, Ray, Sidman & Klinger (1966)	mouse
Dalton (1968) R	primates
Bragg & Dreher (1969) R, A	capuchin
Dalton, Taylor, Henton & Allen (1969)	rhesus monkey
Heffner, Ravizza & Masterton (1969a) A	tree shrew
Heffner, Ravizza & Masterton (1969b)	bushbaby
Ravizza, Heffner & Masterton (1969a)	opossum
Ravizza, Heffner & Masterton (1969b)	hedgehog
Heffner & Masterton (1970)	slow loris, potto
Ray (1970) U	mouse
Heffner, Heffner & Masterton (1971)	guinea-pig
Heffner, Van Oeveren & Masterton (1971)	mouse, cotton rat
Vocalization	
Cowles & Pennington (1943)	rat
Eye-blink	
Markl & Ehret (1973) Rw	mouse

Letters after the date of a study indicate which other procedures were used in the study (see Table I).

animal's behaviour, as indicated by the feedback loops in Figs 1c and d. If an act (known as the "operant") produces reinforcement, there is an increased probability that the animal will repeat the act. This is the basis of operant conditioning. This effect is strongest if reinforcement follows the act immediately. Therefore, efficient operant procedures employ apparatus that automatically monitors and responds to the animal's behaviour. An example of how such apparatus can be incorporated into a hearing study is given in pp. 273–274 of this paper.

With care, the experimenter can shape a subject's behaviour by selectively reinforcing activity that is increasingly similar to the ultimately desired performance. This technique of "successive approximations" can be used to produce a range of performances, from a simple leg flexion (Culler, Finch, Girden & Brogden, 1935) to a complex behavioural "chain". For example, seals have been trained to move to a designated position, press a lever, wait for a trial, and press one of two other levers, depending on whether or not the trial contained a tone stimulus (Möhl, 1968). Figure 2 illustrates the variety of operant apparatus and performance. All diagrams are stylized representations of the apparatus associated with the indicator responses listed in Tables III and IV.

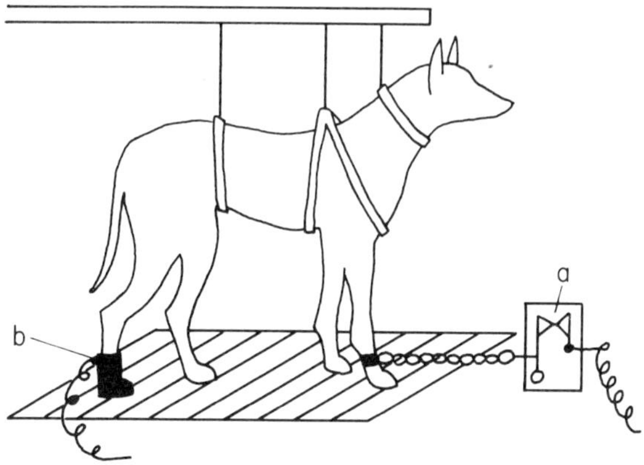

FIG. 2a. Flexion. Used for avoidance only. The subject may avoid shock with a leg flexion, which may either open a mercury switch (a), or open a circuit between its leg and the electrifiable floor grid. All electric contacts are applied directly to the subject's leg or through a boot (b).

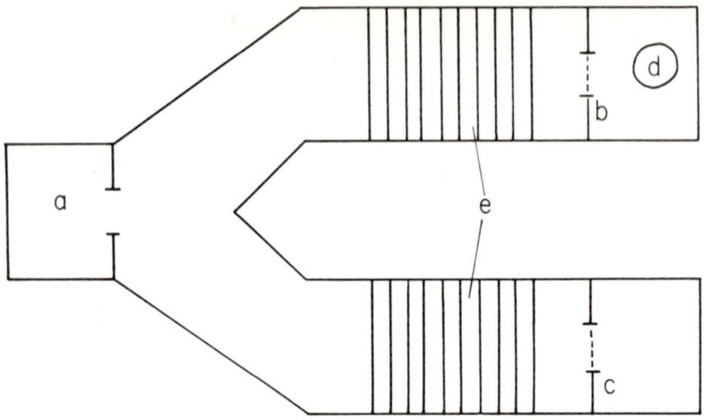

FIG. 2b. Alley choice. May employ both a positive and a negative reinforcer. The subject moves from the starting compartment (a) to one of two goal compartments (b) or (c). A food reward (d) may await the subject at the correct compartment; electric shock may be delivered through a floor grid (e) in the incorrect alley.

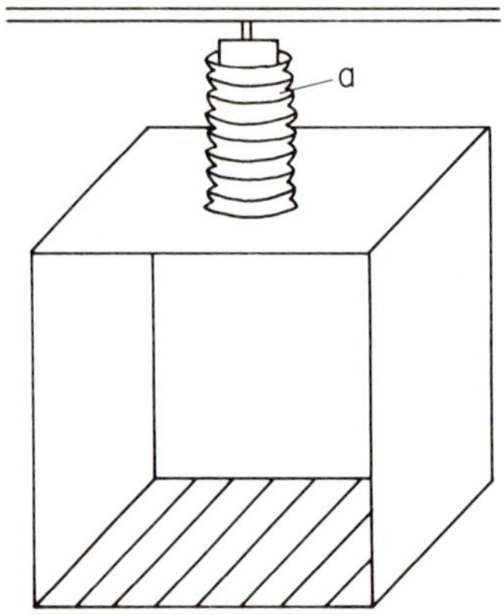

FIG. 2c. Stabilimeter. Used for avoidance only. Any gross body movement which moves the cage serves as the indicator response. The cage is suspended by a spring-like metal bellows (a). Electric shock is delivered between the floor grid and a harness worn by the subject.

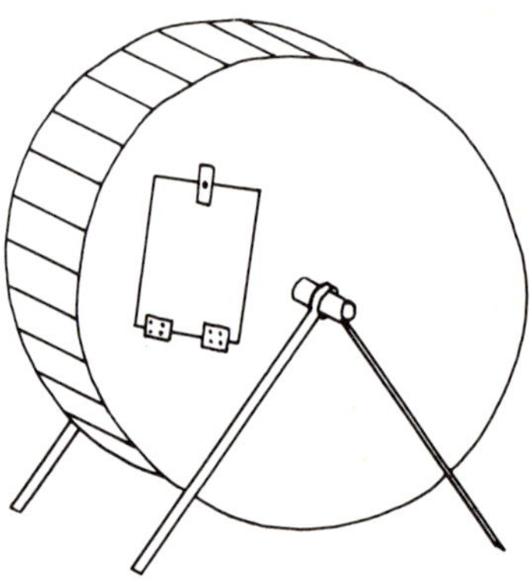

FIG. 2d. Rotator. Used for avoidance only. The indicator response is a walk forward which rotates the cage. Electric shock is delivered through the cross-bars. The cage in this figure rotates around a horizontal axis; a similar model may rotate around a vertical axis. The friction of rotation may be varied to control false positive responses.

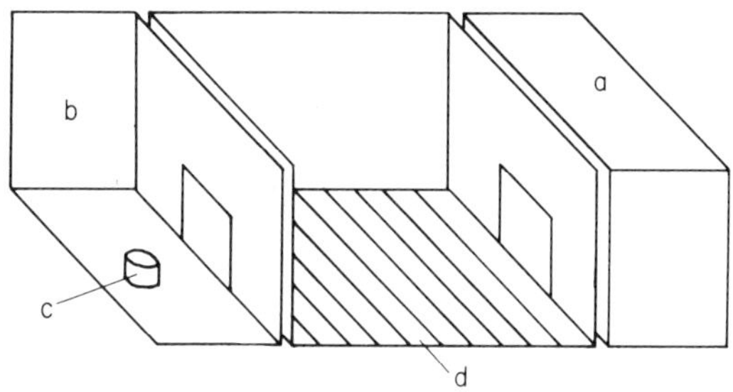

FIG. 2e. One-way alley. Employs both a negative and a positive reinforcer. The subject moves from box (a) to the other box (b), where a food reward (c) is available. The boxes are then detached and reversed in position for the next trial. The floor grid (d) is electrified when an auditory signal is not present.

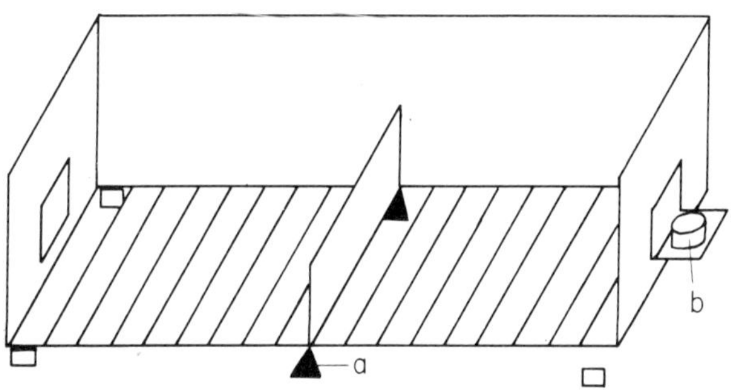

FIG. 2f. Shuttle-box (also known as double-grille cage or tilt-box). Used with negative and/or positive reinforcer. The indicator response is a movement from one side of the cage to the other. The central barrier is optional, as is the fulcrum (a) which permits a mechanical record of the tilt of the cage resulting from the subject's movement. Shock may be delivered independently to either half of the floor grid, and/or food reward (b) is available for correct responses.

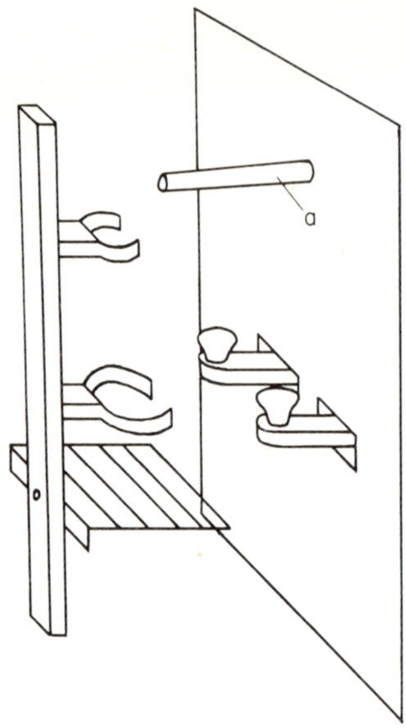

FIG. 2g. Lever- or key-press manipulanda. Used for reward or avoidance. The figure illustrates a two-lever apparatus for primates. One lever is pressed in the presence of a test-tone, the other in its absence. The subject is restrained at the neck and mid-section. Electric shock can be delivered through the chair grid, and/or food reward is delivered through a spout (a) near the subject's mouth.

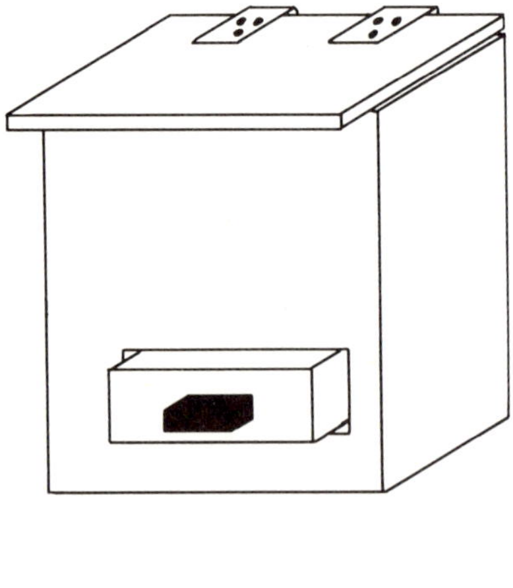

FIG. 2h. Reward-container manipulanda. Used for reward only. The subject may lift the lid or pull out a reward-bearing tray as the indicator response.

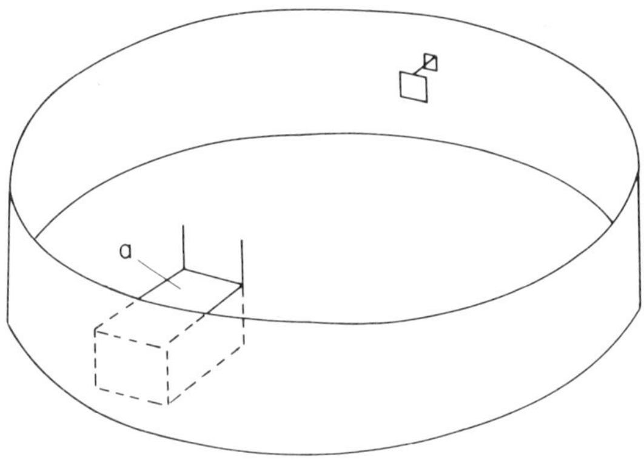

FIG. 2i. One-way movement and wait-in-place observing response. Used for reward only. Figure shows a tank for aquatic mammals. The subject may position itself on or in the structure (a) as the observing response. The indicator response is a lever-press at the opposite side of the tank.

The subject is trained to produce indicator responses during the auditory signal presentation if reinforcement is given only at this time. One example of this "discrimination training" is described in pp. 273–274.

Another feature of operant procedures is that the subject can be alerted to a trial by preceding the trial with a visual or auditory cue. This is commonly used in all types of reward procedures, and some types of avoidance procedures.

In operant procedures, reinforcement can only follow the indicator response. Therefore the animal is motivated to perform the response in the absence of the auditory stimulus. These "false positive" responses may occur very frequently if discrimination training is incomplete. Obviously, these responses weaken the validity of threshold data. There are several methods for treating false positives. (a) Most commonly, false positives are punished with an aversive stimulus, such as an electric shock or a burst of air in the subject's face (Möhl, 1968); many reward studies also employ a brief interruption of the experimental proceedings, known as a "time-out", as punishment. (b) Discrimination training is extended, but false positives are otherwise ignored. (c) A "choice" procedure is used, as described on pp. 254–257 below. (d) By determining the rates of false positive response and true positive response for several conditions of equal auditory sensitivity but unequal motivation, it may be possible to outline the distribution of auditory sensations produced by an identical stimulus. This permits a relatively pure assessment of sensitivity, uncomplicated by non-auditory influences on behaviour. Although established in human psychophysics, "signal detection" analysis such as this is not widely used with infrahuman subjects, and as yet it is not a practical tool for assessing audition. Testing requires very precise performance, and many trials are needed to determine the sensory response to only one intensity of one frequency (e.g. Hack, 1966; Clopton, 1972). Among others, D'Amato (1970) provides a good introduction to this potentially important field.

Avoidance. This (Fig. 1c), also known as aversive or negative reinforcement conditioning, is actually a combination of the operant and respondent principles. As in respondent conditioning, an aversive reinforcer and a sound stimulus are repeatedly paired. Electric shock again has been the only reinforcer used, except for several recent studies in which cats were immersed in water (Sokolovski, 1972, 1973a,b). Avoidance is classified as an operant procedure because the subject can escape or avoid the reinforcer by performing the correct response at the correct time. Once the avoidance is learned, the intensity or the frequency of shock delivered may be reduced, or shock may be replaced with a substitute reinforcer, such as the sound of a buzzer. Table III lists the avoidance studies reviewed. Figure 2 explains each indicator response.

TABLE III

Avoidance procedures

Study	Subject
FLEXION:	
Culler, Finch, Girden & Brogden (1935)	dog
Brogden & Culler (1937)	dog
Lipman & Grassi (1942)	dog
Wollack (1963)	sheep
MOVEMENT:	
Rotator, horizontal axis	
Ades, Mettler & Culler (1937, 1939)	cat
Neff (1940)	cat
Harris (1943) A	primates
Kryter & Ades (1943)	cat

continued

Study	Subject
Neff (1947)	cat
Gross (1952) A	guinea-pig
Schuknecht (1953)	cat
Lindquist, Neff & Schuknecht (1954)	cat
Neff & Hind (1955)	cat
Elliott, Stein & Harrison (1960)	cat
Elliott & McGee (1965)	cat
Rotator, vertical axis	
Gould & Morgan (1942)	rat
Gross (1952) A	guinea-pig
Stabilimeter	
Harris (1943) A	primates
One-way alley (with reward)	
Blackwell & Schlosberg (1943)	rat
Alley choice (with reward)	
Upton (1929a) R	guinea-pig
Henry, F. M. (1935, 1938)	rat
Shuttle (with reward)	
Sokolovski (1972, 1973a,b)	cat
Shuttle without barrier	
Seiden (1958)	marmoset
Landes (1962)	rat
Shuttle with barrier	
Eccher (1942)	rat
Simmons & Beatty (1962)	cat
Miller (1963)	chinchilla
Miller, Watson & Covell (1963)	cat
Miller (1964)	chinchilla
Strother, Parker, Rahm & Crump (1964)	cat
Fujita & Elliott (1965) A,Rw	primates
Peters (1965)	chinchilla
Birch, Warfield, Ruben & Mikaelian (1968) A,Rw	mouse
Heffner, Ravizza & Masterton (1969a) R	tree shrew
Henderson (1969)	chinchilla
Henderson, Onishi, Eldredge & Davis (1969)	chinchilla
Miller (1970)	chinchilla
Ward & Duvall (1971)	chinchilla
Webster & Webster (1972)	kangaroo rat
Eldredge (1973)	chinchilla
MANIPULANDA:	
Dice & Barto (1952) U,U	deermice
Martin, Romba & Gates (1962)	rhesus monkey

continued

TABLE III (*continued*)

Study	Subject
Clack & Herman (1963)	rhesus monkey
Behar, Cronholm & Loeb (1965)	rhesus monkey
Farrer & Prim (1965)	chimpanzee
Fujita & Elliott (1965) A,Rw	primates
Clack (1966a)	rhesus monkey
Birch *et al.* (1968) A,Rw	mouse
Bragg & Dreher (1969) R,R	capuchin
Mitchell, Vernon & Herman (1971)	lemur
UNSPECIFIED:	
Anderson & Wedenberg (1965) U,R,R,Rw	guinea-pig
Wilpizeski & Young (1971)	squirrel monkey

Letters after the date of a study indicate which other procedures were used in the study (see Table I). Figure 2 illustrates each response.

Reward. As implied by the name, this procedure, also known as appetitive or positive reinforcement conditioning, is one in which the subject works to obtain the reinforcer (Fig. 1d). The reinforcer is typically some liquid or food which the animal favours or lives on. In most studies the subject is maintained at some fraction of its normal body weight in order to increase its responsiveness to the reinforcer. Johnson (1966) describes a food deprivation programme designed to ensure tight control of the subject's motivation. As in avoidance procedures, once the animal has mastered the task, it may be possible to reduce motivation, in this case by reducing food-deprivation. Table IV lists the studies which used the indicator responses pictured in Fig. 2.

One-way movement has been more popular in reward than in avoidance procedures, and in a number of cases movement combined with lever pressing is the operant. A particularly novel technique was the use of vocalization as an operant in the sea lion. During the course of a visual discrimination, Schusterman, Balliet & Nixon (1972) discovered that the animal emitted a "click" vocalization when expected stimuli were not presented. The experimenters were then able to control this "clicking" with standard reinforcement techniques. The sea lion is the only known nonhuman mammal to have used a "voluntary" vocal response in audiometry.

As well as making the subject perform the operant response at the correct time, which is known as "rate" or "go, no-go" procedure, reward procedures offer the alternative of giving the subject

TABLE IV

Reward procedures. Studies using rate procedures are listed according to both the indicator response and the observing response. Studies using choice procedures are listed according to the response choice and the observing response

Study	Subject
Indicator responses (rate)	
MOVEMENT:	
One-way	
McGill (1959)	cat
Dalland (1965a,b)	bats
Dalland, Vernon & Peterson (1967)	bats
Anderson (1970)	harbour porpoise
Dalland (1970)	bats
Shuttle	
Mikaelian & Warfield (1970)	cat
MANIPULANDA:	
Reward-container	
Dworkin, Sutherland & Seymour (1934)	cat
Wendt (1934)	primates
Dworkin, Katzman, Hutchison & McCabe (1940)	dog, cat
Wollack (1965)	raccoon
Lever or key	
Elder (1933, 1934)	chimpanzee
Gourevitch, Hack & Hawkins (1960)	rat
Clack & Harris (1963)	rat
Fujita & Elliott (1965) A,A	primates
Gourevitch (1965)	rat
Clack (1966b)	rat
Gourevitch & Hack (1966)	rat
Stebbins, Green & Miller (1966)	macaques
Birch, Warfield, Ruben & Mikaelian (1968) A,A	mouse
Stebbins, Miller, Johnson & Hawkins (1969)	macaques
Gourevitch (1970)	rat, pigtailed monkey
Mitchell, Gillette, Vernon & Herman (1970)	lemurs
Stebbins (1970b)	primates
Stebbins, Pearson & Moody (1970)	macaques
Beecher (1972)	owl monkey
Jacobs & Hall (1972)	Amazon dolphin
Markl & Ehret (1973) R	mouse
Stebbins (1973)	primates
UNSPECIFIED:	
Anderson & Wedenberg (1965) U,R,R,A	guinea-pig

continued

TABLE IV (*continued*)

Study	Subject
Indicator responses (choice)	
(correct response to trials with auditory stimulus/correct response to trials without auditory stimulus)	
Vocalize/remain silent	
Schusterman, Balliet & Nixon (1972)	California sea lion
"Yes" lever/wait-in-place	
Johnson (1966, 1967, 1968a)	bottlenosed porpoise
Hall & Johnson (1972)	killer whale
Jacobs & Hall (1972)	Amazon dolphin
"Yes" lever/"No" lever	
Möhl (1968)	common seal
Terhune & Ronald (1970, 1971, 1972)	harp seal
Observing responses	
(responses performed in the absence of auditory stimulus)	
Wait-in-place	
McGill (1959)	cat
Dalland (1965a,b)	bats
Wollack (1965)	raccoon
Johnson (1966)	bottlenosed porpoise
Dalland *et al.* (1967)	bats
Johnson (1967, 1968a)	bottlenosed porpoise
Birch *et al.* (1968)	mouse
Anderson (1970)	harbour porpoise
Dalland (1970)	bats
Hall & Johnson (1972)	killer whale
Jacobs & Hall (1972)	Amazon dolphin
Schusterman *et al.* (1972)	California sea lion
Lever- or key-press	
Gourevitch *et al.* (1960)	rat
Clack & Harris (1963)	rat
Gourevitch (1965)	rat
Clack (1966b)	rat
Stebbins *et al.* (1966)	macaques
Möhl (1968)	common seal
Gourevitch (1970)	rat, pigtailed monkey
Mitchell *et al.* (1970)	lemurs
Stebbins (1970b)	primates
Stebbins *et al.* (1970)	macaques
Terhune & Ronald (1970, 1971, 1972)	harp seal
Stebbins (1973)	primates

continued

Study	Subject
Target-press with nose	
Gourevitch & Hack (1966)	rat
Spout-lick	
Beecher (1972)	owl monkey

Letters after the date of a study indicate which other procedures were used in the study (see Table I). Figure 2 illustrates each response.

a choice of two responses: one is correct for trials with tone, and the other is correct for trials without tone. Both produce reinforcement. This is referred to as a "choice" procedure. Table IV categorizes choice procedures according to the pair of responses employed. Typically, the interval between cue and trial onset is the same for every trial (occasionally referred to as "single-interval" procedure). In rate procedures this is a variable interval because response timing, not choice, is of primary importance.

Reward procedures may also employ learned behaviour that is performed in the absence of the auditory stimulus and is reinforced by its presentation. For instance, a subject may be trained to wait in a specific position for stimulus presentation. This is termed an "observing response". Table IV lists specific observing responses and the studies which have used them.

Historical trends

In one of the first audiometric studies employing an electrical oscillator, Upton (1929a) observed that "the general impression seems to be that many of the mammals are able to hear noises but are deaf to tones". Clearly, knowledge of mammalian hearing has progressed tremendously within the working lifetime of senior contributors to this symposium. A major breakthrough in research on mammalian audition resulted from experiments on digestion. Pavlov's discovery of the conditioned reflex provided the first method of effectively "instructing" infrahuman animals how to respond in hearing tests. This technique, combined with the development of electrical sound sources, finally permitted accurate, standardized measurement of auditory function to be made in non-verbal animals.

Since Pavlov, the trends in behavioural methodology have been towards the emergence of new techniques and the elimination or increased efficiency of old techniques. For example, Igarashi & Hoyt (1963) and Sokolovski (1972) have developed methods of forecasting the quality of avoidance performance, while the classic Yerkes–Watson alley choice apparatus has not been used for nearly 40 years.

The most significant development, not only in audiometry but in all branches of animal psychophysics, has been that of operant conditioning (Stebbins, 1970a). Most of this development has been achieved by experimental psychologists exploring operant behaviour for its own sake (*J. exper. Analysis Behav.* is devoted entirely to operant research; see Honig (1966) and Catania (1968) for excellent reviews of this field). Nevertheless, application of their findings has been very important to auditory research. For example, the function of ultrasound emitted by Cetacea remained uncertain until hearing in this Order could be assessed; however, in the 1950s, "intensity thresholds (were) out of the question altogether, by any method except the measurement of cochlear potentials" (Kellogg, 1953). Today, operant methods have been adapted to a range of marine mammals for threshold determination and a variety of other precise measurements (e.g. frequency difference limens, Möhl, 1967; masked thresholds, Johnson, 1968b; and sets of acoustic discriminations, Herman & Arbeit, 1973). Another example is the clearer understanding of the evolution of human hearing resulting from a recent audiometric survey of several very varied infrahuman species (Masterton, Heffner & Ravizza, 1969). Operant suppression was used throughout the survey, which would have been impossible with a less flexible technique.

EVALUATION OF TECHNIQUES

Much of this section is necessarily speculative. Understandably, many studies do not report details such as training time, false positive rate, threshold reliability or performance difficulties encountered. But these are crucial to a complete methodological evaluation. Even when these details are reported, it is still difficult to make valid judgements about the techniques because the skill, experience and purpose of the individual experimenter largely determine the output of a particular technique. The studies which are most valuable in comparing techniques are those in which threshold is determined by more than one method. Such studies, however, are rare, and even then are not free of bias (Fujita & Elliott, 1965).

Nevertheless, any researcher about to undertake an audiometric study must have some reason for choosing the technique he does. Largely with this choice in mind, an attempt has been made to appraise the four behavioural procedures on the basis of the quality and efficiency of the measurements they permit. Table V is a summary of this appraisal. The text of this section elaborates most of the statements in the table. In the discussion, preference has been given to statements relevant to physiological experimentation because of the relatively physiological orientation of this symposium. On the other hand, attention has also been given to factors that may possibly influence threshold to a much smaller degree than do various physiological treatments. The disconcerting but exciting implication is that threshold measurements are influenced by procedural biases that have not yet been explored.

TABLE V

*Evaluation of techniques: relative merits of features that may occur**

Potential advantages	Potential disadvantages
I. UNCONDITIONED PROCEDURES (see Table I).	
Response: not influenced by strength of conditioning or magnitude of reinforcement; observable in natural settings.	Response threshold greater than sensation threshold.
Procedure: very fast; simple.	Habituation: rapid; long lasting; affected differently by different sound frequencies.
	Detection and recording of response only visual, not automatic.
No practice:	
sensitivity: as in the wild; relatively stable between sessions.	stimulus unfamiliar.
False positive rate not increased by reinforcement with subthreshold stimuli.	
† The converse of the advantageous factors listed above (e.g. Procedure: *not* very fast), are disadvantages for all conditioned procedures (II, III, IV below). Similarly, the disadvantageous features above become advantageous for conditioned procedures.	
Response graded:	
response proportional to auditory sensation.	response defined by experimenter and apparatus.

continued

* Statements opposite each other in the two columns are not necessarily related. A statement which straddles the two columns refers to a single characteristic which has both advantages and disadvantages, and these are listed immediately below the statement.

TABLE V (*continued*)

Potential advantages	Potential disadvantages
No response motivation:	
response to auditory sensation only.	attention not maximized.
Very many subjects tested.	Response: emitted spontaneously; elicited by many environmental stimuli; influenced by arousal state; unshapeable; variable between individuals.
No cueing:	
no external influences on attention.	attention uncontrolled.
Fixed body position:	
precise sound-field calibration.	restraint of subject influences responsiveness.
II. RESPONDENT PROCEDURES (see Table II). See also note† in I above.	
Response graded:	
response proportional to auditory sensation.	response defined by experimenter and apparatus.
Minimum response motivation:	
response not influenced by its probable consequences.	attention not maximized.
	Response magnitude proportional to reinforcement magnitude.
Response elicited at experimenter's will.	Response: unshapeable; spontaneously emitted; ambiguous at threshold; variable between individuals.
Fixed body position.	Baseline restabilizes slowly following reinforcement (i.e. long intertrial interval necessary; few trials per session).
Relatively fast procedure.	
Simple conditioning apparatus sufficient.	Stable conditioned response not formed (subject rejected).
Many subjects tested.	
	No cueing.
Fixed body position:	
precise sound-field calibration.	restraint of subject influences responsiveness.
	Imprecise threshold assessment?

continued

Potential advantages	Potential disadvantages
III. AVOIDANCE PROCEDURES (see Table III). See also note† in I above.	
Response motivated:	
attention maximized.	response: must be emitted; influenced by its probable consequences and other motivational variables.
Response: shapeable; elicited reflexively by reinforcer; specific to auditory stimulus after a few trials.	Habituation to electric shock.
	Stable response: not formed, or breaks down during testing (subject rejected during or after training).
Reinforcer effect: immediate; controllable over a wide range.	Performance unstable within and between sessions, becoming more so with time.
Cueing and rapid trial presentation with manipulanda.	Random body movement at threshold.
	False positives: rate increases at threshold; require time to eliminate; rate increased if punished; self-reinforcing.
Fixed body position with flexion and manipulanda:	
precise sound-field calibration.	restraint of subject influences responsiveness.
	Unfixed body position with movement: i.e. imprecise sound-field; subject's threshold variability proportional to sound-field variability; search for intensity maxima disrupts performance.
	No cueing with movement and flexion: i.e. attention uncontrolled; subject under stress of concentration for entire session.
	Sophisticated conditioning apparatus for electric shock delivery and programming desirable.
	Imprecise threshold assessment?
IV. REWARD PROCEDURES (see Table IV). See also note† in I above.	
Response motivated:	
attention maximized.	response: must be emitted; influenced by its probable consequences and other motivational variables.
Very shapeable performance.	Reinforcer not immediately effective.
Observing response.	Satiation.

continued

TABLE V (*continued*)

Potential advantages	Potential disadvantages
Stable performance. Least frequent subject rejection. Time-out punishment of false positives. No false positives with choice. Fixed body position.	Deprivation: weakens subject; requires weight-control programme; unsuitable for some species. Lapses of co-operation: increased at threshold; accompanied by emotional display.
Cueing. Rapid trial presentation. Most precise threshold assessment?	False positives: frequency increases at threshold; require time to eliminate; decreases responsiveness to auditory stimulus. "Superstitious" development of undesirable intertrial behaviour. Complex performance; requires time; not practical with all species. Stimulus schedule controls responding. Programming apparatus desirable. Few subjects used.

Procedures with unconditioned responses

Unconditioned procedures yield data more quickly than conditioned procedures since no training is involved. Therefore, they can also accommodate more subjects in a given time. With the rat, for example, Herman & Clack (1963) obtained a 20-frequency audiogram for each of 18 subjects in 6 h using the Preyer reflex, while Clack & Harris (1963) took 2 years to obtain a 12-frequency audiogram for each of five subjects using a complex operant procedure. The speed of unconditioned procedures makes them valuable in the following situations: (a) quick assessment of hearing; (b) situations where conditioning is not feasible or difficult, for example in studies of hearing development in neonates; and (c) studies under natural conditions. Also, Ash (1951) and other workers have suggested that practice greatly improves performance, and thereby distorts the "true (i.e. typical) estimate of (the animal's) auditory capacity" (Ash, 1951). This implies that unconditioned tests more accurately represent the hearing of animals in natural conditions.

With unconditioned procedures, the link between sensation and response is relatively direct. There is no intervention of variables such as strength of conditioning, sensitivity to reinforcer, or strength of motivation. The importance of these variables is discussed below. Conversely, the absence of practice, motivation and cueing could all be expected to decrease the probability that an animal would attend to small signals.

It is uncertain whether unconditioned responses can provide a valid assessment of threshold. Thresholds of several of the responses, including Preyer reflex, startle, and vibrissae twitch, are at least 50 dB above sensory threshold as determined by other means. However, these high threshold values vary with sound frequency, and Murphy & Harris (1961) found that the Preyer reflex audiogram roughly paralleled the conditioned-response audiogram determined by Blackwell & Schlosberg (1943). In other words, thresholds of these responses, which may be part of defence reflexes, might permit indirect calculation of sensory thresholds. However, there are conflicting opinions. Anderson & Wedenberg (1965) and Ralls (1967) discuss the controversy over the Preyer reflex.

Thresholds of orientation-reflex components, such as respiration and immobility, may be near sensory thresholds, but apparently unconditioned-response thresholds have not yet been compared with conditioned-response thresholds under the same acoustic conditions. Bradford *et al.* (1973) reported that human auditory thresholds determined with the unconditioned GSR agree well with verbally-reported thresholds. In any case, unconditioned responses may be used to assess the effects of physiological treatments. For example, Clack (1966b) found that Preyer reflex and conditioned-response thresholds were temporarily raised by approximately the same amount after exposing the subject to loud noise. Furthermore, the Preyer reflex data for recovery from noise seemed less variably dispersed around the estimated course of sensitivity recovery.

Interestingly, Sokolov (1960) found that a subject could not see a weak visual stimulus until the orientating reflex was elicited by a supraliminal auditory stimulus; the visual stimulus was then detectable until the sound-induced orientating reflex ceased. This shows that threshold is a function of the experimental conditions under which it is determined, and suggests that some form of trial cueing may be feasible in unconditioned procedures.

Habituation, a decrease of response magnitude with successive responses, is characteristic of the orientating reflex (Sokolov, 1960). Thus, the same auditory sensation may produce a different response on two separate trials, and responses may cease entirely after a few trials. Early trials are more likely to give positive results. Testing is further invalidated because habituation does not affect responses to different frequencies equally (Miller & Murray, 1966; Berlin, Gill & Leffler, 1968). These effects may linger for weeks after the original habituation (File, 1973). In short, unconditioned responses that may assess threshold directly are unreliable.

There are, however, several possible ways of avoiding habituation:

(a) It can be partially averted if, starting below threshold, stimulus intensity is slightly increased between trials until the animal responds. Although this "ascending" series locates the threshold range with only one response, subsequent responses may be contaminated by habituation. Bradford *et al.* (1973) achieved good test-retest reliability with this method.

(b) Sokolov (1960) presented evidence that the habituation mechanism does not function with very weak, near-threshold stimulation. Therefore, careful stimulus presentation in the near-threshold range might impede habituation.

(c) Pavlov discovered that decorticate animals never habituate (Sokolov, 1960). Sokolov further suggested that habituation and orientation are mediated by different neural pathways. Similarly, File (1974) reported that orientation and habituation react independently to drug treatment; one system can be depressed without disturbing the other. These last two findings indicate the interesting possibility of artificial prevention of habituation.

(d) High threshold, defence-reflex responses habituate much more slowly than orientating responses (Sokolov, 1960). For example, Herman & Clack (1963) obtained reliable threshold measurements with the Preyer reflex, even though they used test trials of "descending" intensity.

Typically, both unconditioned and respondent responses are graded in proportion to stimulus intensity. Therefore, responses to stimuli of threshold intensity are small and variable. The crucial selection of responses which are sound-evoked is based on an arbitrary criterion of response magnitude. Detection of responses depends on the sensitivity of the experimental apparatus. Visual observation is, of course, potentially faulty, and cross-checks, such as the experimenter's formal self-evaluation of his observations (Miller & Murray, 1966), are advisable. Simultaneous records of two or more components of the orientating reflex might be another useful cross-check on the validity of unconditioned responses, although apparently such a procedure has yet to be employed in auditory studies.

The baseline against which responses are measured may be easily disturbed. This is advantageous because small auditory sensations are revealed. However, it is disadvantageous because even small environmental events and spontaneous physiological activity may produce many false indicator responses, thereby increasing the difficulty in selecting true indicator responses evoked by the auditory stimulus. Isolation of the subject and the use of procedures to stabilize the baseline help to solve this problem. Most of these procedures also increase threshold reliability by fixing the subject in a relatively invariable, accurately calibrated sound-field. Miller (1970) found that the consistency of threshold measurements, both within and between subjects, decreased as the variability of stimulus intensity in the sound-field increased.

Unfortunately, these stabilization procedures may influence a subject's responsiveness or attention. For example, some anaesthetics impede the GSR (Soliman, 1969), and physical restraint may put the subject in a state of fright (W. R. Webster & Dunlop, 1965). The experimenter may effectively eliminate false responses by setting the criterion of response acceptance well above the "noise" level, but this may also eliminate many true indicator responses. Semenoff & Young (1964), using the GSR, rcported unusually high thresholds for macaques. These and other authors (e.g. Behar, Cronholm & Loeb, 1965) have attributed the results to the high criterion of response acceptance employed.

The magnitude of an unconditioned response is presumably proportional in some way to the magnitude of the auditory sensation, since the response is generally proportional to stimulus intensity. This suggests that unconditioned responses may be useful in the study of loudness, which is still at a primitive stage in infrahuman subjects (Stebbins, 1966). Graded respondent responses would probably be even better suited to loudness assessment because they are more reliable than unconditioned responses.

Finally, unconditioned responses are fairly fixed, "innate" reflexes, so that they, more than the experimenter, may determine the particular apparatus, treatment and sound-field that can be used with a given species.

Procedures with conditioned responses

Conditioning increases the reliability of responses, the precision of threshold measurement, and the flexibility of experimental design. In return, time and effort must be invested, and the animal's behaviour is directly controlled by the reinforcer, rather than by the auditory stimulus. These effects seem to be graded according to the behavioural complexity of the procedure employed, although the large diversity of contexts of these procedures easily produces exceptions to this statement.

Respondent procedures tend to be simple and relatively fast. Responses are graded, fixed, and unmotivated (i.e. the probable consequences of a response do not determine whether or not the response is emitted; this is further discussed below). However, the magnitude of the unconditioned response is related to the magnitude of the reinforcer (Gormenzo & Moore, 1969), which may influence the thresholds of some individuals more than others.

The reliability of respondent-measured thresholds can be very good (e.g. Jamison, 1951; Dalton, 1968; Heffner, Heffner & Masterton, 1971), but these thresholds may be less precise than thresholds determined with operant methods, particularly reward procedures. An interstudy comparison of sensitivity, whether it be that of the subjects or the techniques, based on reported threshold values is dangerous; key variables, such as the sound sources, the sound-fields, the threshold definitions and the performance criteria (i.e. the levels at which performance is consistent enough to start and stop the collection of data) are all inconsistent between studies. A comparison based on intrastudy thresholds is safer because these intervening variables are more likely to be held constant. *Macaca* monkeys are considerably less sensitive to 4 kHz than to 2 or 8 kHz, at least when tested with reward procedures (Wendt, 1934; Fujita & Elliott, 1965; Gourevitch, 1970; Stebbins, 1973). Avoidance procedures sometimes show this "dip" (Harris, 1943), but sometimes they do not (Clack & Herman, 1963; Behar *et al.*, 1965; Fujita & Elliott, 1965). Respondent procedures have not revealed this dip (Semenoff & Young, 1964; Dalton, 1968; Dalton, Taylor, Henton & Allen, 1969). Dalton (1968) did not detect the dip with two different respondent techniques, GSR and operant suppression, but he did detect it with an electrophysiological procedure, the averaged evoked cortical response. Likewise, Fujita & Elliott (1965) found that a single-lever reward procedure revealed a dip, whereas two avoidance procedures did not; Markl &

Ehret (1973) found a peak of sensitivity in the audiogram of mice with a reward procedure, but not with a respondent procedure. All of these findings suggest that some procedures may be inherently more sensitive than others.

The respondent procedure of operant suppression may be more precise than some operant procedures. Heffner, Ravizza & Masterton (1969a) found that operant suppression revealed a sharp peak of sensitivity in the tree shrew's audiogram, while shuttle-box avoidance did not. Operant suppression, although introduced to animal psychophysics only eight years ago, could possibly be the most popular technique in the field today (Stebbins, 1971). It has a number of advantages which may account for its wide and immediate acceptance. The experimenter strictly controls the behavioural baseline; procedures for establishing and maintaining this baseline have been well investigated; spontaneous or false indicator responses are rare; the procedure is easily adapted to different species and sensory problems; and the subject's body position is fixed but not physically held (Smith, 1970, reviews the advantages and applications more completely). However, it has not always been easy or possible to establish the behavioural baseline (Dalton, 1968; Bragg & Dreher, 1969), and because the subject is engaged in a "food-getting" behaviour when test-tones are presented, weak tones may be masked or the subject's attention may be diverted from them. This latter suggestion implies that increased hunger motivation might elevate threshold values, and presents a possible explanation for a recent, puzzling finding that old guinea-pigs were more sensitive than young ones (Heffner, Heffner *et al.*, 1971).

Operant procedures allow maximum flexibility in experimental design because behaviour, particularly in appetitive situations, can be shaped. A subject can be brought to peak attentiveness because reinforcement consequences are tied to its attentiveness. High reliability of results can be achieved by training an animal until its performance is satisfactory, although avoidance procedures may be more limited in this respect than reward procedures, for reasons discussed below.

Several features of operant procedures may increase a subject's attention to test trials. The use of cues maximizes the subject's attention at the appropriate time, gives it a rest from concentration at other times, and increases response reliability by reducing variation in the subject's attention before trials (Elder, 1934). Unfortunately, cues seem to disrupt performance in some types of avoidance procedures (Elliott, Stein & Harrison, 1960). The subject's observing responses function as cues which it presents itself, and they offer a number of further advantages: (a) behaviour between trials is controlled and automatically recorded; (b) distracting or noisy activity, such as grooming, is prevented; (c) the subject is made to work, which sharpens discrimination performance (Blough, 1966); (d) the subject's body position is fixed so that the subject's "search" of a variable sound-field for regions of maximum sound intensity (Wendt, 1934; Lipman & Grassi, 1942; Johnson, 1966) is restricted, and therefore (e) performance is not disrupted when a change in test-tone frequency changes the position of the intensity maxima in the sound-field (Clack & Harris, 1963); and (f) thresholds are more consistent between sessions because practice effects resulting from this "search" are reduced.

However, because operant methods work on behaviour that the animal emits, rather than on reactions which the experimenter elicits, training and testing procedures become complicated in three related ways.

First, the subject must learn how to perform the indicator response. The success of training depends on the subject's motivation, the choice and presentation of the reinforcer, and the choice of the response to be trained. If motivation is weak or not affected by the reinforcer, or if the reinforcer does not immediately follow the desired behaviour, this behaviour will not be strengthened. It may be a long time before the subject emits even a faint approximation to the desired behaviour. Excess motivation may produce such weakness, fear or preoccupation with reinforcer, that the subject's overt behaviour is excessively restricted. All these possibilities impair learning.

Some responses may be very difficult, if not impossible, for a particular species to acquire. For example, personal observation suggests that a rat will touch a bar much more readily with a paw than with its nose. In avoidance situations, the animal is put in a state of stress. Bolles (1970) suggested that in this condition an animal's behaviour is limited to "species-specific defence reactions". The ease with which an animal can learn a particular response depends on its compatability with these innate responses. This may explain why the rotator and shuttle-box are relatively inefficient procedures with the rat (Bolles, 1970). In some cases experimental conditions may determine the response or motivating agent to be used. For example, Sidman *et al.* (1966) wanted to test mice with severe motor disturbances that precluded all but an "effortless" indicator response. This unusual restriction resulted in the experimenters' original adaptation of conditioned suppression to sensory research. A less beneficial restriction arose midway through the study of Gourevitch, Hack & Hawkins (1960), when the experimenters discovered that antibiotic treatment produced a nephrotoxic condition which prevented further use of water deprivation as a motivator.

Second, after the response has been learned, the animal's efforts and attention are presumably focused on reinforcement. Control of behaviour must be shifted from this powerful stimulus to an almost powerless one, the auditory stimulus. Prior to 1930 it was thought that most mammals were tone-deaf because they were tested in ill-suited forms of the alley-choice apparatus. Tests with Pavlovian procedures brought different results, leading a number of workers to conclude that these alley-choice procedures incorporated a number of strong, competing influences which prevented the control of behaviour by pure tones (e.g. Anrep, 1920; Upton, 1929a). In hindsight, it is now possible to attribute this failure to at least two specific factors. First, it is clear that the degree of stimulus control is a function of stimulus salience (Terrance, 1966). For example, Landes (1962) found that rats could not be trained to avoid a 4 kHz signal, but responded well to 8 kHz, a frequency to which they are more sensitive (Gourevitch, 1965). Once trained, the animals also responded to 4 kHz. Similarly, a conflict in reports of rats' sensitivity to ultrasonic frequencies (cf. Peacock & Williams, 1962; Gourevitch & Hack, 1966) may possibly be due to a difference in the training stimuli employed. This effect is suggested by alley-choice studies in which broad-band noises, but not pure tones, were effective training stimuli (e.g. Hunter, 1915, 1927). Secondly, in the alley-choice procedure random choice alone produces reinforcement in half the trials. Thus, the animal is often reinforced without paying any attention to the auditory stimulus. Such adventitious reinforcement severely retards operant studies because subjects may respond to this reinforcement more readily than to the intended stimuli (Blough, 1966). One of the first operant procedures used to measure threshold was a reward procedure (Dworkin, Sutherland & Seymour, 1934). Unfortunately, the team of researchers involved did not realize this, and they modelled their study on Pavlovian principles: a tone was presented, food was dropped into a trough, and the animal obtained the food on lifting the lid of this trough. Here again, the subject could obtain reinforcement without attending to the auditory stimulus. Eventually threshold measurements were made with this procedure, but many months were required, and false positive responses consistently marred the subject's performance (Dworkin, Katzman, Hutchison & McCabe, 1940). Two similarly faulty reward studies with primates appeared at about the same time (Elder, 1934; Wendt, 1934). The naïve inefficiency of these experiments was undoubtedly a major reason why positive reinforcement was not used in a hearing study for the next 20 years. Avoidance techniques, however, were extensively used in this period, mostly in experiments on hearing physiology. Such experiments still rely almost exclusively on aversive procedures, while purely psychophysical experiments more commonly incorporate appetitive procedures (Stebbins, 1970a).

Third, during threshold tests the animal's performance is still under motivational control, and therefore the validity of threshold data may suffer. For example, the rate of false positive responses depends on the preprogrammed consequences of the responses. If there is a small risk for wrong responses, as there would be if those responses incurred no punishment, and a potentially high pay-off for correct guesses, as there would be if the subject were very hungry, the subject will obviously make more guesses. This adventitiously increases the probability that a response will coincide with a test-tone, and thus lowers the measure "threshold". Conversely, stiffer penalties, or any other intervention which decreases response motivation, would artificially raise the measured "threshold". Various ways of treating false positive responses may have individual drawbacks. In certain avoidance situations punishment may paradoxically increase the false positive rate (Clack & Herman, 1963).

In choice procedures there are no false positive responses, but approximately half the trials must not contain a tone stimulus, and therefore tests are inefficient. Further, the additional complexity of two responses requires additional training, and it may be too difficult to be practical with some individuals or species (Clack & Harris, 1963; Mitchell, Gillette *et al.*, 1970).

Other factors, such as hormonal state, may affect an animal's motivation. Wendt (1934), for example, suggested that the rate of false positive responses may be related to a female's oestrus cycle. In studies of auditory physiology, insults intended only for the hearing mechanism may also affect motivation. This is seldom considered, and none of the studies reviewed included control procedures for this factor. However, Henderson (1969) suggested the possibility of a motivational shift to explain differences between man and chinchilla in the effect of loud noise on temporal integration. Another example may emerge from a conflict in reports of the effects of X-radiation on audition. Murphy & Harris (1961) used the unconditioned Preyer reflex, and reported small, inconsistent effects, while Brogden & Culler (1937) and Landes (1962) both used shock avoidance and produced greater effects. Motivational intervention is further suggested by Landes' (1962) report of large differences in reaction to conditioning between irradiated and non-irradiated groups. This is consistent with a number of non-audiometric studies in which X-radiation significantly affected avoidance performance (Sherrod, 1973).

Avoidance conditioning does not seem to provide the high degree of behaviour control obtained by reward conditioning. Failures to condition the desired response occur with both reward and avoidance techniques, but more commonly with avoidance. Several workers have found reward-controlled behaviour to be more stable than avoidance-controlled behaviour (e.g. Birch, Warfield, Ruben & Mikaelian, 1968; Stebbins, 1970a, 1971; Gourevitch, personal communication). In time, avoidance-controlled behaviour may break down entirely (Hoffman, 1966; Elliott, 1967), in which case the subject must be rejected after a substantial effort to train it. Random movement between avoidance responses has been reported for cats (Elliott *et al.*, 1960) and rats (Landes, 1962). This detracts from the subject's attention, and increases variance in the sound field and threshold measurement. Elliott *et al.* (1960) also reported that the level of threshold was related to the amount of such "neurotic" movement. Fujita & Elliott (1965) used both avoidance and reward with monkeys, but reported thresholds for high test-frequencies with the reward procedure only. The authors give no reason for this, but possibly the "neurotic" movement associated with avoidance procedures, combined with the normally-enhanced variability of high-frequency sound-fields, prevented reliable threshold measurements.

An avoidance response is presumably reinforcing because it removes the threat of noxious stimulation. Thus, false positive avoidance responses are "self-reinforcing", in that they provide reinforcement whether they actually prevent electric shock or not. False responses are therefore difficult to eliminate. For example, Hoffman (1966) reported more rapid and complete discrimination training with reward than with avoidance conditioning.

Reward procedures, however, may incorporate their own disadvantages. A counterpart of the "neurotic" movement of avoidance procedures is "superstitious" intertrial behaviour (Skinner, 1948). Tonal stimuli are reinforcing because they are associated with the opportunities to obtain reward. Consequently, any behaviour immediately followed by a tone will be strengthened. Such behaviour will then have a greater probability of occurrence and therefore a greater probability of preceding the tone, which further strengthens this behaviour. Although unrelated to tone presentation, this behaviour may become an activity pattern which prevents controlled test conditions. A disadvantage of "wait-in-place" observing responses is that the animal may develop such undesirable intertrial behaviour. Active observing responses, such as a spout-lick or a key-press, may partially mask test tones, but the animal's behaviour is more controlled.

A disadvantage of observing responses is that they are reinforced with the auditory stimulus according to a predetermined schedule. Clack & Harris (1963), Blough (1966), and Stebbins (1970b) give examples of animals which "cracked the system", and were ultimately responding to the schedule, rather than to the auditory stimulus.

Motivation to work is reduced in reward procedures when the subject is satiated with the reinforcer, and in avoidance procedures when the subject becomes habituated to electric shock (Azrin & Holz, 1966). Habituation to shock may be more serious than satiation because it is more permanent. The use of brief pulses of shock, rather than continuous shock, counteracts this effect (Clack & Herman, 1963).

Reward-controlled behaviour is occasionally disrupted by emotional displays, such as cage-gnawing or trains of false positive responses, following difficult discriminations or reductions in reinforcement frequency. Elder (1934) reported that tantrums during threshold tests interrupted experimentation for days at a time. Such disruption may be evoked by excessive delay of reinforcement if the experimenter demands too great an increase in the subject's performance. On the the other hand, if training progresses too slowly the subject becomes so accustomed to one level of performance that no increases in performance level can be accomplished smoothly (e.g. Gourevitch, 1970).

Conclusion

Even this necessarily limited analysis suggests that each technique has its own particular merits. Furthermore, various forms of all these techniques have yielded satisfactory threshold assessments. In these respects, the choice of procedure seems irrelevant. On the other hand, procedures differ in characteristics such as speed, potential for behavioural control, and sensitivity to threshold. Different attributes are desirable for different purposes. Ideally, an experimenter should choose a technique to match both the purpose of his study and the behavioural characteristics of his subjects. However, at this stage it is impossible to do this because there is insufficient understanding of the inherent differences in the four basic procedures. This essay has made a preliminary identification of a number of these differences. The list is probably not complete, and a truly valid assessment of these differences can only be a practical one. An experimental comparison of at least six of the most popular behavioural techniques conducted by a small team, possibly composed of experienced and inexperienced researchers, would be an invaluable aid to sensory research.

TECHNIQUE FOR SMALL MAMMALS

Figure 3 illustrates a technique currently being developed to determine the thresholds of small mammals which are thought to be sensitive to ultrasounds (Brown, 1973).

Response A, the observing response, is the interruption of a photocell beam when the subject positions itself properly on a small stand. A lamp indicates that the subject is positioned correctly. If it remains in this position for the required time (a variable interval from 1 to 12 sec), a 2-sec test-tone is presented. Reinforcement is given if the subject performs the indicator response (response B), which is the interruption of a second photocell beam, within 4 sec of tone onset. The reward substance is a mash of the animal's food pellets, water and a sweetened vitamin-syrup.

This performance is trained in the following manner. Before conditioning commences, a subject in its home cage is introduced to the reinforcer mash and to the loud sound of the reinforcer dipper. In the first training session the animal is placed in the empty experimental box for one hour. As it investigates the experimental box, it emits response B a number of times, although the only consequence of this response is the sound of a small relay. Discrimination training begins early in the next session; a salient auditory stimulus, broad-band noise, is presented. During its

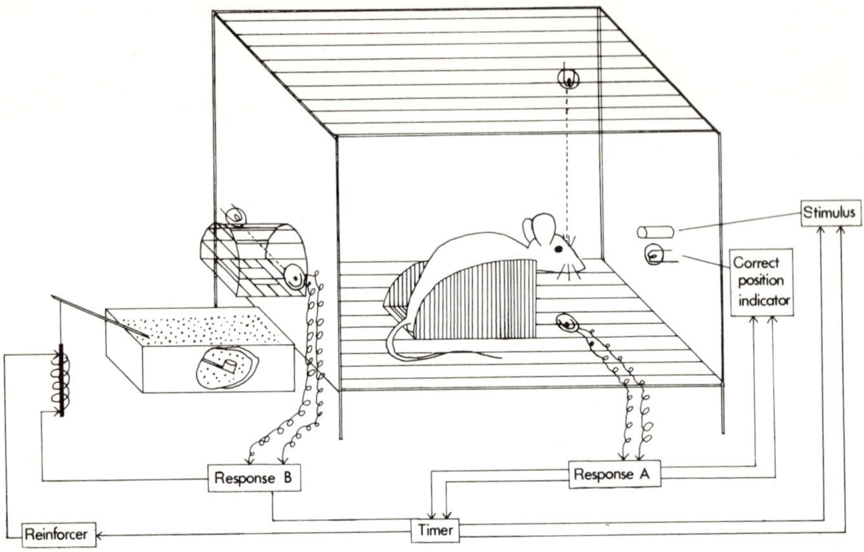

FIG. 3. Apparatus and programme of a reward procedure for use with small mammals. The electrical circuit which operates the reinforcer dipper is closed by the timer, after response A has been performed for a pre-set duration, and by response B. See the text for a further description of this procedure.

presence each performance of response B is reinforced. The broad-band noise is turned off after three or four reinforcements. The animal emits response B repeatedly after this, although none of these responses is reinforced. When the response rate subsides to the unreinforced operant level determined in the first session, the noise is turned on again until the first response is made; this response produces reinforcement and turns off the broad-band noise. As before, subsequent responses go unreinforced until the operant level is reached. Then the broad-band noise is turned on again and the next response produces reinforcement. Within an hour, this regime produces a sharp responsiveness to the sound. The positioning stand for response A is then introduced, and response A is shaped using the broad-band noise to reward successive approximations. It is then fairly straightforward, but tedious to increment the required duration of response A from 0 to 12 sec. The auditory stimulus is changed from broad-band noise to pure tone without disrupting the subject's performance.

SUMMARY

Behavioural measurements of intensity thresholds rely on stimulus control of behaviour: when an auditory stimulus is presented, the subject performs a specific behavioural act. An auditory sensation is thus inferred. Pavlov's discovery of conditioned reflexes provided the first means of achieving stimulus control powerful enough to produce reliable responses to very weak pure-tones, and thereby enabled valid threshold assessment to be made.

Four behavioural procedures are now used to assess threshold: unconditioned, respondent, avoidance operant and reward operant procedures. Each achieves stimulus control in a particular way, and each is associated with a variety of indicator responses.

Each has its own characteristics, which may be advantageous or disadvantageous, depending on the purpose for which they are employed.

Unconditioned procedures tend to be fast, imprecise and dependent on unmalleable responses. Conditioned procedures are the opposite, to an extent which is generally proportional to the complexity of the subject's performance. All behavioural procedures are subject to influences which weaken both stimulus control and the validity of threshold data.

Acknowledgements

I am indebted to Professor J. D. Pye for his consistent encouragement and very helpful advice on many aspects of this research and to Mr Y. M. Au, Mrs E. M. Jones, Miss S. C. Oldfield, Mrs V. E. Osman, Dr G. D. Sales and Dr J. C. Smith for their kind and important assistance. Above all, I wish to thank my family for their invaluable support of this work.

References

Ades, H. W., Mettler, F. A. & Culler, E. A. (1937). Distribution of acoustic pathways in the medial geniculate bodies. *Am. J. Physiol.* **119**: 257–258.

Ades, H. W., Mettler, F. A. & Culler, E. A. (1939). Effect of lesions in the medial geniculate bodies upon hearing in the cat. *Am. J. Physiol.* **125**: 15–23.

Alexander, I. E., Githler, F. J., Diess, A. & Rodwin, H. M. (1955). Selective cochlear destruction, cochlear potentials and hearing loss. *Archs Otolar.* **61**: 46–53.

Alford, B. R. & Ruben, R. J. (1963). Physiological, behavioral and anatomical correlates of the development of hearing in the mouse. *Ann. Otol. Rhinol. Lar.* **72**: 237–247.

Ames, D. R. & Arehart, L. A. (1972). Physiological response of lambs to auditory stimuli. *J. anim. Sci.* **34**: 994–998.

Anderson, H. & Wedenberg, E. (1965). A new method for hearing testing in the guinea-pig. *Acta otolar.* **60**: 375–393.

Anderson, S. (1970). Auditory sensitivity of the harbour porpoise, *Phocoena phocoena*. In *Investigations on Cetacea.* **2**: 255–259. Pilleri, G. (ed.). Berne, Switzerland: Bentili AG Berne-Bümpliz.

Anrep, G. V. (1920). Pitch discrimination in the dog. *J. Physiol., Lond.* **53**: 367–385.

Ash, P. (1951). The sensory capacities of infrahuman mammals: vision, audition, gustation. *Psychol. Bull.* **48**: 289–326.

Azrin, N. H. & Holz, W. C. (1966). Punishment. In *Operant behavior: areas of research and application*: 380–447. Honig, W. K. (ed.). New York: Appleton-Century-Crofts.

Beecher, M. D. (1972). Auditory sensitivity of the owl monkey. *J. acoust. Soc. Am.* **52**: 170.

Behar, I., Cronholm, J. N. & Loeb, M. (1965). Auditory sensitivity of the rhesus monkey. *J. comp. physiol. Psychol.* **59**: 426–428.

Berlin, C. I. (1963). Hearing in mice via GSR audiometry. *J. Speech Hear. Res.* **6**: 359–368.

Berlin, C. I., Gill, A. & Leffler, M. (1968). Hearing in mice by GSR audiometry: I. Magnitude of unconditioned GSR as an index of frequency sensitivity. *J. Speech Hear. Res.* **11**: 159–168.

Birch, L. M., Warfield, D., Ruben, R. J. & Mikaelian, D. O. (1968). Behavioral measurements of pure tone thresholds in normal CBA-J mice. *J. aud. Res.* **8**: 459–468.

Blackwell, H. R. & Schlosberg, H. (1943). Octave generalization, pitch discrimination and loudness thresholds in the white rat. *J. exp. Psychol.* **33**: 407–419.

Blough, D. S. (1966). The study of animal sensory processes by operant methods. In *Operant behavior: areas of research and application*: 345–379. Honig, W. K. (ed.). New York: Appleton-Century-Crofts.

Bolles, R. C. (1970). Species-specific defense reactions and avoidance learning. *Psychol. Rev.* **77**: 32–48.

Bradford, L. J., McKinley, J. H., Rousey, C. L. & Klein, D. E. (1973). Measurement of hearing in dogs by respiration audiometry. *Am. J. vet. Res.* **34**: 1183–1187.

Bragg, V. C. & Dreher, D. E. (1969). A shock-avoidance technique for determining audiologic thresholds in the cebus monkey. *J. aud. Res.* **9**: 270–277.

Brogden, W. J. & Culler, E. A. (1937). Increased acoustic sensitivity in dogs following roentgen-radiation of the hypophysis. *Am. J. Physiol.* **119**: 13–23.

Brown, A. M. (1973). High levels of responsiveness from the inferior colliculus of rodents at ultrasonic frequencies. *J. comp. Physiol.* **83**: 393–406.

Catania, A. C. (ed.) (1968). *Contemporary research in operant behavior.* Glenview, Ill.: Scott, Foresman and Company.

Chang, H.-T. (1936). An auditory reflex of the hedgehog. *Chin. J. Physiol.* **10**: 119–124.

Clack, T. D. (1966a). Effect of signal duration on the auditory sensitivity of humans and monkeys (*Macaca mulatta*). *J. acoust. Soc. Am.* **40**: 1140–1146.

Clack, T. D. (1966b). Behavioral and pinna-reflex thresholds in the rat: audiometric and TTS covariations. *J. acoust. Soc. Am.* **40**: 1186–1187.

Clack, T. D. & Harris, J. D. (1963). Auditory thresholds in the rat by a two-lever technique. *J. aud. Res.* **3**: 53–63.

Clack, T. D. & Herman, P. N. (1963). A single-lever psychophysiological adjustment procedure for measuring auditory thresholds in the monkey. *J. aud. Res.* **3**: 175–183.

Clopton, B. C. (1972). Detection of increments in noise intensity by monkeys. *J. exp. Analysis Behav.* **17**: 473–481.

Corbeille, C. & Baldes, E. J. (1929a). I. Respiratory responses to acoustic stimulation in intact and decerebrate animals. *Am. J. Physiol.* **88**: 481–490.

Corbielle, C. & Baldes, E. J. (1929b). II. Cardiac responses to acoustic stimulation in intact and decerebrate rabbits. *Am. J. Physiol.* **88**: 491–494.

Corbielle, C. & Baldes, E. J. (1929c). III. The effect of acoustic stimulation in the blood pressure of urethanized dogs. *Am. J. Physiol.* **88**: 495–497.

Cowles, J. T. & Pennington, L. A. (1943). An improved conditioning technique for determining auditory acuity of the rat. *J. Psychol.* **15**: 41–47.

Criborn, C. (1970). Respiratory response to acoustic excitation of mice. *Life Sciences* **9**: 13–20.

Crifo, S. (1973). Shiver-audiometry in the conditioned guinea-pig (simplified Anderson–Wedenberg test). *Acta otolar.* **75**: 38-44.

Crifo, S. (1974). Identification of ototoxic drugs by guinea-pig shiver audiometry. *Audiology* **13**: 302–310.

Culler, E., Finch, G., Girden, E. & Brogden, W. (1935). Measurements of acuity by the conditioned-response technique. *J. gen. Psychol.* **12**: 223–227.

Dalland, J. I. (1965a). Auditory thresholds in the bat: a behavioural technique. *J. aud. Res.* **5**: 95–108.

Dalland, J. I. (1965b). Hearing sensitivity in bats. *Science, N. Y.* **150**: 1185–1186.

Dalland, J. I. (1970). The measurement of ultrasonic hearing. In *Animal psychophysics: the design and conduct of sensory experiments*: 21–40. Stebbins, W. C. (ed.). New York: Appleton-Century-Crofts.

Dalland, J. I., Vernon, J. A. & Peterson, E. A. (1967). Hearing and cochlear microphonic potentials in the bat *Eptesicus fuscus*. *J. Neurophysiol.* **30**: 697–709.

Dalton, L. W., Jr. (1968), Auditory sensitivity in the rhesus (*Macaca mulatta*) and the white-throated capuchin (*Cebus capuchinus*) monkeys: a comparison of three techniques. *U.S. Air Force 6571 Aeromed. Res. Lab. Tech. Rep.* No. 68–14.

Dalton, L. W., Jr., Taylor, H., Henton, W. & Allen, J. N. (1969). Auditory thresholds in the rhesus monkey using a closed-system helmet. *J. aud. Res.* **9**: 178–182.

D'Amato, M. R. (1970). *Experimental psychology: methodology, psychophysics, and learning*. New York & London: McGraw-Hill Book Company.

Denes, P. & Kocher, W. (1961). The reliability of Preyer's ear reflex in the mouse (*Mus musculus*). *Naturwissenschaften* **48**: 82–83.

Deol, M. S. & Kocher, W. (1958). A new gene for deafness in the mouse. *Heredity* **12**: 463–466.

Dice, L. R. & Barto, E. (1952). Ability of mice of the genus *Peromyscus* to hear ultrasonic sounds. *Science, N.Y.* **116**: 110–111.

Dworkin, S., Katzman, J., Hutchison, G. A. & McCabe, J. R. (1940). Hearing acuity of animals as measured by conditioning methods. *J. exp. Psychol.* **26**: 281–298.

Dworkin, S., Sutherland, G. F. & Seymour, S. L. (1934). A conditioned reflex method of testing hearing in cats. *Q. Jl. exp. Physiol.* **24**: 23–30.

Eccher, W. (1942). *Comparative thresholds of pitch and intensity in rats.* Paper read at E. Psychol. Ass., 1942. Reported in Blackwell & Schlosberg (1943).

Elder, J. H. (1933). Audiometric studies with the chimpanzee. *Psychol. Bull.* **30**: 547–548.

Elder, J. H. (1934). Auditory acuity of the chimpanzee. *J. comp. Psychol.* **17**: 157–183.

Eldredge, D. H. (1973). Anatomical and physiological correlates of threshold shifts in the chinchilla after exposure to noise. In *Disorders of auditory function*: 103–114. Taylor, W. (ed.). New York & London: Academic Press.

Elliott, D. N. (1967). Effect of peripheral lesions on acuity and discrimination in animals. In *Sensorineural hearing processes and disorders*: 179–189. Graham, A. B. (ed.). Boston: Little, Brown and Company.

Elliott, D. N. & McGee, T. M. (1965). Effect of cochlear lesions upon audiograms and intensity discrimination in cats. *Ann. Otol. Rhinol. Lar.* **74**: 386–408.

Elliott, D. N., Stein, L. & Harrison, M. J. (1960). Determination of absolute-intensity thresholds and frequency-difference thresholds in cats. *J. acoust. Soc. Am.* **32**: 380–384.

Farrer, D. N. & Prim, M. M. (1965). A preliminary report on auditory frequency threshold comparisons of humans and pre-adolescent chimpanzees. *U.S. Air Force 6571 Aeromed. Res. Lab. Tech. Rep.* No. 65–6.

File, S. E. (1973). Long-term retention of behavioural habituation in the rat. *Anim. Behav.* **21**: 585–589.

File, S. E. (1974). Changes in orienting and habituation following administration of chlorpromazine. *Q. Jl. exp. Psychol.* **26**: 125–130.

Fleschler, M. (1965). Adequate acoustic stimulus for startle reaction in the rat. *J. comp. physiol. Psychol.* **60**: 200–207.

Fujita, S. & Elliott, D. N. (1965). Thresholds of audition for three species of monkey. *J. acoust. Soc. Am.* **37**: 139–144.

Gautier, H. (1972). Respiratory and heart rate responses to auditory stimulations. *Physiol. Behav.* **8**: 327–332.

Gill, A. & Berlin, C. I. (1968). Hearing in mice by GSR audiometry: II. Magnitude of unconditioned GSR as a function of intensity and frequency interactions. *J. Speech Hear. Res.* **11**: 169–178.

Gormenzo, I. & Moore, J. W. (1969). Classical conditioning. In *Learning: processes*: 119–203. Marx, M. H. (ed.). New York & London: The Macmillan Company.

Gould, J. & Morgan, C. T. (1942). Auditory sensitivity in the rat. *J. comp. Psychol.* **34**: 321–329.

Gourevitch, G. (1965). Auditory masking in the rat. *J. acoust. Soc. Am.* **37**: 439–443.

Gourevitch, G. (1970). Detectability of tones in quiet and in noise by rats and monkeys. In *Animal psychophysics: the design and conduct of sensory experiments*: 67–97. Stebbins, W. C. (ed.). New York: Appleton-Century-Crofts.

Gourevitch, G. & Hack, M. H. (1966). Audibility in the rat. *J. comp. physiol. Psychol.* **62**: 289–291.

Gourevitch, G., Hack, M. H. & Hawkins, J. E., Jr. (1960). Auditory thresholds in the rat measured by an operant technique. *Science, N.Y.* **131**: 1046–1047.

Graham, F. K. & Clifton, R. K. (1966). Heart-rate change as a component of the orienting response. *Psychol. Bull.* **65**: 305–320.

Gross, N. B. (1952). The effects of cochlear lesions on the auditory response of the guinea-pig. *J. comp. physiol. Psychol.* **45**: 127–139.

Guilford, J. P. (1954). *Psychometric methods.* New York & London: McGraw-Hill Book Company, Inc.

Hack, M. H. (1966). Receiver operating characteristics in the rat. *J. aud. Res.* **6**: 229–234.

Hack, M. H. (1968). The developmental Preyer reflex in the sh-1 mouse. *J. aud. Res.* **8**: 449–457.

Hall, J. D. & Johnson, C. S. (1972). Auditory thresholds of a killer whale *Orcinus orca* Linnaeus. *J. acoust. Soc. Am.* **51**: 515–517.

Harris, J. D. (1943). The auditory acuity of pre-adolescent monkeys. *J. comp. Psychol.* **35**: 255–265.

Heffner, H. E. & Masterton, B. (1970). Hearing in primitive primates: slow loris (*Nycticebus coucang*) and potto (*Perodicticus potto*). *J. comp. physiol. Psychol.* **71**: 175–182.

Heffner, H. E., Ravizza, R. J. & Masterton, B. (1969a). Hearing in primitive mammals, III: tree shrews (*Tupaia glis*). *J. aud. Res.* **9**: 12–18.

Heffner, H. E., Ravizza, R. J. & Masterton, B. (1969b). Hearing in primitive mammals, IV: bushbaby (*Galago senegalensis*). *J. aud. Res.* **9**: 19–23.

Heffner, H. E., Van Oeveren, R. & Masterton, B. (1971). Hearing in two species of wild rodents. *J. acoust. Soc. Am.* **50**: 148.

Heffner, R., Heffner, H. E. & Masterton, B. (1971). Behavioral measurements of absolute and frequency-difference thresholds in guinea-pig. *J. acoust. Soc. Am.* **49**: 1888–1895.

Henderson, D. (1969). Temporal summation of acoustic signals by the chinchilla. *J. acoust. Soc. Am.* **46**: 474–475.

Henderson, D., Onishi, S., Eldredge, D. H. & Davis, H. (1969). A comparison of chinchilla auditory evoked response and behavioral response thresholds. *Percept. Psychophys.* **5**: 41–45.

Henry, F. M. (1935). Comparative thresholds of audition in the white rat. *Psychol. Bull.* **32**: 536.

Henry, F. M. (1938). Audition in the white rat. III. Absolute and relative intensity thresholds. *J. comp. Psychol.* **26**: 45–62.

Henry, K. R. (1972). Pinna reflex thresholds and audiogenic seizures: developmental changes after acoustic priming. *J. comp. physiol. Psychol.* **79**: 77–81.

Henry, K. R., Wallick, M. & Davis, M. (1972). Inferior collicular lesions: effects on audiogenic seizure and Preyer reflex. *Physiol. Behav.* **9**: 885–887.

Herman, L. M. & Arbeit, W. R. (1973). Stimulus control and auditory discrimination learning sets in the bottlenose dolphin. *J. exp. Analysis Behav.* **19**: 374–394.

Herman, P. N. & Clack, T. D. (1963). Post-stimulatory shift of the Preyer reflex threshold in the white rat. *J. aud. Res.* **3**: 189–204.

Hoffman, H. S. (1966). The analysis of discriminated avoidance. In *Operant behavior: areas of research and application*: 499–530. Honig, W. K. (ed.). New York: Appleton-Century-Crofts.

Honig, W. K. (ed.) (1966). *Operant behavior: areas of research and application.* New York: Appleton-Century-Crofts.

Horton, G. P. (1933). A quantitative study of hearing in the guinea-pig. (*Cavia cobaye*). *J. comp. Psychol.* **15**: 59–73.

Horton, G. P. (1934). The effect of intense and prolonged acoustical stimulation on the auditory sensitivity of guinea-pigs. *J. comp. Psychol.* **18**: 405–417.

Hunter, W. S. (1915). The auditory sensitivity of the white rat. *J. Anim. Behav.* **5**: 312–329.

Hunter, W. S. (1927). Further data on the auditory sensitivity of the white rat. *Pedag. Semin.* **34**: 177–186.

Igarashi, M. & Hoyt, R. F. (1963). Selection of feline subjects for auditory testing. An examination of behavior during basic conditioning of the response. *J. aud. Res.* **3**: 169–174.

Jacobs, D. W. & Hall, J. D. (1972). Auditory thresholds of a fresh water dolphin, *Inia geoffrensis* Blainville. *J. acoust. Soc. Am.* **51**: 530–533.

Jamison, J. H. (1951). Measurement of auditory intensity thresholds in the rat by conditioning of an autonomic response. *J. comp. physiol. Psychol.* **44**: 118–125.

Johnson, C. S. (1966). Auditory thresholds of the bottlenosed porpoise (*Tursiops truncatus*, Montagu). *U.S. Nav. Ordnance Test Stn Tech. Publ.* No. 4178.

Johnson, C. S. (1967). Sound detection thresholds in marine mammals. In *Marine bio-acoustics* **2**: 247–260. Tavolga, W. N. (ed.). New York: Pergamon Press.

Johnson, C. S. (1968a). Relation between absolute threshold and duration-of-tone pulse in the bottlenosed porpoise. *J. acoust. Soc. Am.* **43**: 757–763.

Johnson, C. S. (1968b). Masked tonal thresholds in the bottlenosed porpoise. *J. acoust. Soc. Am.* **44**: 965–967.

Kellogg, W. N. (1953). Ultrasonic hearing in the porpoise. *J. comp. physiol. Psychol.* **46**: 446–450.

Kellogg, W. N. & Kohler, R. (1952). Responses of the porpoise to ultrasonic frequencies. *Science, N.Y.* **166**: 250–252.

Koyama, Y. (1964). Changes of the responsiveness of the respiration to the repetitive sound stimulation. *J. oto-rhino-lar. Soc. Japan* **67**: 974–981.

Kryter, K. D. & Ades, H. W. (1943). Studies on the function of the higher acoustic nervous centers in the cat. *Am. J. Psychol.* **56**: 501–536.

Landes, A. B. (1962). The effect of radiation on the auditory threshold of the white rat. *J. aud. Res.* **2**: 208–219.

Lindquist, N. E., Neff, W. D. & Schuknecht, H. F. (1954). Stimulation deafness: a study of hearing losses resulting from noise or blast impulses. *J. comp. physiol. Psychol.* **47**: 406–411.

Lipman, E. & Grassi, J. R. (1942). Comparative auditory sensitivity of man and dog. *Am. J. Psychol.* **55**: 84–89.

Markl, V. H. & Ehret, G. (1973). Die Hörschwelle der Maus (*Mus musculus*). *Z. Tierpsychol.* **33**: 274–286.

Martin, P., Romba, J. J. & Gates, H. W. (1962). A method for the study of hearing loss and recovery in rhesus monkeys. *U.S. Army Hum. Engng. Lab. Tech. Memo.* No. 11–62. Maryland: Aberdeen Proving Ground.

Masterton, B., Heffner, H. E. & Ravizza, R. (1969). The evolution of human hearing. *J. acoust. Soc. Am.* **45**: 966–985.

McGill, T. E. (1959). Auditory sensitivity and the magnitude of the cochlear potential. *Ann. Otol. Rhinol. Lar.* **68**: 193–207.

Mikaelian, D. & Ruben, R. J. (1964). Hearing degeneration in Shaker-1 mouse. *Archs Otolar.* **80**: 418–430.

Mikaelian, D. & Ruben, R. J. (1965). Development of hearing in the normal CBA-J mouse: correlation of physiological observations with behavioral responses and with cochlear anatomy. *Acta otolar.* **59**: 451–461.

Mikaelian, D. & Warfield, D. (1970). Long-term partial eighth nerve section in cats. *Archs Otolar.* **91**: 371–375.

Miller, J. D. (1963). Audibility curve of chinchilla. *J. acoust. Soc. Am.* **35**: 1907.

Miller, J. D. (1964). Auditory sensitivity of the chinchilla in quiet and in noise. *J. acoust. Soc. Am.* **36**: 2010.

Miller, J. D. (1970). Audibility curve of the chinchilla. *J. acoust. Soc. Am.* **48**: 513–523.

Miller, J. D. & Murray, F. S. (1966). Guinea-pig's immobility response to sound: threshold and habituation. *J. comp. physiol. Psychol.* **61**: 227–233.

Miller, J. D., Watson, C. S. & Covell, W. P. (1963). Deafening effects of noise on the cat. *Acta otolar.* Suppl. **176**.

Mitchell, C., Gillette, R., Vernon, J. A. & Herman, P. (1970). Pure-tone auditory behavioral thresholds in three species of lemurs. *J. acoust. Soc. Am.* **48**: 531–535.

Mitchell, C., Vernon, J. A. & Herman, P. (1971). What does the lemur really hear? *J. acoust. Soc. Am.* **50**: 710–711.

Möhl, B. (1967). Frequency discrimination in the common seal. In *Underwater acoustics.* **2**: 43–54. Albers, V. M. (ed.). New York: Plenum Publishing Corporation.

Möhl, B. (1968). Auditory sensitivity of the common seal in air and water. *J. aud. Res.* **8**: 28–38.

Morgan, C. & Gould, J. (1941). Acoustical determinants of the "neurotic pattern" in rats. *Psychol. Rec.* **4**: 258–268.

Murphy, J. & Harris, J. D. (1961). Negligible effects of X-radiation of the head upon hearing in the rat. *J. aud. Res.* **1**: 117–132.

Neff, W. D. (1940). An experimental investigation of hearing following partial section of the VIIIth nerve. *Psychol. Bull.* **37**: 527–528.

Neff, W. D. (1947). The effects of partial section of the auditory nerve. *J. comp. physiol. Psychol.* **40**: 203–215.

Neff, W. E. & Hind, J. E. (1955). Auditory thresholds of the cat. *J. acoust. Soc. Am.* **27**: 480–483.

Peacock, L. J. & Williams, M. (1962). An ultrasonic device for recording activity. *Am. J. Psychol.* **75**: 648–652.

Peters, E. N. (1965). Temporary shifts in auditory thresholds of chinchillas after exposure to noise. *J. acoust. Soc. Am.* **37**: 831–833.

Poduschka, W. (1968). Über die Wahrnehmung von Ultraschall beim Igel, *Erinaceus europaeus romanicus. Z. vergl. Physiol.* **61**: 420-426.

Powers, B. M., Warfield, D. & Ruben, R. J. (1966). Development of the Preyer response to different frequencies in normal CBA-J mice. *J. aud. Res.* **6**: 425–435.

Ralls, K. (1967). Auditory sensitivity in mice, *Peromyscus* and *Mus musculus. Anim. Behav.* **15**: 123–128.

Ravizza, R. J., Heffner, H. E. & Masterton, B. (1969a). Hearing in primitive mammals, I: opossum (*Didelphis virginianus*). *J. aud. Res.* **9**: 1–7.

Ravizza, R. J., Heffner, H. E. & Masterton, B. (1969b). Hearing in primitive mammals, II: hedgehog (*Hemiechinus auritus*). *J. aud. Res.* **9**: 8–11.

Ray, B. A. (1970). Psychophysical testing of neurologic mutant mice. In *Animal psychophysics: the design and conduct of sensory experiments*: 99–124. Stebbins, W. C. (ed.). New York: Appleton-Century-Crofts.

Schleidt, W. M. (1948). Töne hoher Frequenz bei Mäusen. *Experientia* **4**: 145–146.

Schleidt, W. M. (1951). Töne hoher Frequenz bei Mäusen. *Experientia* **7**: 65–66.

Schleidt, W. M. (1952). Reactionen auf Töne hoher Frequenz bei Nagern. *Naturwissenschaften* **39**: 69–70.

Schuknecht, H. F. (1953). Techniques for study of cochlear function and pathology in experimental animals. Development of the anatomical frequency scale for the cat. *Archs Otolar.* **58**: 377–397.

Schusterman, R. J., Balliet, R. F. & Nixon, J. (1972). Underwater audiogram of the California sea lion by the conditioned vocalization technique. *J. exp. Analysis Behav.* **17**: 339–350.

Seiden, H. R. (1958). *Auditory acuity of the marmoset monkey* (Hapale jacchus). Unpublished Ph.D. Thesis: Princeton University.

Semenoff, W. A. & Young, F. A. (1964). Comparison of the auditory acuity of man and monkey. *J. comp. physiol. Psychol.* **57**: 89–93.

Sherrick, C. & Bilger, R. (1959). Auditory sensitivity of the guinea-pig to low frequency tones. *Percept. Mot. Skills* **9**: 339–344.

Sherrod, K. B. (1973). Effects of irradiation and/or handling on shuttle box performance of C57B1/6 mice. *Physiol. Behav.* **10**: 721–724.

Sidman, M., Ray, B. A., Sidman, R. L. & Klinger, J. (1966). Hearing and vision in neurological mutant mice. *Expl Neurol.* **16**: 377–402.

Simmons, F. B. & Beatty, D. L. (1962): The significance of round window recorded cochlear potentials in hearing. *Ann. Otol. Rhinol. Lar.* **71**: 767–800.

Simpson, G. G. (1945). The principles of classification and a classification of mammals. *Bull. Am. Mus. nat, Hist.* **85**: i–xvi, 1–350.

Skinner, B. F. (1948). "Superstition" in the pigeon. *J. exp. Psychol.* **38**: 168–172.

Smith, J. C. (1970). Conditioned suppression as an animal psychophysical technique. In *Animal psychophysics: the design and conduct of sensory experiments*: 125–159. Stebbins, W. C. (ed.). New York: Appleton-Century-Crofts.

Sokolov, E. N. (1960). A neuronal model of stimulus and the orienting reflex. *Probl. Psychol.* **3**: 217–219.

Sokolovski, A. (1972). "Push-pull" conditioning technique for measuring auditory thresholds in cats. *Arch. klin. exp. Ohr. Nas.-Kehlk. Heilk.* **203**: 87–108.

Sokolovski, A. (1973a). Normal threshold of hearing for cat for free-field listening. *Arch. klin. exp. Ohr. Nas.-Kehlk. Heilk.* **203**: 232–240.

Sokolovski, A. (1973b). The protective action of the stapedius muscle in noise-induced hearing loss in cats. *Arch. klin. exp. Ohr. Nas.-Kehlk. Heilk.* **203**: 289–309.

Soliman, S. M. (1969). The GSR and conditionability of mice with different brain weights. *J. aud. Res.* **9**: 145–158.

Stebbins, W. C. (1966). Auditory reaction time and the derivation of equal loudness contours for the monkey. *J. exp. Analysis Behav.* **9**: 135–142.

Stebbins, W. C. (1970a). Principles of animal psychophysics. In *Animal psychophysics: the design and conduct of sensory experiments*: 1–19. Stebbins, W. C. (ed.). New York: Appleton-Century-Crofts.

Stebbins, W. C. (1970b). Studies of hearing and hearing loss in the monkey. In *Animal psychophysics: the design and conduct of sensory experiments*: 41–66. Stebbins, W. C. (ed.). New York: Appleton-Century-Crofts.

Stebbins, W. C. (1971). Hearing. In *Behavior of nonhuman primates.* **3**: 159–192. Schrier, A. M. & Stollnitz, F. (eds). New York & London: Academic Press.

Stebbins, W. C. (1973). Hearing of old world monkeys (*Cercopithecinae*). *Am. J. phys. Anthrop.* **38**: 357–364.

Stebbins, W. C., Green, S. & Miller, F. L. (1966). Auditory sensitivity of the monkey. *Science, N.Y.* **153**: 1646–1647.

Stebbins, W. C., Miller, J. M., Johnson, L.-G. & Hawkins, J. E., Jr. (1969). Ototoxic hearing loss and cochlear pathology in the monkey. *Ann. Otol. Rhinol. Lar.* **78**: 1007–1026.

Stebbins, W. C., Pearson, R. D. & Moody, D. B. (1970). Hearing in the monkey (*Macaca*): absolute and differential sensitivity. *J. acoust. Soc. Am.* **47**: 67.

Strother, W. F., Parker, D. E., Rahm, W. E., Jr. & Crump, J. F. (1964). The effects of anaesthetics upon the ear. V. Cochlear potentials and behavioral thresholds. *Ann. Otol. Rhinol. Lar.* **73**: 141–152.

Terhune, J. M. & Ronald, K. (1970). The audiogram and calls of the harp seal (*Pagophilus groenlandicus*) in air. *Proc. Seventh A. Conf. Sonar. Diving Mammals.* Menlo Park, Calif.: Stanford Res. Inst.

Terhune, J. M. & Ronald, K. (1971). The harp seal, *Pagophilus groenlandicus* (Erxleben, 1777). X. The air audiogram. *Can. J. Zool.* **49**: 385–390.

Terhune, J. M. & Ronald, K. (1972). The harp seal, *Pagophilus groenlandicus* (Erxleben, 1777). III. The underwater audiogram. *Can. J. Zool.* **50**: 565–569.

Terrace, H. S. (1966). Stimulus control. In *Operant behavior: areas of research and application*: 271–344. Honig, W. K. (ed.). New York: Appleton-Century-Crofts.

Upton, M. (1929a). The auditory sensitivity of guinea-pigs. *Am. J. Psychol.* **41**: 412–421.

Upton, M. (1929b). Functional disturbances of hearing in guinea-pigs after long exposure to an intense tone. *J. gen. Psychol.* **2**: 397–412.

Ward, W. D. & Duvall, A. J. (1971). Behavioral and ultrastructural correlates of acoustic trauma. *Ann. Otol. Rhinol. Lar.* **80**: 881–897.

Webster, D. B. & Webster, M. (1972). Kangaroo rat auditory thresholds before and after middle ear reduction. *Brain Behav. Evolution* **5**: 41–53.

Webster, W. R. & Dunlop, C. W. (1965). Unanaesthetized cats, auditory evoked potentials and test cage acoustics: a methodological study. *J. aud. Res.* **5**: 11–25.

Wendt, G. R. (1934). Auditory acuity of monkeys. *Comp. Psychol. Monogr.* **10**: No. 3.

Wever, E. G. (1930). The upper limit of hearing in the cat. *J. comp. Psychol.* **10**: 221–233.

Wilpizeski, C. R. & Young, I. M. (1971). Air and bone conduction pure tone detection thresholds for squirrel monkey. *J. acoust. Soc. Am.* **50**: 147–148.

Wollack, C. H. (1963). The auditory acuity of the sheep (*Ovis aries*). *J. aud. Res.* **3**: 121–132.

Wollack, C. H. (1965). Auditory thresholds in the raccoon (*Procyon lotor*). *J. aud. Res.* **5**: 139–144.

Appendix

Studies classified by species

Order Marsupialia

Didelphis virginianus (opossum)

Ravizza *et al.* (1969a) R

Order Insectivora
Family Erinaceidae (hedgehogs)
Chang (1936) U
Erinaceus europaeus romanicus
Poduschka (1968) U
Hemiechinus auritus
Ravizza *et al.* (1969b) R
Order Chiroptera (bats)
Myotis lucifugus
Dalland (1965b, 1970) Rw
Eptesicus fuscus
Dalland (1965a,b, 1970) Rw
Dalland *et al.* (1967) Rw
Order Primates
Tupaia glis (tree shrew)
Heffner, Ravizza *et al.* (1969a) R,A
Stebbins (1970b) Rw
Lemur spp. (lemurs)
L. catta
Mitchell, Gillette *et al.* (1970) Rw
Mitchell, Vernon *et al.* (1971) A
L. macaco fulvus
Mitchell, Gillette *et al.* (1970) Rw
L. m. macaco
Mitchell, Gillette *et al.* (1970) Rw
Nycticebus coucang (slow loris)
Heffner & Masterton (1970) R
Perodicticus potto (potto)
Heffner & Masterton (1970) R
Galago senegalensis (bushbaby)
Heffner, Ravizza *et al.* (1969b) R
Aotus trivirgatus (owl monkey)
Beecher (1972) Rw
Cebus capuchinus (capuchin)
Dalton (1968) R,R
Bragg & Dreher (1969) R,R,A
Saimiri sciureus (squirrel monkey)
Fujita & Elliott (1965) A,A,Rw
Wilpizeski & Young (1971) A
Ateleus (=*Ateles*) *ater* (black spider monkey)
Wendt (1934) Rw
Hapale (=*Callithrix*) *jacchus* (marmoset)
Seiden (1958) A
Cercopithecus aethiops
Stebbins (1970b, 1973) Rw
Macaca spp. (macaques)
Stebbins, Pearson *et al.* (1970) Rw
M. arctoides
Stebbins (1970b, 1973) Rw

Order Primates—(*continued*)
- *M. fascicularis* (cynomologus monkey)
  - Fujita & Elliott (1965) A,Rw
  - Stebbins, Green *et al.* (1966) Rw
  - Stebbins (1970b, 1973) Rw
- *M. irus*
  - Stebbins, Miller *et al.* (1969) Rw
- *M. mulatta* (rhesus monkey)
  - Wendt (1934) Rw
  - Harris (1943) A,A
  - Martin *et al.* (1962) A
  - Clack & Herman (1963) A
  - Behar *et al.* (1965) A
  - Fujita & Elliott (1965) A,Rw
  - Clack (1966a) A
  - Dalton (1968) R,R
  - Dalton *et al.* (1969) R
  - Stebbins (1970b, 1973) Rw
- *M. nemestrina* (pigtailed monkey)
  - Semenoff & Young (1964) R
  - Stebbins, Green *et al.* (1966) Rw
  - Stebbins, Miller *et al.* (1969) Rw
  - Gourevitch (1970) Rw
  - Stebbins (1970b) Rw
  - Stebbins, Pearson *et al.* (1970) Rw
  - Stebbins (1973) Rw
- *M. rhesus*
  - Harris (1943) A,A
- *M. sinica*
  - Harris (1943) A,A
- *Cercocebus torquatus*
  - Wendt (1934) Rw
- *Papio* spp. (baboons)
- *P. anubis*
  - Wendt (1934) Rw
- *P. papio*
  - Stebbins (1970b, 1973) Rw
- *Erythrocebus patas*
  - Stebbins (1970b, 1973) Rw
- *Pan* sp. (chimpanzee)
  - Elder (1933, 1934) Rw
  - Farrer & Prim (1965) A

Order Lagomorpha
- *Oryctolagus* sp. (rabbit)
  - Corbielle & Baldes (1929a,b) U,U
  - Koyama (1964) U

Order Rodentia (rodents)
- *Dipodomys merriami* (kangaroo rat)
  - Webster & Webster (1972) A

Order Rodentia (rodents)—(*continued*)

Peromyscus spp. (deermice)

Dice & Barto (1952) U,U,A

Sigmodon hispidus (cotton rat)

Heffner, Van Oeveren *et al.* (1971) R

Cricetus cricetus

Schleidt (1952) U

Mesocricetus auratus

Schleidt (1952) U

Clethrionomys (=*Evotomys*) *glareolus* (red-backed vole)

Schleidt (1948, 1951, 1952) U,U

Micromys minutus

Schleidt (1952) U

Rattus norvegicus (rat)

Henry, F. M. (1935, 1938) A

Morgan & Gould (1941) U

Eccher (1942) A

Gould & Morgan (1942) A

Blackwell & Schlosberg (1943) A

Cowles & Pennington (1943) R

Jamison (1951) R

Schleidt (1952) U

Gourevitch *et al.* (1960) Rw

Murphy & Harris (1961) U

Landes (1962) A

Clack & Harris (1963) Rw

Herman & Clack (1963) U

Fleschler (1965) U

Gourevitch (1965) Rw

Clack (1966b) U, Rw

Gourevitch & Hack (1966) Rw

Gourevitch (1970) Rw

File (1973) U

R. rattus (=*Mus rattus*) (house rat)

Schleidt (1952) U

Mus musculus (mouse)

(listed by strain)

CBA-J

Alford & Ruben (1963) U

Berlin (1963) U,R

Mikaelian & Ruben (1965) U

Powers *et al.* (1966) U

Berlin *et al.* (1968) U

Birch *et al.* (1968) A,A,Rw

Gill & Berlin (1968) U

C57BL/6J

Sidman *et al.* (1966) R

Ray (1970) U,R

Henry, K. R. (1972) U,U

Henry, K. R. *et al.* (1972) U,U

Order Rodentia (rodents)—(*continued*)
CBA/Gr X C57BL/Gr
Denes & Kocher (1961) U
DBA/2J
Henry, K. R. *et al.* (1972) U,U
sh-1
Berlin (1963) U
Mikaelian & Ruben (1964) U
Hack (1968) U
qk, pi, rl, c^h, kr, BALB/cGn
Sidman *et al.* (1966) R
Ray (1970) U,R
dn
Deol & Kocher (1958) U
BALB/cj, MA/j, C57BR/cdj
Soliman (1969) U
Wild or unspecified
Schleidt (1952) U
Criborn (1970) U
Heffner, Van Oeveren *et al.* (1971) R
NMRI
Markl & Ehret (1973) R,Rw
Glis glis
Schleidt (1952) U
Muscardinus avellanarinus
Schleidt (1952) U
Eliomys quercinus
Schleidt (1952) U
Cavia spp. (guinea-pig)
Upton (1929a) R,A
Upton (1929b) R
Horton (1933, 1934) R
Gross (1952) A,A
Schleidt (1952) U
Alexander *et al.* (1955) R
Sherrick & Bilger (1959) R
Anderson & Wedenberg (1965) U,R,R,A,Rw
Miller & Murray (1966) U
Heffner, Heffner *et al.* (1971) R
Crifo (1973, 1974) U,R
Chinchilla chinchilla (chinchilla)
Miller (1963, 1964) A
Peters (1965) A
Henderson (1969) A
Henderson *et al.* (1969) A
Miller (1970) A
Ward & Duvall (1971) A
Eldredge (1973) A
Order Cetacea
Inia geoffrensis (Amazon dolphin)
Jacobs & Hall (1972) Rw

Order Cetacea—(*continued*)

Stennella (=*Stenella*) *plagiodon* (long-snouted dolphin)

Kellogg & Kohler (1952) U

Kellogg (1953) U

Tursiops truncatus (bottlenosed porpoise)

Kellogg & Kohler (1952) U

Kellogg (1953) U

Johnson (1966, 1967, 1968a) Rw

Orcinus orca (killer whale)

Hall & Johnson (1972) Rw

Phocoena phocoena (harbour porpoise)

Anderson (1970) Rw

Order Carnivora

Canis familiaris (dog)

Corbielle & Baldes (1929a,c) U,U

Culler *et al.* (1935) A

Brogden & Culler (1937) A

Dworkin, Katzman *et al.* (1940) Rw

Lipmann & Grassi (1942) A

Bradford *et al.* (1973) U

Procyon lotor (raccoon)

Wollack (1965) Rw

Felis catus (cat)

Corbielle & Baldes (1929a) U

Wever (1930) R

Dworkin, Sutherland *et al.* (1934) Rw

Ades *et al.* (1937, 1939) A

Dworkin, Katzman *et al.* (1940) Rw

Neff (1940) A

Kryter & Ades (1943) A

Neff (1947) A

Schuknecht (1953) A

Lindquist *et al.* (1954) A

Neff & Hind (1955) A

McGill (1959) Rw

Elliott *et al.* (1960) A

Simmons & Beatty (1962) A

Miller *et al.* (1963) A

Strother *et al.* (1964) A

Elliott & McGee (1965) A

Mikaelian & Warfield (1970) Rw

Sokolovski (1972, 1973a,b) A

Zalophus californianus (California sea lion)

Schusterman *et al.* (1972) Rw

Pagophilus groenlandicus (harp seal)

Terhune & Ronald (1970, 1971, 1972) Rw

Phoca vitulina vitulina (common seal)

Möhl (1968) Rw

Order Artiodactyla
Ovis aries (sheep)
Wollack (1963) A
Ames & Arehart (1972) U

Letters following the date of each study indicate the procedures used in the study. U = unconditioned; R = respondent; A = avoidance; Rw = reward (see Table I). The studies of each species are listed chronologically and then alphabetically as in the list of references. The taxonomic system employed is that of Simpson (1945). When used, the common names of species or orders are those given in the studies. In cases of conflict, or to facilitate cross-referencing in Tables I–IV, a common name provided by Simpson (1945) or by one of the studies is used.

Symp. zool. Soc. Lond. (1975) No. 37, 291–305

THE AUDITORY RESPONSE CRADLE: A DEVICE FOR THE OBJECTIVE ASSESSMENT OF AUDITORY STATE IN THE NEONATE

M. J. BENNETT

Department of Mechanical Engineering, Brunel University, Middlesex, England

SYNOPSIS

A system for recording physiological response measures from the human newborn has been developed. This apparatus, termed the "auditory response cradle", provides a means of assessing the auditory state of the infant during the neonatal period. The requirement for an examination at this age was analysed and found to be based upon three points: the accessibility of the population, the advantages of early treatment and/or habilitation, and the need to review constantly possible drug ototoxicity. The existing methods were found generally to involve subjective visual observations, frequently carried out using poorly defined stimuli in unsuitable environments. The auditory response cradle was designed to improve upon this unsatisfactory situation. A multiparameter approach to the problem was adopted for monitoring both the responses and the arousal state of the subject. The equipment incorporates a series of non-contacting transducers in a fully adjustable body and head support system. The neonate lies supine in a foam padded bed, the head being supported in a separate foam rest which can pivot on the spinal axis. Acoustic stimuli are presented from miniature loudspeakers via waveguides and ear inserts. Secondary waveguides enable the sound pressure level set up in the meatus to be measured. Bed mounted instrumentation permits auditory responses in the form of head rotation, head jerk and general limb movements to be polygraphically recorded. Other transducers, fitted to bands around the child, add cardiac and respiratory rate signals to the body activity parameters. All five channels may be related directly on the polygraph to the measured stimulus level. In this way, an accurate assessment may be made of the acoustic responses of the neonatal subject. Preliminary trials have produced promising results. Full validation of the technique awaits the completion of an extensive programme of work which includes follow-up examinations.

INTRODUCTION

The desirability of an early assessment of auditory state in the infant has been stressed by many workers (Downs & Sterritt, 1964; Gerber, 1972; Poole, Goetzinger & Rousey, 1966). The case for carrying out these examinations in the neonatal period comprises three main points. Firstly, a large percentage of the population is readily accessible in the hospital nursery, a situation not again achieved until the eighth month (health visitor) screening for "at risk" infants. Next, detection of auditory disfunction at this time will make possible remedial action and/or an early start to a habilitation programme. Finally, a point often overlooked is raised by Morgon, Charachon & Bringuier (1971), namely that there is an urgent need to monitor any ototoxic effects caused by drugs taken during pregnancy.

One method of assessing the auditory state of the neonate is to use the so-called auditory reflexes. Using this approach it is necessary to examine the nature of the relationship between the responses and the hearing capability of the subject, using "hearing" in the sense implying cognitive reaction to a stimulus. This is only possible by linking neonatal data to developmental examinations of the child over a period of years. Attempts to do this have not, as yet, proved successful. Rosenblith (1970) attempted to answer the question and concluded "the data obtained gave little support to the theory that those children who fail to respond to auditory stimuli as neonates will show hearing defects in later assessments". Similarly Ling (unpublished) "supports the view that results of mass screening programmes for newborn infants are of limited validity". A closer examination of such studies indicates that it is possibly not the premise of neonatal screening that is at fault, but it is instead the inadequate techniques available with which to carry out the tests. For example, Rosenblith (1970) used a rattle and a bell to provide the auditory stimulus, both of which are inherently unrepeatable and ill-defined. In addition, the order in which these were presented was determined by the arousal state of the infant thus confusing the state variable with the variable under test. Although Ling (unpublished) defined his stimuli with great care, his conclusion was based on the use of narrow-band white noise presented in 50 msec bursts and he himself comments that "a stimulus duration of 50 msec is probably less than optimal". In addition, these studies and others such as those of Eisenberg, Griffin, Coursin & Hunter (1964) and Ling, Ling & Doehring (1970) rely entirely on subjective assessments of a response by one or more observers. This adds to the difficulties as inter-observer reliability has been found to be low in areas where marked responses are not present (Downs & Sterritt, 1964; Ling *et al.*, 1970). If these factors are all taken together, it becomes evident that the true status of neonatal response assessment is by no means clear. It was with this in mind that the present study was commenced with the aim of providing an objective recording system for the response measures together with carefully controlled and monitored presentations of the auditory stimuli.

THE SELECTION OF RESPONSE MEASURES

The choice between a single or a multi-parameter approach

There are a number of physiological measures that may be employed for the assessment of neonatal auditory responsivity. Parameters such as respiration and cardiac rates, limb, head and eye movements have all been used individually as response measures. Reliance on a single index, however, may presuppose that all responses are equivalent indicators of stimulus effectiveness. Turkewitz, Moreau, Birch & Davis (1971) have argued that this is not the case and that there are different thresholds for different responses. In addition, a single parameter may reflect merely a change in another variable and not a direct response; for example, heart rate may, in some cases, be regarded as secondary to changes in respiration and body movement (Bench & Smyth, 1967; Hutt, Lenard & Prechtl, 1969).

The need to assess arousal state

A further case for a multi-parameter approach is made by the need to assess accurately the arousal state of the subject. This arises from the premise that activity or arousal state exerts a significant effect upon the response ratio, the response pattern and the response strength (Eisenberg, Hunter, Griffin & Coursin 1965). Escalona (1962) points out that variables such as state may be dealt with in any of five ways. "They may be ignored, they may be circumvented, they may be controlled, they may be representatively included, and they may be studied in their own right". Since this statement, so much evidence has accumulated (Brackbill, 1970; Ashton, 1971; Eisenberg, Griffin *et al.* 1964; and many others), that we cannot ignore the state variable. Of the alternatives, that of representatively including state as a variable is the method to be adopted.

Having chosen this course of action, a major difficulty arises, namely that of defining "state". Hutt *et al.* (1969) conclude that states should "not be placed on a continuum. Rather they may be better regarded as qualitatively different neurophysiological conditions with specific functional organization". It is evident that the greater the number of state categories defined, the more difficult it is to distinguish accurately between them.

In consequence, it was decided to use a relatively simple scale adapted from that of Prechtl & Beintema (1964).

State I: eyes closed, regular respiration, regular heart rate, no movements except sudden generalized startles.

State II: eyes closed, irregular respiration, irregular heart rate with large fluctuations, small muscular twitches, no gross movements.

State III: eyes open, small fluctuations in respiration and heart rates, no gross movements.

State IV: eyes open, movements of head, limbs and trunk.

State V: crying.

The physiological measurements made on the subject must therefore act both as response indices and as means of obtaining an arousal state assessment.

The chosen response measures

Head turn

Work by Papousek (1961) showed that repeatable head rotation reflexes in response to an auditory stimulus could be obtained from infants in the first months of life by suitable conditioning. Siqueland & Lipsitt (1966) further found that this relatively consistently elicited reflex permitted a study of auditory discrimination in newborns. These conclusions led Bench & Murphy (1970) to develop the "Papousek cradle" and thus to investigate the unconditioned head turn reaction to sound. Their apparatus produced promising results and led to the present study at Brunel University.

Startle

A startle reaction has frequently been reported as an auditory response (Downs & Sterritt, 1964; Eisenberg, Griffin *et al.*, 1964; and many others). The startle response has been shown to involve both head and limb movements (Bench, Collyer, Langford & Toms, 1972). In a prototype apparatus, it was found that the most reliable measure was the occurrence of a sharp backward jerk of the head. This movement was selected for instrumentation in the auditory response cradle.

General trunk and limb activity

The preceding two parameters provide little information from which arousal state assessments can be made. General trunk and limb movements were chosen as being both response measures—as part of the startle reaction—and as indicators of the overall level of bodily activity.

Respiratory rate

A respiratory response to an auditory stimulus has been noted subjectively since the reactions to sound were first studied. Specific investigations have confirmed that marked changes occur at both stimulus onset and offset, and with variation of the sound intensity (Poole, Goetzinger & Rousey, 1966; Steinschneider, 1968; Sameroff, 1971; Gerber, 1972). In addition, a knowledge of the respiratory pattern has been found to be of assistance in determining the arousal state of the infant (Wolff, 1966), although it is, in itself, insufficient as an index (Hutt *et al.*, 1969).

Cardiac rate

The cardiac rate is similarly applicable as an indicator of both auditory response and arousal state. Changes seen after the presentation of an acoustic stimulus occur in adult subjects (Lang & Hnatiow, 1962; Meyers & Gullickson, 1967), and with the neonate (Turkewitz *et al.*, 1971). Of particular interest to this study is the finding of Bartoshuk (1964), that an approximately linear relationship exists between the cardiac responses plotted on a logarithmic scale and the stimulus intensity in decibels.

THE DESIGN OF THE AUDITORY RESPONSE CRADLE

The main components

The auditory response cradle system is shown in Fig. 1. It is constructed in three basic units: the cradle, the sound source trolley and the polygraph trolley. Of primary importance are the body and head support systems. These require to be fully adjustable to accommodate the wide variations in size between pre- and postmature newborns. Figure 2 shows the design finally evolved. The neonate lies supine in the bed section which is shaped and foam padded to hold a wrapped baby firmly. The position of the bed section relative to the head rest may be altered to accommodate different neck lengths by means of an internal rack and pinion.

The head rest is made from a flexible foam (ICI Limited) with a density comparable with the latex used for upholstery. Moulding techniques were developed for this material and a series of trials led to the current design. Recesses at the side permit access to the ears and a further cut-out at the rear allows the foam to deform and accept different head sizes. The head rest is fitted to a carriage which is free to rotate through an included angle of 15 degrees

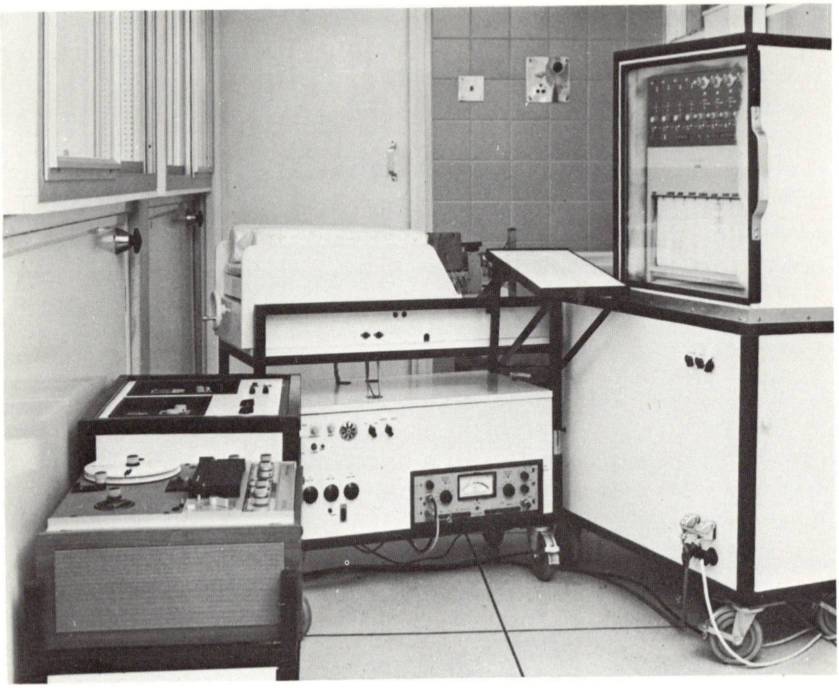

FIG. 1. Overall view of the apparatus: from left to right the sound source, the cradle and the polygraph units.

about an axis in line with the spine. Spring stops are provided at the ends of the free movement: these provide a gradually increasing torque which restrains the head turn and prevents sharp impulses from occurring at the extremities of motion. Two adjustments are provided to allow for variations both in the overall size of the infant and in individual cranial depth.

The head rest also houses the miniature loudspeakers that are close-coupled to the subject's ears by means of waveguides and ear inserts. The loudspeakers were adapted from Medresco hearing aid earpieces. The ear inserts were moulded from a cold cure silastomer rubber to a design obtained in a previous study (Bennett, in preparation). The sound source trolley contains a standard Madsen TBN60 audiometer, a white noise generator (Dawe Instruments, type 419C) and a Ferrograph Series Seven tape recorder from which stimuli, such as voice sounds, may be replayed.

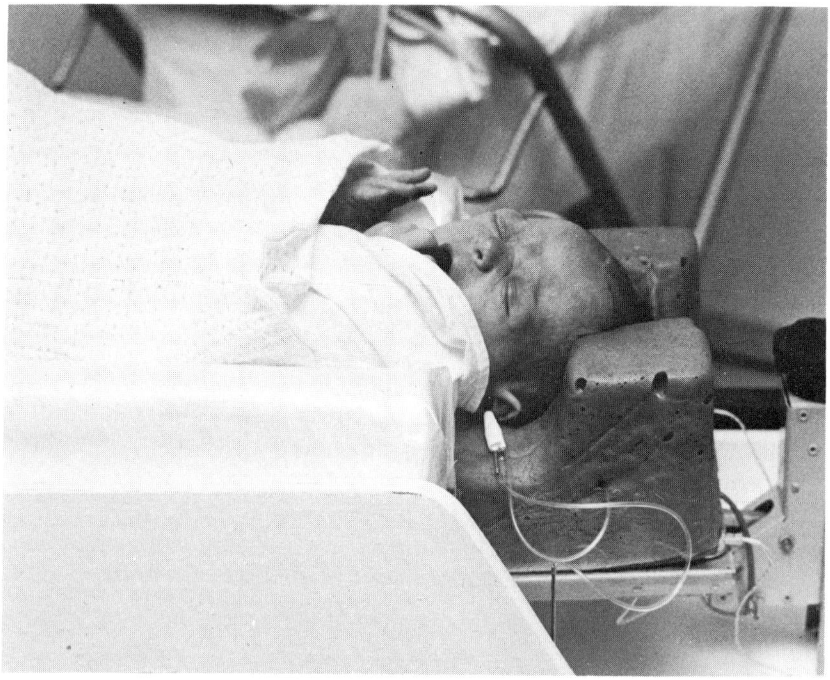

FIG. 2. The head support system showing the ear inserts and waveguides.

The operation of the auditory response cradle is sequenced by means of a master control unit in the main trolley. The sequence is initiated by a manual push button. The polygraph is started for a period of 15 sec to allow for the transducers to be balanced, after which time a prestimulus timer operates. This switches on an indicator light for 5 sec to allow a visual assessment of the arousal state of the child to be made. Simultaneously a record of this interval is made on an additional polygraph channel. As this light is extinguished, the stimulus timer applies the selected sound to the left or the right ear. This operation is again recorded on the polygraph. After the stimulus period, a poststimulus light is illuminated for a second visual state assessment to be made. The recorder is then stopped and the system automatically reset. A manual reset button is also provided should the operator wish to halt the sequence at any time.

The third trolley (Fig. 1) houses the polygraph during transit and contains further signal conditioning equipment. The overall concept is proving acceptable in the hospital environment and is readily transportable in a purpose-fitted estate car.

The instrumentation

When examining the feasibility of any particular form of instrumentation, the primary consideration was for the safety of the subject. To this end, it was decided to reject all forms of transducer that required physical contact with the child. Although this was disadvantageous in ruling out electrophysiological measurement, it had the inherent advantage of obviating the necessity for the preparation of electrode sites.

The choice of recorder was based on the need for accurate temporal analysis which meant that a good transient response was a prerequisite. The Allcoscript eight channel polygraph is used, having a response up to 40 Hz and a rise time of less than 10 msec. The amplifiers chosen require input levels of the order of 500 mV thus requiring relatively high output transducer systems.

Cradle mounted instrumentation

The pivoted arrangement for the head support system enabled the head turn to be recorded directly from a low friction rotary potentiometer which controlled the output from an associated bridge network. A belt drive from the bearing was used in order to obtain an increase in sensitivity from the pulley ratio.

The startle head jerk signal was obtained from a transducer mounted beneath the subject's cranium in the foam of the head support. This consisted of a one-inch-diameter pad of carbon impregnated foam sandwiched between two metal plates. As the foam is compressed, the resistance varies in the range 5000 to 500 ohms. Although the foam is nonlinear in compression, and does not offer repeatability, it is sensitive, cheap and cannot be overloaded.

Overall body movement detection again used the carbon impregnated foam as a sensing medium. This time, thin sheets of the foam were made into a mattress, the top and bottom layers having wire stitched into them to provide electrical contact areas. Changes in compression occurring anywhere on the mattress could thus be recorded after suitable signal conditioning. The variations in resistance were also used to trigger a chain of pulses to operate an apnoea alarm with a variable delay.

Respiratory and cardiac rate measurement

The selection of techniques for respiratory and cardiac rate measurement was somewhat more difficult than for the other parameters. For respiration, the decision not to use contacting methods eliminated electrophysiological methods such as impedance pneumography which had been used previously with success on the neonate (Pallett & Scopes, 1965). Also rejected was the placement of a thermistor adjacent to the nares because of the difficulty of finding a suitable location (Heron & Jacobs, 1968), and because of the possibility of mouth breathing. Trials were carried out with a pressure sensitive system. A small plastic funnel was located by means of a wire support in the region of the face and connected to an electronic micromanometer with a sensitivity of 0·001 mm of water gauge. Even with this apparatus, the pressure variations present during state I were too low to be detected.

Possible use of the information from the cradle mattress was suggested by the apnoea alarm work of Gundersen & Dahlin (1971). Experiments with both direct resistance change and the triggered pulses showed that although regular patterns were present in the traces obtained from the system, these did not relate to the directly observed respiratory excursions. In addition, even slight body movement completely masked the signals for a considerable period.

The methods left open were those requiring a band around the child. Of these, the mercury-in-rubber strain gauge is probably the most common. The transducer, together with the signal conditioning apparatus, is somewhat expensive, and sensitive to general body movement. In consequence, it was decided to develop a similar system more suited to our needs. After several trials, it was found possible to use again the one-inch-diameter, carbon impregnated, foam pad. The transducer was fitted to a band placed around the abdomen of the child over the clothes. A simple low voltage bridge circuit converts the resistance changes and produces clear respiratory signals on the polygraph unmasked by normal movements.

Several methods for cardiac rate measurement were also tried. Eventually it was found that heart beats could be picked up by an accelerometer fitted to a band around the thorax. A piezo-resistive type was determined to be the most suitable from the points of view of both sensitivity and size. The signal from the associated carrier amplifier required high-pass filtering to remove the low frequency

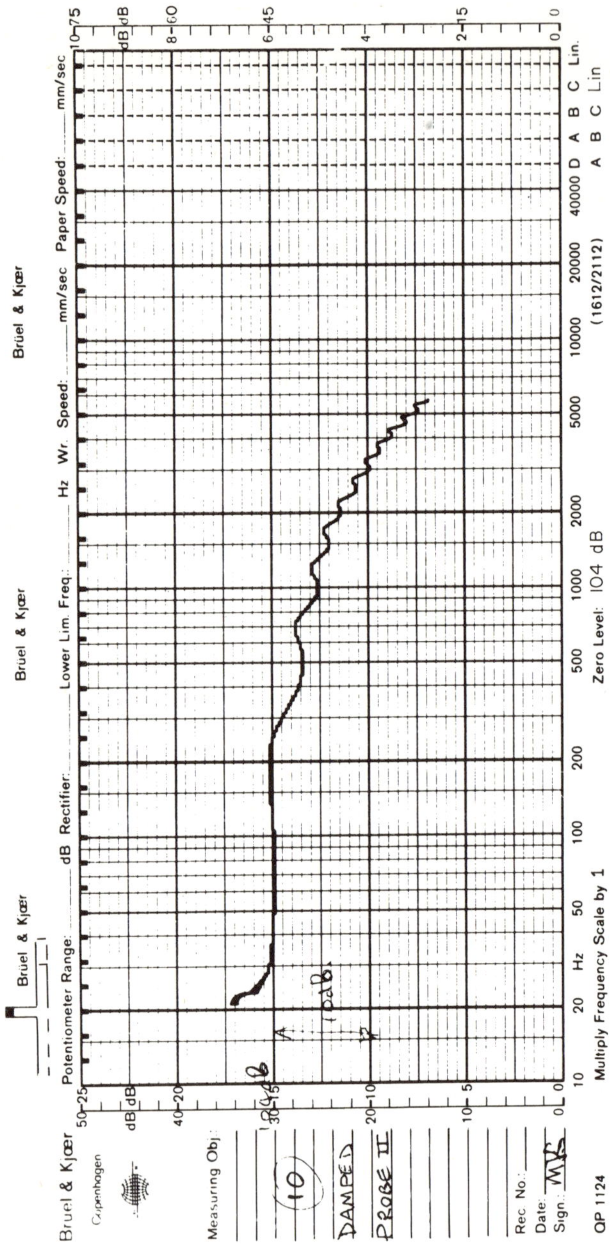

FIG. 3. A probe microphone calibration curve.

respiratory signals. Masking of the signal by body movements resulting from a generalized startle response can occur. To improve upon this, a cardiotachometer is under development that will be sensitive only to the heart beat waveform.

The stimulus level

For any objective testing, it is essential that the stimulus sound pressure level be known in the ear. This is particularly so in this case since, to the author's knowledge, no data exist on the acoustic properties of the ear of the neonate. This means that the conventional calibration of the sound system is not possible as no suitable artificial ear is available. In addition, prior calibration makes no allowance for meatal size, the presence of vernix, or the likely event of a poorly fitting ear insert. In consequence, it was decided to employ a probe microphone technique for direct measurement.

Second waveguides were fitted to each of the ear inserts and connected to individual half-inch condenser microphones (Bruel and Kjaer Ltd., type 4134). The outputs from these are switched to a Bruel and Kjaer type 2707 measuring amplifier. The signal is rectified and a d.c. level produced proportional to the sound pressure level in decibels. This is taken to the polygraph to give a continuous display of the meatal sound pressure level.

The recorded levels require correction for the characteristics of the condenser microphone and probe tube. Calibration curves were determined by placing the open end of the probe tube in a hard wall cavity in which a known, constant, sound pressure level was established. The frequency of the sound was varied from 20 to 5000 Hz and the output from the microphone plotted directly, producing a frequency response curve. Fundamental resonances were damped by adding wire wool to the tubes to act as an acoustic resistance. Figure 3 shows a typical curve.

EVALUATION OF THE AUDITORY RESPONSE CRADLE

The proximity of Brunel University to Hillingdon Hospital has enabled close co-operation to be established with the clinical staff. In particular, a laboratory has been established in the Maternity Unit for evaluation trials on the cradle. Although located in an extremely quiet area, measurements indicated that the ambient sound level required reduction for the purposes of the tests. The provision of double glazing and sound resistant doors reduced the sound level to around 36 dB ("A" scale). No significant

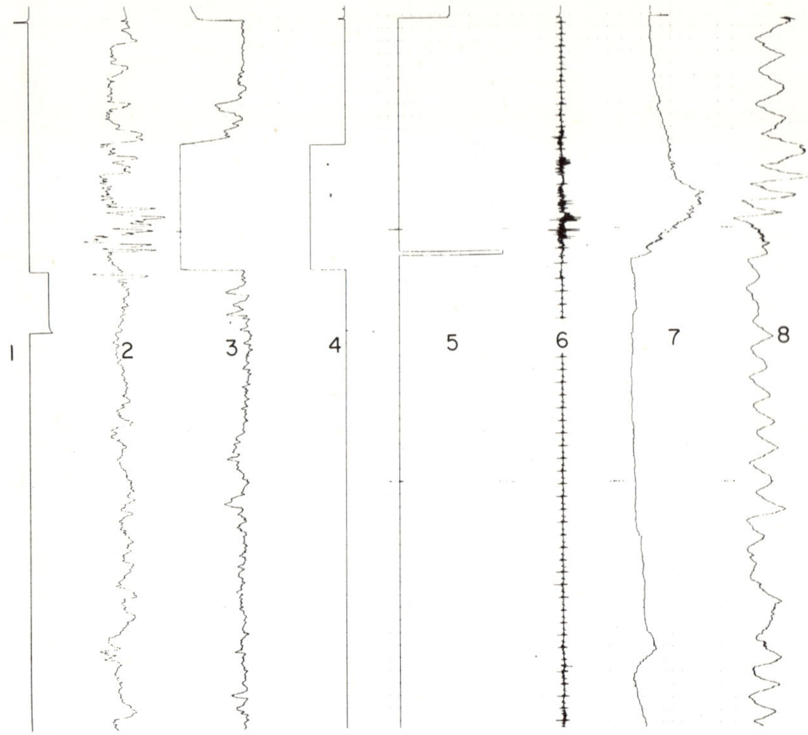

FIG. 4. Specimen test result. Channel 1: prestimulus interval, 2: trunk and limb activity, 3: meatal sound pressure level, 4: stimulus interval, 5: head turn, 6: cardiac beat, 7: startle, 8: respiration.

structure-borne vibrations could be measured by accelerometers fitted to the walls and floor. It was therefore decided that no further modifications to the laboratory were needed.

A pilot study has been carried out to assist the planning of the main programme. This has produced most promising results. Figure 4 shows a typical polygraph chart. The prestimulus period is indicated by channel 1 and the sound-on interval by channel 4, the sound level and rise time being shown by trace 3. The stimulus used in this case was broad spectrum noise at 90 dB sound pressure level. At other times, the level recorded is indicative of the background, physiologically generated, sound pressure in the meatus which is of the order of 50 dB re 0·0002 dynes/cm^2. After the stimulus onset a clear startle reaction is seen on channel 7, the latency being 0·4 sec. The adjacent right hand trace also shows a response in the respiratory pattern: first a pause, then an acceleration before a deceleration to the resting rate. The activity change, seen on the mattress trace on channel 2, somewhat masks the cardiac signals, channel 6. This will be improved after the addition of the cardiotachometer mentioned previously. Channel 5 records the output from the head turn transducer. During this test, the head was hard over against the left hand stop. Additional head pressure occurred in the same direction 0·5 sec after stimulus onset. This activated a secondary switch which gave the pulse seen on the trace.

The relatively loud stimulus used in the test was designed to show overall response. With lower level sounds, particularly pure tones, reactions are frequently much less general. On occasions, detailed temporal analysis of heart and respiratory signals is required to confirm a response. This illustrates the discriminatory nature of the auditory response cradle as, in these cases, no visually observable responses may occur. Additional benefits accrue from the time-linking of all the traces. Spontaneous movements occurring say 0·2 sec prior to a stimulus may be recorded visually as a response; from the chart the lack of a recognized latency period would register a "no test" condition.

The main test programme, involving over 1000 neonates, is about to begin. A data handling facility is being established to enable the results of the tests to be linked to maternal and neonatal data using the techniques of multivariate analysis.

Full evaluation of the auditory response cradle as a means of assessing the auditory state of the infant at the neonatal state can only be carried out by reviewing the test population at a later stage. This is to be achieved by carrying out an epidemiological survey using information from the Local Authority Computer records. Individual cases will be seen as required.

From this work, it is hoped that a method for the routine audiological screening of the newborn will be developed. In particular it is anticipated that a greatly simplified version of the cradle will be evolved which could become standard, low-cost, ward equipment for use by the nursing staff.

Acknowledgements

The author gratefully acknowledges the assistance of the staffs of the Regional Audiology Research Unit, Reading, Berkshire, and Hillingdon Hospital, Uxbridge, Middlesex.

REFERENCES

Ashton, R. (1971). State and the auditory reactivity of the human neonate. *J. exp. Child Psychol.* **12**: 339–346.

Bartoshuk, A. K. (1964). Human neonatal cardiac responses to sound: a power function. *Psychon. Sci.* **1**: 151–152.

Bench, R. J., Collyer, Y., Langford, C. & Toms, R. (1972). A comparison between the neonatal sound evoked startle response and the head-drop (Moro) reflex. *Devl Med. Child Neurol.* **14**: 308–317.

Bench, R. J. & Murphy, K. P. (1970). The Papousek cradle: a device for measuring babies' head movement responses to auditory stimulation. *J. Lar. Otol.* **84**: 521–523.

Bench, R. J. & Smyth, C. N. (1967). The choice of response index in psychophysiological studies of the human neonate. *Digest 7th Int. Conf. Med. & Biol. Eng.*: 244–245.

Bennett, M. J. (in preparation). *Acoustic impedance bridge measurements with the neonate.*

Brackbill, Y. (1970). Acoustic variation and arousal level in infants. *Psychophysiology* **6**: 517–526.

Downs, M. P. & Sterritt, G. M. (1964). Identification audiometry for neonates: a preliminary report. *J. aud. Res.* **4**: 69–80.

Eisenberg, R. B., Griffin, E. J., Coursin, D. B. & Hunter, M. A. (1965). Auditory behavior in the human neonate: a preliminary report. *J. Speech Hear. Res.* **7**: 245–269.

Eisenberg, R. B., Hunter, M. A., Griffin, E. J. & Coursin, D. B. (1965). Auditory behavior in the human neonate: a preliminary report. *Int. Audiol.* **4**: 65–68.

Escalona, S. K. (1962). The study of individual differences and the problem of state. *J. Am. Acad. Child Psychiat.* **1**: 11–37.

Gerber, S. E. (1972). Biomedical technology and the detection of birth defects. *Rehabil. Lit.* **33**: 322–325.

Gundersen, J. & Dahlin, K. (1971). Monitoring of breathing with a segmental air-filled mattress. *Med. Biol. Eng.* **9**: 541–547.

Heron, T. G. & Jacobs, R. (1968). A physiological response of the neonate to auditory stimulation. *Int. Audiol.* **7**: 41–47.

Hutt, S. J., Lenard, H. G. & Prechtl, H. F. R. (1969). Psychophysiological studies in newborn infants. In *Advances in child development and behavior.* **4**: 128–172. Reese, H. W. & Lipsitt, L. P. (eds). New York & London: Academic Press.

Lang, P. H. & Hnatiow, M. (1962). Stimulus repetition and the heart rate response. *J. comp. physiol. Psychol.* **55**: 781–785.

Ling, D. (unpublished). *Response validity in auditory tests of newborn infants.*

Ling, D., Ling, A. H. & Doehring, D. G. (1970). Stimulus, response and observer variables in the auditory screening of newborn infants. *J. Speech Hear. Res.* **13**: 9–18.

Meyers, W. J. & Gullickson, G. R. (1967). The evoked heart rate response: the influence of auditory stimulus repetition, pattern reversal and autonomic arousal level. *Psychophysiol.* **4**: 56–66.

Morgon, A., Charachon, D. & Bringuier, N. (1971). Disorders of the auditory apparatus caused by embryopathy or foetopathy: prophylaxis and treatment. *Acta otolar.* Suppl. No. 291.

Pallet, J. E. & Scopes, J. W. (1965). Recording respiration in newborn babies by measuring impedance of the chest. *Med. Electron. Biol. Eng.* **3**: 161–168.

Papousek, H. (1961). Conditioned head rotation reflexes in infants in the first months of life. *Acta paediat.* **50:** 565–576.

Poole, R., Goetzinger, C. P. & Rousey, C. L. (1966). A study of the effects of auditory stimuli on respiration. *Acta otolar.* **61**: 143–152.

Prechtl, H. F. R. & Beintema, D. J. (1964). The neurological examination of the full-term newborn infant. *Clinics in devl Med.* No. 12.

Rosenblith, J. F. (1970). Are newborn auditory responses prognostic of deafness? *Trans. Am. Acad. Ophthal. Otolar.*, Symp. Conservation of Hearing through Neonatal Testing **74**: 1215–1228.

Sameroff, A. J. (1971). Respiration and sucking as components of the orientating reaction in newborns. *Psychophysiol.* **7**: 213–222.

Siqueland, E. R. & Lipsitt, L. P. (1966). Conditioned head-turning in human newborns. *J. exp. Child Psychol.* **3**: 356–376.

Steinschneider, A. (1968). Sound intensity and respiratory responses in the neonate. *Psychosom. Med.* **30**: 534–541.

Turkewitz, G., Moreau, T., Birch, H. G. & Davis, L. (1971). Relationships among responses in the human newborn: the non-association and non-equivalence among different indicators of responsiveness. *Psychophysiol.* **7**: 233–247.

Wolff, P. H. (1966). The causes, controls and organization of behaviour in the neonate. *Psychol. Issues.* **5** (1, whole No. 17).

Symp. zool. Soc. Lond. (1975) No. 37, 307–315.

OPERANT AUDIOMETRY WITH CHILDREN

I. ST JAMES-ROBERTS

Department of Psychology, University of Liverpool, Liverpool, England

SYNOPSIS

Five different types of audiometric tests are discriminated and the implications of the operant model for the use of some of these are considered. Attention is then confined to tests which employ purposive behavioural response and to the application of operant methodologies to such tests.

Operant audiometric tests are characterized by the use of formal, programmed methodologies, controlled response strategies and reward of appropriate response. These attempt to control extraneous experimenter and test variables and hence to optimize response attainment. Principal differences between tests concern the response and reward techniques employed. These are evaluated and the relative merits of visual, food and token reinforcers and of TROCA, SROCA and play techniques are considered.

INTRODUCTION

The last few years have seen a considerable increase in the number of types of audiometric tests. There are now at least five separate types. Tests (1) to (4) are "objective"; tests (3) to (5) "subjective".

1. Tests which attempt to assess the conductivity of some aspect of the external or middle ear e.g. acoustic impedance tests.
2. Tests which measure the electrical activity of the cochlea, auditory pathway or central nervous system. Stapedial reflex tests may be included in this category, since the index used is presumed to reflect a transduced record of pathway activity.
3. Tests which measure the activity of some visceral organ or structure innervated by the sympathetic branch of the autonomic nervous system e.g. psychogalvanic tests.
4. Tests which depend on an involuntary behavioural response from the testee e.g. startle or reflex orientation audiometry.
5. Tests which require an active voluntary response; the majority of tests which fall into the category of "subjective audiometry" would be included here.

Each type of test has its shortcomings. The first three provide limited information only on aspects which may be necessary, but are certainly not sufficient, for co-ordinated response to sound. For example, even an averaged evoked electroencephalographic response obtained from electrodes positioned over the temporal cortex cannot be taken as direct evidence of capability of co-ordinated response to auditory stimuli in the testee. In addition, "objective" audiometric tests—those which are presumed to be independent of the voluntary response of the testee—are often so conducted as to depend for their results on interpretation of test data by a tester or observer, a deficiency which might be expected to preclude the use of the term "objective".

As the classification above suggests, the distinction between "objective" and "subjective" tests has been further obscured by recent findings that the supposedly "reflex" mechanisms used by tests of types 3 and 4 may be modified by both classical and operant conditioning methods. In fact, the finding that, for example, autonomic response is susceptible to classical (or Pavlovian) conditioning techniques is not a new one. Bordley, Hardy & Richter have used the approach to produce more reliable psychogalvanic response to sound as early as 1948. The method used in this, and similar studies, is to pair auditory and electric shock stimuli prior to testing, with the result that during testing itself the auditory stimuli elicit the marked and habituation-resistant galvanic response originally attributed to the shock stimuli. In practice, however, the test is unreliable and insensitive and not suited to routine use with, for example, child testees.

OPERANT METHODS

Biofeedback and operant research findings

Operant methods for modification of reflex responses are the province of the branch of psychophysiology known as biofeedback. Although biofeedback itself does not have immediate relevance for audiometric testing, the finding that autonomic and other reflex responses are susceptible to voluntary control is an important one. For example, as used in audiometry, the term "objective" presupposes the proven resistance of the response concerned to modification. However, the findings of biofeedback research have implications beyond those of questioning the appropriateness of the term "objective" for tests of types 3 and 4. These implications derive principally from three well established operant research findings: (a) where modification of response can occur, the result may be either enhancement or deterioration of response, depending on situational contingencies; (b) modification need not involve the active purposive participation of the individual concerned, that is, he does not need to "consciously" change his response; (c) previous modification, or learning, may affect present response.

The first of these three points is of most obvious interest to the audiometrician employing tests of type 3 or 4, since the possibility that some test contingency may interfere with test response is important. However, the latter two points, though more subtle, are relevant too. They argue that the tester employing a test which, for example, uses a startle or more especially an orientation as its response, needs to bear in mind not only the tendency of such responses to habituate, but also the previous experience of the testee with respect to response of the sort employed. If the contingencies of such experience have failed to encourage, for example, orientation to sound—which is perhaps likely to be the case with an individual who has a hearing disorder—then it is reasonable to expect the testee to be less likely to orient to sound than if such orientation had previously been rewarding.

It must be pointed out that these considerations affect tests of type 5 as well, and that they are likely to be differentially influential on the various responses used by tests of types 3 and 4. The startle response of the neonate, for example, is presumably less likely to be subject to modification than the orientation response of the seven-year-old. However, the main value of tests of this type has been their presumed ability to involve the complete response repertoire of the testee without the need of his co-operation. If, as seems to be the case, this ability is suspect, then tests such as "conditioned reflex audiometry" are both misnamed and incapable of distinguishing response deficit due to hearing pathology from response deficit due to modified response ability.

Although the discussion so far has been concerned with test situations where operant modification of response may interfere with testing, the term operant audiometry is normally used to refer only to test situations where deliberate attempts are made to modify, and particularly to optimize, a response which is regarded as under testee control. Although tests concerned are those which conform to type 5, it will be clear from the foregoing discussion that some of the tests concerned employ response techniques appropriate to type 4: orientation toward a sound source, for example, is frequently employed. Before turning to these tests, it should be noted that the term "operant audiometry" has been used here to refer to the integration of operant conditioning and audiometric methodologies in preference to the synonyms "instrumental" and "behavioural" audiometry. The latter term has particularly been avoided, since it is in frequent use in the literature in reference to any audiometric test which requires a demonstrative response from its testees. The phrase "behavioural observation audiometry" for example, is currently in popular use to describe the procedures of orthodox clinical pure-tone audiometry for infants and young children.

Characterization of operant audiometric tests

Tests which contain some operant features existed as long ago as 1947 (Dix & Hallpike), although it is only the recent popularity of operant methodologies in general which has given rise to the formalized procedures of operant audiometry proper. Although considerable differences exist between tests, they may be said to be typified by their adherence to the following general principles.

(a) It is possible to employ systematic manipulation of test contingencies to optimize testee response.
(b) Test instructions and procedures employ a carefully programmed series of stages, each with its own performance criterion. Successful performance at each stage has to be achieved before the next may be attempted. In this way, gradual and controlled incrementation of response is possible, a procedure which allows a complex terminal level of performance to be achieved with minimum error and hence minimal frustration in the testee. The method is known as reinforcement of successive approximations. It has the additional advantage of allowing each testee to proceed at his own pace while his success or failure at each stage of the test provides useful general information on his capabilities.
(c) A reward, or reinforcer, is supplied contingent upon appropriate response. In audiometry, a reinforcer is normally provided for every appropriate response, although in other areas a variable ratio of reward to correct response may sometimes be used, particularly where many responses are forthcoming.
(d) A simple and formalized response strategy is used, with accompanying response criteria. This, in juxtaposition with the formalized methodology, enables the control of experimenter variables and hence precludes both testee response to cues unintentionally provided by the experimenter and possible misinterpretation of the testee's behaviour.

Since the use of a programmed methodology and reinforcement of response are intended to offset both the novel and possibly alarming nature of the audiometric test situation and the potentially sterile and boring nature of the testing itself, it is perhaps not surprising that operant audiometry has achieved notable successes with difficult-to-test children. Lloyd (1965) and Bricker & Bricker (1969), for example, have used the method successfully with retardates, Meyerson & Michael (1960) with "exceptional children" and St James-Roberts (1974) with children characterized by "inconsistent auditory response". The majority of such studies have employed pure-tone auditory stimuli and descending method-of-limits threshold tracking techniques. Suprathreshold stimuli are used initially to establish response, the intensity of the stimuli then being successively reduced. Provided that a proportion of suprathreshold stimuli is maintained throughout testing, the effect of this method is to produce intermittent reinforcement, since some false positive response (response when no stimulus is presented) and, particularly, false negative response (no response when a stimulus is presented) will occur and will be unrewarded. Intermittent reinforcement has been found to be most effective in maintaining response (Skinner, 1969). In use, most authors have reported this method to be efficient in eliciting and sustaining interest in test stimuli (Lloyd, 1966).

Differences between operant audiometric tests

Differences between the various operant audiometric tests available stem principally from the different response and reward techniques employed. Providing the response used is simple enough to assure the testee's ability to give it and providing it is readily quantifiable, the exact nature of the response used is not of great significance. The most frequently used are button press and lever pull, although "play audiometry", for example, often uses a specific manipulation of a toy for its response. With responses of the latter kind, considerable care must be taken to ensure that a predetermined criterion is selected to define response, and that the criterion is adhered to throughout testing. Otherwise there is a danger, particularly where a relatively ill-defined response such as an orientation toward the sound source is used, that random movement will be registered as response.

Selection of an effective reward is most important for operant audiometry, as for all techniques which employ operant methodologies, since achievement of reward constitutes the primary test incentive. This paper will conclude by examining the advantages and disadvantages of four of the techniques most commonly used: social reinforcers, token reinforcers, food reinforcers and visual reinforcers.

Social reinforcement operant conditioning audiometry (SROCA)

Social reinforcement is probably the most commonly used of the reinforcer types, particularly in routine assessment audiometry. Normally, such reinforcement is simply the applause or approval provided by the tester for appropriate response, although post-test applause may also be given (Frisina, 1963; O'Neill, Oyer & Hillis, 1961). The approval is presumed both to inform the testee of the correctness of his response (knowledge of results) and to motivate response toward acquisition of approval.

Although social reinforcement has the obvious advantage of being easy to use, it has severe limitations. In particular, it is difficult to formalize approval so that it is provided only immediately after correct response—the testee will often seek approval before making a response he believes to be correct, for instance, with the result that confusion may occur as to what is being approved. Where a relatively informal response is used, as in play audiometry (Barr, 1955; Lowell, Rushford, Hoverstein & Stoner, 1956) consisting of inserting a peg in a hole or dropping a block into a box, the tester may often find that he has reinforced a response quite unconnected with the response-to-sound he had in mind. In addition, approval is not reinforcing for all children, while even those for whom it is rewarding will often lose interest if extended or difficult testing is required.

Token reinforcers

These, together with food reinforcers, are often known as "tangible reinforcers", the term tangible reinforcement operant conditioning audiometry (TROCA) having been coined by Lloyd (1966) and others to suit tests of this type.

Token reinforcers, as the name implies, constitute a sort of currency which may later be exchanged for food, toys, visits, or a range of other activities. Once their value has been acquired they are very effective, particularly with older children. Their disadvantage is that the association between the tokens and the rewards has to be learned, normally necessitating pretraining, while post-test sessions, during which the tokens are exchanged, occupy further time. In addition, the separation of reception-of-token from exchange-of-token-for-reward poses problems where young or impatient children are concerned.

Food reinforcers

The ubiquitous "Smartie" is the food reinforcer most frequently used, although other edible items, such as popcorn, biscuits and so on, have also been employed (Meyerson & Michael, 1960; Spradlin & Lloyd, 1965). Small sweets, in particular, have the advantage that they can be awarded using automated dispensers, which help to avoid extraneous social reinforcers inadvertently provided by the experimenter.

Food reinforcers are commonly used in animal work, where the reinforcing value of the food is normally enhanced by a period of deprivation prior to testing. With children, such deprivation is neither desirable nor possible, with the result that even the greediest of individuals can quickly become satiated with an excess of sweets, with resulting decline in performance. Moreover, the consumption of food and the mess which inevitably results tend to interfere with or prevent test response.

Visual reinforcers

A wide variety of visual reinforcers has been used, including pictures and slides (Dix & Hallpike, 1947; Miller, 1963; Shimizu & Nakamura, 1957), animated animals and puppets (Green, 1958; Sullivan, Miller & Polisor, 1962), toy trains (Ewing, 1930; Ishisawa, 1960) and mechanical toys (MacPherson, 1960). As with food reinforcers, completely automated presentation of visual reinforcers is often possible, producing a more controlled test situation.

The apparent attractiveness of visual reinforcers has been qualified by their tendency to satiate, that is, they quickly become boring and repetitive and, as a result, the co-operation of the testee is lost. Development of visual reinforcers which change or progress in some way which reflects the testee's performance may overcome this deficiency.

Combined, and positive versus negative, reinforcers

Although the various reinforcer types in contemporary use have been described separately, in practice two or more reinforcers are often used simultaneously. Social reinforcers, particularly, are often used to supplement food or visual ones. The use of negative reinforcement (punishment) for incorrect response has also been investigated with the aim of reducing false positive and negative response. However, in the absence of negative reinforcers which are strongly aversive, for example the electric shock stimuli often used with animals, it is frequently found that a bored testee will try to introduce some variety into the test situation by acquiring negative reinforcers, a strategy which tends to conflict with test requirements. In consequence, tests in use in assessment centres in this country and North America depend almost exclusively on positive reinforcement techniques.

References

Barr, B. (1955). Pure-tone audiometry for pre-school children. *Acta otolar.* Suppl. **121.**

Bordley, J. E., Hardy, W. G. & Richter, C. P. (1948). Audiometry with the use of galvanic skin resistance response: a preliminary report. *Johns Hopk. Hosp. Bull.* **82**: 569.

Bricker, W. A. & Bricker, D. D. (1969). Four operant procedures for establishing auditory stimulus control with low-functioning children. *Am. J. ment. Defic.* **73**: 981–987.

Dix, M. R. & Hallpike, C. (1947). The peep-show: a new technique for pure-tone audiometry in young children. *Br. med. J.* **1947(2)**: 719–723.

Ewing, A. W. G. (1930). *Aphasia in children.* London: Oxford Medical Publications.

Frisina, D. R. (1963). Measurement of hearing in children. In *Modern developments in audiology*: 155–174. Jerger, J. (ed.). New York and London: Academic Press.

Green, D. S. (1958). The pup-show: a simple, inexpensive modification of the peep-show. *J. Speech Hear. Disord.* **23**: 118–120.

Ishisawa, A. (1960). A study on play audiometry (Japanese text). *Otol. Fukuoka* **6** (suppl. 7): 397–415. From *dsh.Abst.* (1962). **2**: 201.

Lloyd, L. L. (1965). Use of the slide-show audiometric technique with mentally retarded children. *Except. Child.* **32**: 93–98.

Lloyd, L. L. (1966). Behavioural audiometry viewed as an operant procedure. *J. Speech Hear. Disord.* **31**: 128–136.

Lowell, E., Rushford, G., Hoverstein, G. & Stoner, M. (1956). Evaluation of pure-tone audiometry with pre-school age children. *J. Speech Hear. Disord.* **21**: 292–302.

MacPherson, J. B. (1960). *The evaluation and development of techniques for testing the auditory acuity of trainable mentally retarded children.* Unpublished Doct. Dissertion: University of Texas.

Meyerson, L. & Michael, J. L. (1960). The measurement of sensory thresholds in exceptional children: an experimental approach to some problems of differential diagnosis and education with special reference to hearing. *U.S. Office of Ed. Cooperative Research Project* No. 418. University of Houston.

Miller, A. L. (1963). The use of slide projectors in pure-tone audiometric testing. *J. Speech Hear. Disord.* **28**: 94–96.

O'Neill, J., Oyer, H. & Hillis, J. (1961). Audiometric procedures used with children. *J. Speech Hear. Disord.* **26**: 61–66.

Shimizu, H. & Nakamura, F. (1957). Pure-tone audiometry in children: lantern-slides test. *Ann. Otol. Rhinol. Lar.* **66**: 292–398.

Skinner, B. F. (1969). *Contingencies of reinforcement*. New York: Appleton-Century-Crofts.

Spradlin, J. E. & Lloyd, L. L. (1965). Operant conditioning audiometry (OCA) with low-level retardates: a preliminary report. In *The audiologic assessment of the mentally retarded: The proceedings of a national conference*: 45–58. Lloyd, L. L. & Frisina, D. R. (eds). Parsons, Kansas: Speech and Hearing Department.

St James-Roberts, I. (1974). An apparatus employing visual and tactile pre-training for use in behavioural audiometry with difficult-to-test children *Brit. J. Audiol.* **8**: 55–60.

Sullivan, R., Miller, M. H. & Polisor, I. A. (1962). The portable peep-show: a further modification of the peep-show. *Arch. Otolar.* **76**: 49–51.

Symp. zool. Soc. Lond. (1975) No. 37, 317–330.

SOUND COMMUNICATION IN RODENTS

JANE C. SMITH

Department of Zoology and Comparative Physiology,
Queen Mary College, London, England

SYNOPSIS

Infant rodents of all the murid and cricetid species that have been studied produce "distress" calls when displaced from the nest. Until recently evidence for the communication function of these sounds was largely circumstantial, but in 1970 (b) Sewell demonstrated that lactating *Apodemus* would leave their own litter in the nest and come to a loudspeaker relaying the prerecorded ultrasounds of a day 5 litter.

A pilot study has now confirmed this phenomenon for *Apodemus* and extended it to seven other myomorph species. In general it was found that virgin females and males showed a lower level of responsiveness than lactating females but this lower level could be enhanced by experience with young (priming). In most species, responses were also elicited by sounds recorded from the young of other species. Responses declined with repeated testing and, for lactating females, with time postpartum.

Studies of the distress calls of infant *Peromyscus maniculatus* revealed that two distinct pulse types, "audible" and ultrasonic, are produced. The occurrence and intensity of these two types appeared to be independently related to cold stress and tactile stimulation. Detailed sound replay experiments on adult *P. maniculatus* showed that both of these pulse types elicited phonotaxis in lactating females.

A signal generator was used to model the sounds of young *P. maniculatus* and mice. Lactating *P. maniculatus* responded more often to tape-recorded model "audible" pulses than to tape-recorded real pulses, but the reverse was true for ultrasounds. Experiments on mice investigated the effects of changes in pulse length, pulse pattern and pulse frequency on the responses of primed females.

INTRODUCTION

Infant myomorph rodents of all the species that have been studied from two families, the Muridae (for example, mice and rats) and the Cricetidae (for example, hamsters and deermice), produce distress calls when removed or scattered from the nest. The "distress" calls of both murid and cricetid species are purely ultrasonic, but some cricetid species also produce calls with audible components. All species produce audible pain squeaks. Studies of these distress calls have mainly been concerned with three aspects: their structure, the motivation for their production and their possible communication function.

Sewell (1969, 1970a) made the first detailed study of the structure of the ultrasounds of infants of several species and recent additions to this field have been published by Brooks & Banks (1973) on the ultrasounds of infant *Dicrostonyx groenlandicus*, the collared lemming, and Colvin (1973) on the ultrasounds of infants from several species of *Microtus*, short-tailed voles. Eisenberg (1968) and Hart & King (1966) have studied the structure of the partly audible calls of infant *Peromyscus*, deermice. In general it seems that the calls of neonatal rodents are longer (usually 50–200 msec), lower in frequency and show a more simple and stereotyped structure than those of older pups. The possible functional importance of this will be mentioned later. This whole subject has recently been reviewed by Sales & Pye (1974).

It was suggested by Zippelius & Schleidt (1956), who first detected ultrasonic distress calls from infant rodents, that these sounds are produced in response to cold stress and tactile stimulation, and this has since been confirmed by other workers. Okon (1970a,b, 1971a,b, 1972) has shown that cold stress and tactile stimulation affect calling independently in five rodent species.

It has commonly been supposed by all those who have studied them, that infant distress calls serve to initiate and/or facilitate maternal behaviour, especially retrieving. Noirot (1964, 1969a, 1970) produced evidence that ultrasounds may affect retrieving and also nest-building in the mouse, and she has also suggested that ultrasounds may reduce maternal aggression in this species (Noirot, 1966, 1969b,c). Evidence for these communication functions was circumstantial, however, until Sewell (1970b) showed that lactating *Apodemus sylvaticus*, woodmice, were attracted to a loudspeaker relaying the prerecorded sounds of young. This indicated that such sounds do have communication value to the females and probably guide a mother to stray young. A similar technique has been used in the present study, the aim of which was to investigate further the communication function of infant rodent distress calls. Since this work began, Allin & Banks (1972) have shown that albino rats will also respond to replayed infant calls, but the response that they obtained was far less marked than that of *A. sylvaticus*.

PRELIMINARY EXPERIMENTS

The test enclosure used to study rodent responses to sound was a small mouse cage with one end divided into two chambers which the animal could enter from the rest of the cage (Fig. 1). A loudspeaker was placed at each of two holes in the side walls opening into these chambers. Signals were relayed into either chamber but only when the animal had been on the nest for at least two minutes. An emergence was recorded if the animal's nose

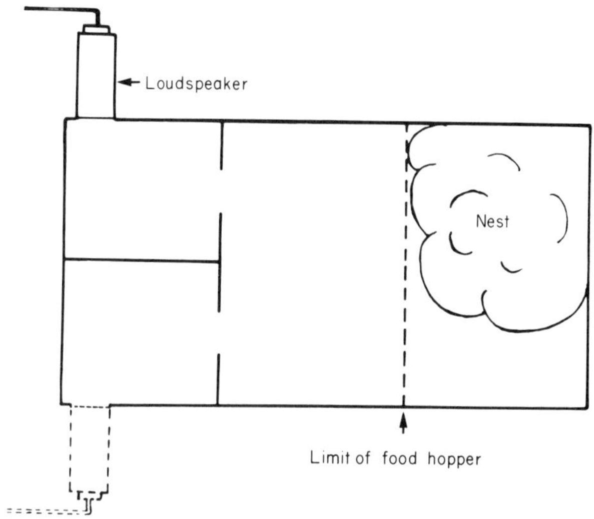

FIG. 1. Plan view of the experimental cage.

protruded from under the food hopper, where the animal nests, otherwise a null response was recorded. A correct response was recorded if the animal entered the chamber into which the sound was played, an incorrect response was recorded if the animal entered the other chamber. Signal presentations lasted for two or for five minutes and careful attention was paid to the replay levels.

A superficial survey tested the responses of animals from eight myomorph species to replayed distress calls of conspecific young. All species showed some responsiveness to sound but *Clethrionomys glareolus* (the bank vole), *Calomys callosus* and *Peromyscus maniculatus* were the most responsive. Albino rats were the least responsive although at least one response was given by most lactating female rats. The findings of this survey were consistent with the difference, between the results of Sewell (1970b) and of Allin & Banks (1972), that was mentioned earlier. Further details will be published elsewhere.

In all species, lactating females were more responsive than other females and males, but the responsiveness of these non-lactating animals could be enhanced by experience with young. The most responsive species also responded to the calls of the newborn young of other species. Responses declined with repeated testing and, for lactating females, with time postpartum.

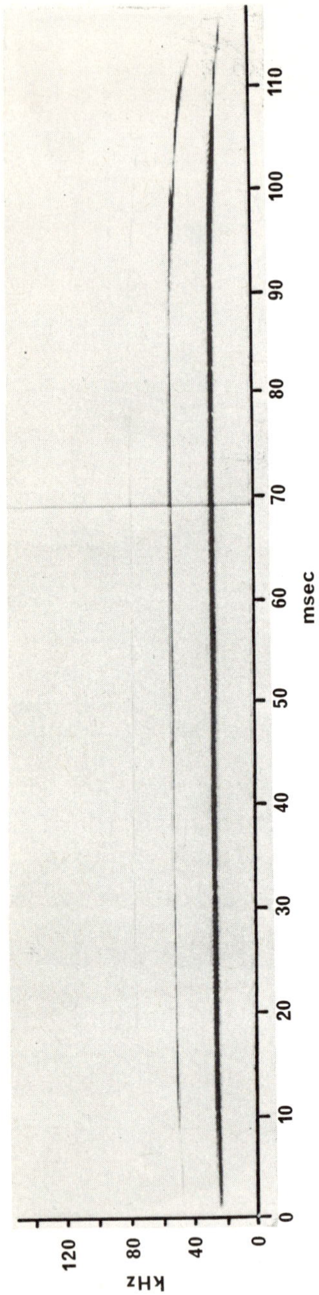

FIG. 2. Sonagram of an "audible" pulse produced by an isolated day 5 *P. maniculatus* pup.

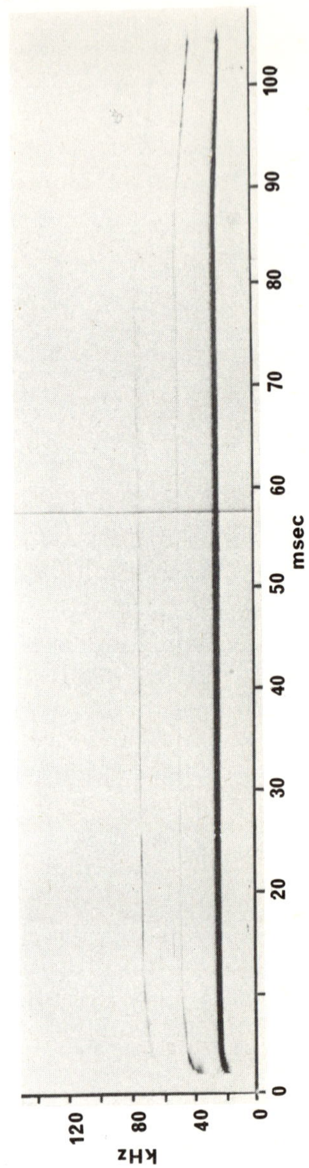

FIG. 3. Sonagram of the model "audible" pulse used in sound replay experiments.

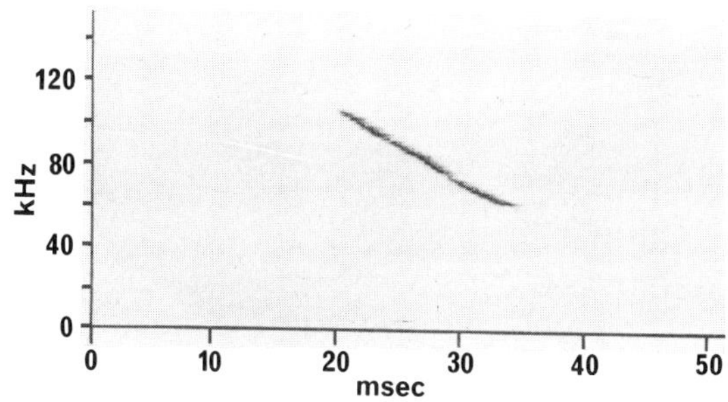

FIG. 4. Sonagram of a rapid downsweep ultrasonic pulse, produced by a day 10 *P. maniculatus* pup.

THE SOUNDS OF INFANT *P. MANICULATUS*

In parallel with the studies of adult responses, it was necessary to collect some background information on the sounds produced by the young of some of the species that were being studied, where such information was incomplete or non-existent.

For *P. maniculatus* it was found that two completely different sounds were produced by the young (Smith, 1972, 1974). These have been termed "audible" and ultrasonic. "Audible" pulses (e.g. Fig. 2) consist of a series of harmonic components with a fundamental component, which is usually also the strongest component, in the range 8–29 kHz. They are usually 50–200 msec long and are almost certainly the same sounds as those described by earlier workers (e.g. Hart & King, 1966). Ultrasonic pulses, which had not been previously described, are single component sounds within the frequency range 60–140 kHz and are 3–45 msec in duration. They

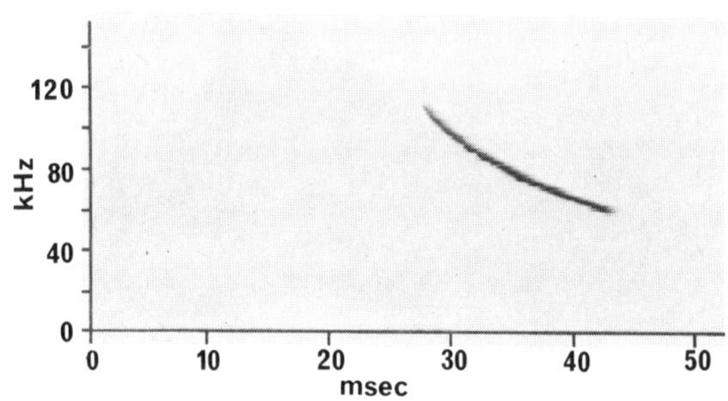

FIG. 5. Sonagram of the model ultrasonic pulse used in sound replay experiments.

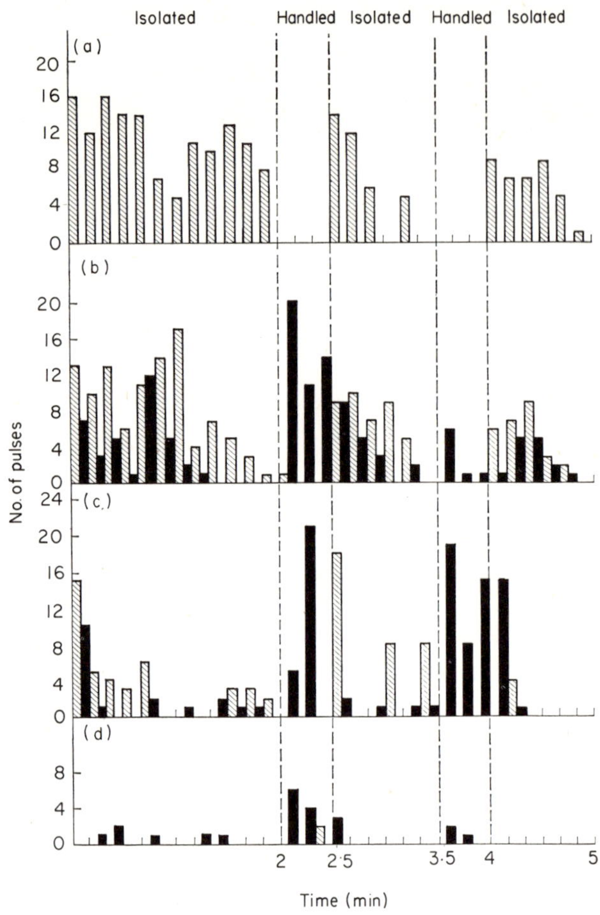

FIG. 6. "Audible" and ultrasonic calling by *P. maniculatus* pups subjected to alternating conditions of isolation and handling. The number of pulses of each day emitted in each 10 sec period is plotted for pups of different ages: (a) day 4; (b) day 7; (c) day 14; (d) day 17. Solid areas, ultrasonic pulses; hatched areas, "audible" pulses. From Smith (1972).

are much more variable in structure than the "audible" pulses; patterns include frequency sweeps, usually downwards but sometimes upwards, or occasionally fluctuating changes of frequency. An example of the simplest pattern that was found, the rapid downsweep, is shown in Fig. 4.

The acoustic structure, the sound pressure level (SPL) and the rate of production of these two types of pulse were studied with respect to age, daily from birth until day 21 (about the time at which the pups are weaned), and also with respect to the conditions of cold stress and tactile stimulation. Cold stress involved isolated exposure at room temperature and tactile stimulation involved handling (rolling gently between finger and thumb) at room temperature.

Typical vocal responses of single pups to alternate isolation and handling over five minutes are shown in Fig. 6. It can be seen that more "audible" calls were produced by young pups than by older ones and that the rate of "audible" calling during isolation was much higher than that during handling. Ultrasonic pulses were only detected from pups older than day 4 and the rate of their production was increased during handling. The effect of handling on the production of each pulse type was largely independent of the effect on the other pulse type. For example, day 4 pups showed cessation of "audible" calling on handling, even though no ultrasounds were detected (see Fig. 6); day 17 pups showed increased

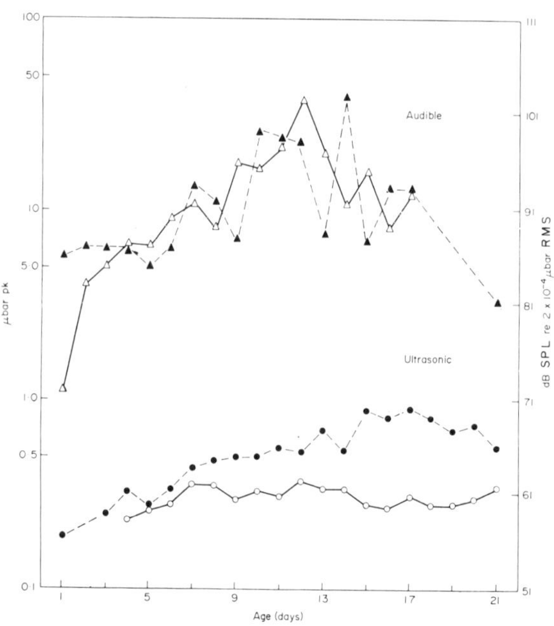

FIG. 7. The average SPL reached by the majority of "audible" and ultrasonic pulses produced by groups of ten *P. maniculatus* pups, subjected either to handling or isolation, against age, in days. ▲, "audible" pulses after handling; △, "audible" pulses, isolation; ●, ultrasonic pulses during handling; ○, ultrasonic pulses, isolation. From Smith (1972).

ultrasound production on handling, although at this age few "audible" pulses were detected.

The SPL measurements obtained for the "audible" pulses and for the ultrasonic pulses produced by ten pups under separate conditions of isolation and handling are shown in Fig. 7. The SPL of "audible" pulses was measured before and just after handling, because, as has been mentioned, few were produced actually during handling. Ultrasonic pulses were measured before and during handling. It was found that "audible" pulses were considerably louder than ultrasonic pulses (Fig. 7) and that they were of similar SPL before and after handling. The SPL of ultrasonic pulses was considerably greater during handling than during isolation and showed an increase with age.

The SPL and occurrence of the two types of pulse produced by infant *P. maniculatus* were therefore found to be independently related to age and motivating conditions. It was considered possible that these two types might have different communication functions: "audible" pulses, which seemed to be typical of young pups subjected to cold stress, might promote maternal behaviour, especially retrieving; and ultrasonic pulses, which appeared to be typical of older pups when handled, might reduce maternal aggression. These functions are similar to those that Noirot suggested for the less varied and purely ultrasonic mouse sounds (see p. 318).

SOUND REPLAY EXPERIMENTS ON *P. MANICULATUS*

Experiments using "real" sounds

Responses of lactating *P. maniculatus* to the two types of pulse produced by the young were tested in order to investigate the ideas that have just been outlined. Tape-recordings of the "audible" cries of a day 5 pup and of the ultrasonic pulses of a day 10 pup were relayed to lactating females during different periods postpartum. Their responses were scored by the methods described on pp. 318–319.

Nine females were tested daily with three presentations of each experimental stimulus and three presentations of the control stimulus (the background noise of a tape-recording), in random order. The experimental signals were relayed at the same SPL and signal-to-noise ratio (SNR), and the control stimulus was relayed at a similar level to the background noise on the experimental recordings. Three "group A" females were tested on days 3–10 postpartum, three "group B" on days 7–14 and three "group C" on days 10–17. All three groups of females showed more responses to experimental than to control stimuli. Group C females showed a greater number of responses to ultrasonic pulses than to "audible" pulses but there was no difference between the responses made to the two pulse types by the other groups. Mothers tested over days 3–10 postpartum responded to a large proportion of the experimental pulse presentations, but fewer responses were made by the other groups. All the animals showed a high incidence (95–100%) of correct responses indicating that they could localize the sound source. These results suggest that both types of pulse were attractive to the females. The fact that group C females were more attracted by ultrasonic pulses than by "audible" calls, and that the period over which they were tested is the period of maximum ultrasonic calling by the young, might indicate some coincidence between maternal responsiveness and infant vocal behaviour in this species. No attempt has yet been made to assess whether ultrasounds reduce maternal aggression in *Peromyscus*.

Experiments using synthesized sounds

The pilot survey (pp. 318–319) had indicated that the responses of many adult rodents to replayed sounds were rather non-specific, so it seemed interesting to try to ascertain which features of the sounds were attractive to the adults. For these experiments a signal generator was used. This instrument had been designed by Pye (1966, 1968) for the production of model bat-pulses. The use of this instrument allowed greater flexibility in the elucidation of the importance of various signal parameters than the use of replayed tape-recordings. This was because the signal parameters could be varied independently, without the cooperation of the original emitter. Although filtering and editing of tape-recorded signals would have been possible, this would have been time-consuming and restrictive.

To test the feasibility of the use of a signal generator to determine the factors important in eliciting phonotaxis, lactating female *P. maniculatus* were tested on days 5–12 postpartum with tape-recordings of real and model sounds of either "audible" or ultrasonic pulse types. Sound spectrograms of the model "audible" and ultrasonic pulses are shown in Figs 3 and 5 (compare with Figs 2 and 4). Only the rapid downsweep type of ultrasonic pulse could be modelled by the generator. Trains of model pulses were unlike trains of real ones in that each pulse was completely identical with the last; the difference was especially marked for the ultrasonic pulses which, as has been mentioned, were very varied in structure.

The results showed that lactating *P. maniculatus* appeared to find the model "audible" pulses more attractive than their real counterparts but that the reverse was true of ultrasonic pulses. It is thought that the stereotyped structure of the model ultrasonic pulses might have been a cause of the relatively lower responsiveness of the females. It may be possible to test this with a new and more flexible generator which has just become available. This stereotypy of structure may, however, be an attractive feature of the sounds of neonatal rodents ("audible" pulses in this species), hence the relatively greater responsiveness of the females towards the models. This idea is supported by the fact, mentioned earlier, that it appears that newborn rodents of all species typically produce fairly long sounds showing a more simple and stereotyped structure than those of older pups. It seems possible that these interspecific similarities may represent important functional features and their presence may explain the lack of specificity of the responses of adults to replayed sounds that was mentioned on pp. 318–319.

EXPERIMENTS WITH MICE

Work has begun on a series of experiments to investigate the importance of some of the parameters of infant mouse sounds. Mice were adopted as subjects because the sounds of the young are easier to model and because they have been more extensively studied than those of *P. maniculatus*. The subjects were virgin female mice which had previously been given experience with young, because earlier experiments (Smith, 1974) had shown that these females responded to replayed sounds as strongly as did lactating females, and the responses of the latter were complicated by possible effects of changing stimuli from the developing litter and postpartum hormonal changes. The signal generator was used to produce a stylized mouse pulse (reference pulse) which was used as a test stimulus. This consisted of a 65 msec constant frequency portion at 65 kHz followed by a downsweep to 45 kHz in 15 msec. Responses to the reference pulse were compared with responses to pulses showing certain modifications of this structure. The experiments are summarized in Table I. In the first experiments on albino mice, pulses of frequency 65–45 kHz were more attractive than "identical" pulses of lower or higher frequency. Sixty-five kHz

TABLE I

Summary of experiments on mice

Animals	Experiment	Pulse 1	Duration (msec)	Frequency (kHz)	Pattern	Pulse 2	Duration (msec)	Frequency (kHz)	Pattern	Preferred pulse*
TO Swiss	A	Reference	80	65–45		A	80	55–35		Reference $P = 0{\cdot}02$
TO Swiss	B	Reference	80	65–45		B	80	75–35		Reference $P > 0{\cdot}05$
TO Swiss	C	C1	80	65		C2	15	65		C1 $P = 0{\cdot}02$
TO Swiss	D	Reference	80	65–45		D	80	65–45		Neither $P \gg 0{\cdot}05$
House mice	E	Reference	80	65–45		A	80	55–35		Reference $P < 0{\cdot}02$
House mice	F	Reference	80	65–45		F	15	65–45		Reference $P < 0{\cdot}05$

* *P* values are for correct responses using the Wilcoxon matched pairs signed-ranks test.

is the middle of the range of the sounds produced by newborn mice. Long (80 msec) pulses of constant frequency at 65 kHz appeared to be more attractive than similar short (15 msec) pulses. There was no significant difference ($P > 0{\cdot}05$) between the responses to the reference pulse and to a slow downsweep pulse of identical length and frequency range.

Albino mice were, however, much less responsive to test signals than were *P. maniculatus* and it was thought that this might be due to sensory or behavioural abnormalities as a result of domestication. To test this, house mice were then studied in a similar experiment, with very similar results (compare experiment E with experiment A in Table I). Further evidence that long pulses are more attractive than shorter ones was provided by experiment F (see Table I).

CONCLUSION

These experiments confirm and extend previous findings on the communication function of infant rodent sounds using tape-recorder playback. It has also been shown that *Peromyscus* and mice respond to synthesized models of conspecific infant sounds. The apparent lack of specificity which was found in the responses of adult rodents to replayed tape-recordings of young, suggests that common structural features of such sounds might be important. As has been mentioned, a new and more flexible signal generator has been developed for use in future research. In particular it is hoped to investigate the responses of rodents in larger arenas than were used in the experiments just described, and also to study the responses of groups of animals in established colonies rather than isolated individuals. Such experiments should indicate more accurately how important infant calls are in the natural situation.

SUMMARY

Adult rodents from eight myomorph species showed phonotaxis to replayed tape-recordings of conspecific infant sounds. Lactating females were the most responsive and experience with young enhanced the otherwise lower responsiveness of non-maternal animals. Responses were not entirely species specific, they declined with repeated testing and, for lactating females, with time postpartum.

Infant *P. maniculatus* were found to produce both "audible" and ultrasonic calls, the occurrence and intensity of which could be independently related to age and motivating conditions. Both types of call elicited phonotaxis in lactating females.

Synthesized models of infant calls were tested on *P. maniculatus* and mice. Model "audible" calls were strong stimuli for phonotaxis in lactating *P. maniculatus*, whereas model ultrasonic calls were relatively weak stimuli. Primed female mice appeared to be more attracted by long calls than short ones, and calls within the range 65–45 kHz were more attractive than two models of different frequency but with otherwise identical structure.

Acknowledgements

This work was carried out at King's College, London, during the tenure of an SRC Research Studentship and the Harold Row Studentship. I am grateful for thc laboratory facilities provided by Professor D. R. Arthur and for the supervision of Professor J. D. Pye. Figure 1 is from a drawing kindly provided by Dr G. D. Sales. Figures 6 and 7 appear by courtesy of The Zoological Society of London.

References

Allin, J. T. & Banks, E. M. (1972). Functional aspects of ultrasound production by infant albino rats (*Rattus norvegicus*). *Anim. Behav.* **20**: 175–185.

Brooks, R. J. & Banks, E. M. (1973). Behavioural biology of the collared lemming (*Dicrostonyx groenlandicus* Traill): an analysis of acoustic communication. *Anim. Behav. Monogr.* **6**: Part 1.

Colvin, M. A. (1973). Analysis of acoustic structure and function in ultrasounds of neonatal *Microtus*. *Behaviour* **44**: 234–263.

Eisenberg, J. F. (1968). Behaviour patterns. In *Biology of* Peromyscus (*Rodentia*): 451–495. King, J. A. (ed.). *Spec. Publs.* **2**, *Am. Soc. Mammal.*

Hart, F. M. & King, J. A. (1966). Distress vocalisations of young in two sub-species of *Peromyscus maniculatus*. *J. Mammal.* **47**: 287–293.

Noirot, E. (1964). Changes in the responsiveness to young in the adult mouse: II The effect of external stimuli. *J. comp. physiol. Psychol.* **57**: 97–99.

Noirot, E. (1966). Ultrasons et comportements maternels chez les petits rongeurs. *Annls Soc. R. zool. Belg.* **95**: 47–56.

Noirot, E. (1969a). Changes in responsiveness to young in the adult mouse: V Priming. *Anim. Behav.* **17**: 542–546.

Noirot, E. (1969b). Serial order of maternal responses in mice. *Anim. Behav.* **17**: 547–550.

Noirot, E. (1969c). Interactions between reproductive and territorial behaviour in female mice. *Int. mental health res. newsl.* **11**: 10–11.

Noirot, E. (1970). Selective priming of maternal responses by auditory and olfactory cues from mouse pups. *Devl Psychobiol.* **2**: 273–276.

Okon, E. E. (1970a). The effect of environmental temperature on the production of ultrasounds by isolated non-handled albino mouse pups. *J. Zool., Lond.* **162**: 71–83.

Okon, E. E. (1970b). The ultrasonic responses of albino mouse pups to tactile stimuli. *J. Zool., Lond.* **162**: 485–492.

Okon, E. E. (1971a). The temperature relations of vocalization in infant golden hamsters and Wistar rats. *J. Zool., Lond.* **164**: 227–237.

Okon, E. E. (1971b). *Motivation for the production of ultrasounds in infant rodents.* Ph.D. Thesis: University of London.

Okon, E. E. (1972). Factors affecting ultrasound production in infant rodents. *J. Zool., Lond.* **168**: 139–148.

Pye, J. D. (1966). Synthesizing the waveforms of bat's pulses. In *Animal sonar systems: biology and bionics*: 43–66. Busnel, R.-G. (ed.). Jouy-en-Josas: Laboratoire Physiologie Acoustique.

Pye, J. D. (1968). Animal sonar in air. *Ultrasonics* **6**: 32–38.

Sales, G. D. & Pye, J. D. (1974). *Ultrasonic communication by animals.* London: Chapman & Hall.

Sewell, G. D. (1969). *Ultrasound in small mammals.* Ph.D. Thesis: University of London.

Sewell, G. D. (1970a). Ultrasonic signals from rodents. *Ultrasonics* **8**: 26–30.

Sewell, G. D. (1970b). Ultrasonic communication in rodents. *Nature, Lond.* **227**: 410.

Smith, J. C. (1972). Sound production by infant *Peromyscus maniculatus* (Rodentia: Myomorpha). *J. Zool., Lond.* **168**: 369–379.

Smith, J. C. (1974). *Sound communication in some myomorph rodents with special reference to adult–infant relationships.* Ph.D. Thesis: University of London.

Zippelius, H. M. & Scheidt, W. M. (1956). Ultraschallaute bei jungen Mäusen. *Naturwissenschaften* **43**: 502.

Symp. zool. Soc. Lond. (1975) No. 37, 331–338.

MATERNAL RECOGNITION OF INFANT SOUNDS

V. VUORENKOSKI

Department of Pediatrics, Keski-Pohjanmaa Central Hospital, Kokkola, Finland and *Department of Speech Communication Royal Institute of Technology (KTH), Stockholm, Sweden*

SYNOPSIS

Crying is an important physiological function for the newborn infant as an aspect of the ventilatory system. It is also a reflection of different physiological states, normal or abnormal, and an expression of emotional states, needs and demands. Finally, the infant sound is an important part of the communication between mother and child.

These early manifestations of communication are psychosomatic. The most important among them may be the role of crying in the establishment of lactation: crying is one of the natural stimuli which trigger a release of oxytocic hormone, causing the milk ejection reflex, "let-down".

A more general quality of mother–child communication is the identification, conscious or unconscious, of different aspects of the vocal utterances of the infant. The first identification problem may be the recognition of characteristics peculiar to the individual baby as soon as the first day after the delivery. This supports the hypothesis that individual vocal differences exist—within unknown limits—at or soon after birth. In a normal mother–child relationship, the mother's recognition of her child's vocalization rises significantly during the first six months of the child's life.

The most common identification problem is the recognition of the different physiological states and needs of the baby. Studies with prerecorded cry sounds, in multiple choice tests, confirm that different groups of adults are able to identify at least some basic neonatal states: "hunger", "pain" and "pleasure". Mothers are especially proficient at this. The data obtained in these studies also indicate that experience and training have a significant effect on the level of identification. A specific sound training is important in recognizing abnormal sound characteristics in different diseases and pathological conditions.

INTRODUCTION—DIFFERENT FUNCTIONS OF THE INFANT SOUND

Sound production is an important physiological function for the human newborn infant as an aspect of the ventilatory system, e.g. grunting in hyaline membrane disease (Yao, Lind & Vuorenkoski, 1971; Vuorenkoski & Vuorenkoski, in prep.). Infant sounds also reflect different physiological states, normal and abnormal. The assessment of normality is one of the most interesting problems in infant vocalization studies (Vuorenkoski, Lind, Wasz-Höckert & Partanen, 1971; Vuorenkoski, Kaunisto, Tjernlund & Vuorenkoski, 1973). The infant sound is also an expression of different states, needs and demands (Illingworth, 1955; Ostwald, 1972). Finally, different aspects of infant sounds must be considered as a part of the communication between the child and the mother (Bernal, 1972).

CRY SOUND AND "LET-DOWN"

Some early manifestations of the infant–mother interaction are psychosomatic. An example is the role of crying in the establishment of lactation: crying is one of the natural stimuli which trigger a release of oxytocic hormone, causing the milk ejection reflex, "let-down". Experimental studies performed with infra-red thermography demonstrated a raised temperature in the area of the mammary glands as a result of a cry stimulus (Vuorenkoski, Wasz-Höckert, Koivisto & Lind, 1969). In the milk release, oxytocic hormone produces contraction of the myoepithelial cells and thus a rise in milk pressure in the ducts and cisterns.

Of 40 normal, primipara mothers tested 3–5 days after delivery (with a Swedish thermograph AGA Thermovision) 36 showed a significant increase in breast temperature at the end of a cry stimulus. The latter was a continuous recording of crying from a normal newborn infant, consisting of two parts: first 4 min of spontaneous crying recorded just before feeding, followed immediately by a recording of pinch-evoked "pain" cries lasting 3 min. Variations were found both in the latency and in the degree of the temperature change. The average latency between the beginning of the cry stimulus and the first change in temperature was about 3 min. On the other hand, no differences were found between the different types of cry stimuli. This may be due to the fact that "hunger" and "pain" cries often become similar, depending on the nature and degree of the stimuli (Vuorenkoski, unpublished data).

INDIVIDUAL CRY CHARACTERISTICS

The most general problem in infant–mother interaction is the identification, conscious or unconscious, of different aspects of the vocal utterances of the baby. The first identification problem may be the recognition of individual sound characteristics. Here we encounter some difficult questions. How early may the human infant cry have significant individual sound characteristics? How are these to be measured? In our experience, the "contour voice print" technique (used sometimes for adult voices, Kersta, 1962), is not very reliable in recording infant sounds and we must use normal sound spectrograms and subjective listening.

An infant does not produce exactly the same sound on two separate occasions and no two infants vocalize in the same way (Lind, personal communication). For practical purposes, we must make appropriate definitions. Some examples of similarities may illustrate the problems. We may have two relatively complex signals with minimal differences, produced by the same infant (Smith–Lemli–Opitz syndrome) after an interval of several months (Wasz-Höckert, Similä, Rosberg, Vuorenkoski & Lind, 1969). But short, low-intensity, monotonous humming "pleasure" sounds from two normal infants may display even smaller differences. In some diseases, the individual voice qualities are so effectively masked by the pathological condition that the recognition of different individuals with the same disease seems to be impossible. A good example of this is the cry sound produced in *maladie du cri du chat* and in some other chromosomal aberrations (Vuorenkoski, Lind, Partanen, Lejeune, Lafourcade & Wasz-Höckert, 1966; Vuorenkoski, Kaunisto, Tjernlund & Vesa, 1970) (Figs 1, 2 and 3). On the other hand, there are also apparently some phonatory qualities (specific types of nasality and hoarseness) related to certain diseases with growth failure, which may aid the possibility of identification of individuals (Vuorenkoski, Tjernlund & Perheentupa, 1972).

Experimental studies of the maternal recognition of normal newborn babies (Formby, 1967; Valanne, Vuorenkoski, Partanen, Lind & Wasz-Höckert, 1967) have shown that some mothers are able to identify the voice of their baby during the first days after the delivery. This supports the hypothesis that individual vocal differences exist—within unknown limits—at or soon after birth. In a normal mother–child relationship maternal recognition of the individual voice characteristics of her child rises significantly during the first six months of life (Vuorenkoski & Rhen, in prep.). Defective recognition seems to be due mainly to an indifferent attitude on the part of the mother towards her child.

RECOGNITION OF STATES

The most common problem in maternal recognition of infant sounds is the identification of the different states and needs of the baby. Studies with prerecorded infant sounds in multiple choice tests confirm that different groups of adults are able to identify at least some situational or emotional states: hunger, pain and pleasure. Mothers are especially proficient at this task. This ability has been demonstrated both with vocalizations containing several consecutive signals (Wasz-Höckert, Partanen, Vuorenkoski, Valanne & Michelsson, 1964) and with isolated signals (Wasz-Höckert,

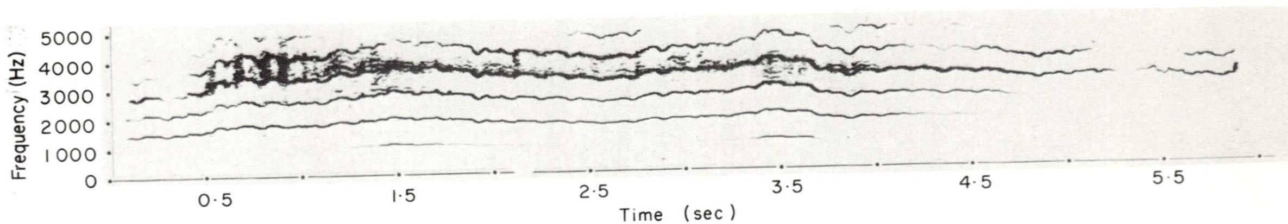

FIG. 1. Sound spectrogram representing a typical spontaneous pain cry signal recorded from a baby aged two months with *maladie du cri du chat*, a chromosomal aberration, karyotype 46, XX, Bp – ; an abnormally long, high-pitched, monotonous cry.

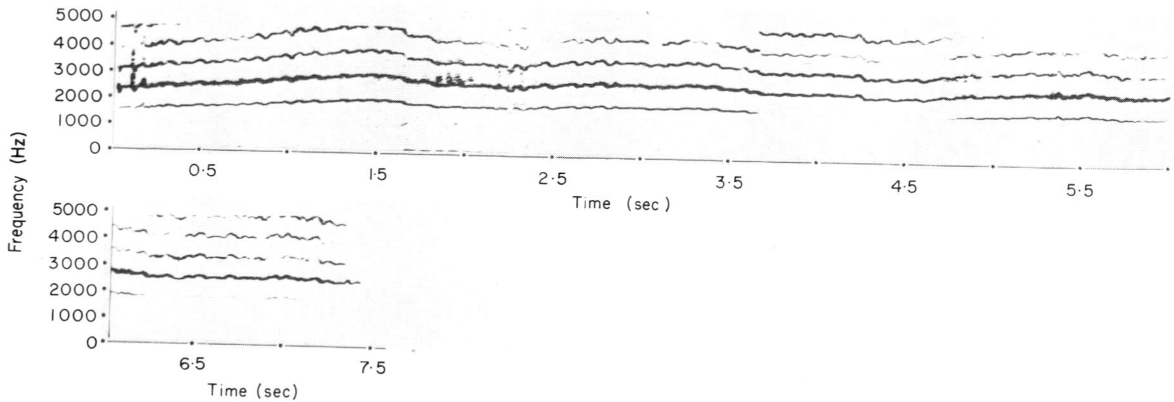

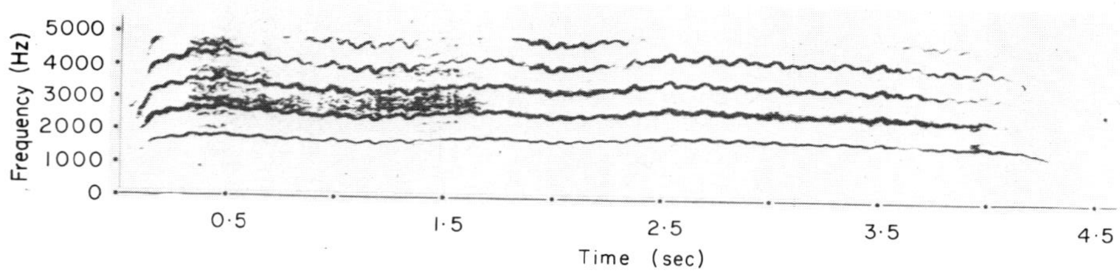

FIGS 2 and 3. Spectrograms of pain cries recorded from a newborn baby (age two days) with an apparently normal phenotype but with a minor chromosomal aberration, karyotype 46, XX, Dp +. During the first year of life deviations were recorded in EEG and in the neurological examination. There is a striking similarity between these cries and the cry in Fig. 1, both in physical analysis and in the auditory observation.

Lind, Vuorenkoski, Partanen & Valanne, 1968). The data obtained in these studies also indicate that experience and training have a significant effect on the level of identification. The effect of experience is also demonstrated in studies with "rooming-in" and other mothers, where rooming-in mothers judged themselves able to attribute more to their babies' cries (Greenberg, Rosenberg & Lind, 1973). Bernal (1972), in her study of crying and maternal response during the first ten days of life, showed that mothers who already had one child were more likely to feed their baby in response to crying than primiparae, but thought it doubtful that this was due to the poor cry interpretation of primiparae.

It is evident that a very poor cry interpretation can lead to a serious disfunction of mother–child interaction and may further contribute to conditions terminating in the *battered child syndrome.*

REAL TIME INSTRUMENTAL IDENTIFICATION OF INFANT SOUNDS

How much and which types of cry sound should a normal and healthy baby emit during a certain time period? Or, putting the question in another way, how much crying over a given time period will the mother or caretaker accept as normal and respond to appropriately? These are some of the problems which we are now studying with the help of the "cry analyser", an apparatus specially developed for real time detection and analysis of infant vocalizations (Vuorenkoski, Kaunisto, Tjernlund & Vesa, 1970; Vuorenkoski, Kaunisto, Tjernlund & Vuorenkoski, 1973). The apparatus (manufactured by AB Special Instrument, Stockholm) consists of an analyser and a strip printer. The analyser contains six cumulative counters for the numbers of low- and high-pitched cries, for the total duration of these cries, and for a count of heart beats and respirations. For recording the cry signal, a specially developed contact microphone is used, attached to the skin of the upper chest of the infant. Heart and respiratory rates are recorded by means of two chest electrodes. For real time visual reading of the data, the positions of the counters, the actual time and the time when the analysis was started, are presented on a digital display. The printer records the measured data automatically on pressure sensitive tape with pre-set intervals, or the operation is manual. We hope that this apparatus, a kind of artificial observer, will help us to understand more about the sounds of the newborn and young infant in both health and disease.

References

Bernal, J. (1972). Crying during the first 10 days of life and the maternal responses. *Devl Med. Child Neurol.* **14:** 362–372.

Formby, D. (1967). Maternal recognition of infant's cry. *Devl Med. Child Neurol.* **9**: 293–298.

Greenberg, M., Rosenberg, I. & Lind, J. (1973). First mothers rooming-in with their newborns: its impact upon the mother. *Am. J. Orthopsychiat.* **43**: 783–788.

Illingworth, R. S. (1955). Crying in infants and children. *Br. Med. J.* **1955(1)**: 75–78.

Kersta, L. G. (1962). Voiceprint identification. *Nature, Lond.* **196**: 1253–1257.

Ostwald, P. (1972). The sounds of infancy. *Devl Med. Child Neurol.* **14**: 350–361.

Valanne, E. H., Vuorenkoski, V., Partanen, T. J., Lind, J. & Wasz-Höckert, O. (1967). The ability of human mothers to identify the hunger cry signals of their own newborn infants during the lying-in period. *Experientia* **23**: 768–769.

Vuorenkoski, V., Kaunisto, M., Tjernlund, P. & Vesa, L. (1970). Cry detector. A clinical apparatus for surveillance of pitch and activity in the crying of a newborn infant. *Acta paediat., Stockh.* Suppl. **206**: 103–104.

Vuorenkoski, V., Kaunisto, M., Tjernlund, P. & Vuorenkoski, L. (1973). An improved model of cry detector: real time analysis of cry, respiratory and heart rates in the newborn infant. *Acta paediat., Stockh.* Suppl. **236**: 84–85.

Vuorenkoski, V., Lind, J., Partanen, T. J., Lejeune, J., Lafourcade, J. & Wasz-Höckert, O. (1966). Spectrographic analysis of cries from children with *maladie du cri du chat. Annl. Paediat. Fenn.* **12**: 174–180.

Vuorenkoski, V., Lind, J., Wasz-Höckert, O. & Partanen, T. J. (1971). Cry score. A method for evaluating the degree of abnormality in the pain cry response of the newborn and young infant. *Speech Transm. Lab. Q. Prog. St. Rep.* **1**: 68–75.

Vuorenkoski, V. & Rhen, K. (in preparation). *The significance of crying in the interaction between the mother and the newborn infant.*

Vuorenkoski, V., Tjernlund, P. & Perheentupa, J. (1972). Auditory perception of voice qualities and speaking fundamental frequency in Mulibrey-nanism and in some other children with growth failure. *Speech Transm. Lab. Q. Prog. St. Rep.* **2–3**: 64–74.

Vuorenkoski, V. & Vuorenkoski, L. (in preparation). *Simultaneous monitoring of grunting, cry, respiratory- and heart rates in Hyaline Membrane Disease.*

Vuorenkoski, V., Wasz-Höckert, O., Koivisto, E. & Lind, J. (1969). The effect of cry stimulus on the temperature of the lactating breast of primiparae. A thermographic study. *Experientia* **25**: 1286–1287.

Vuorenkoski, V., Wasz-Höckert, O., Lind, J., Koivisto, M. & Partanen, T. J. (1971). Training the ability to identify some pathological conditions in newborn infants by listening to the pain cry. *Speech Transm. Lab. Q. Prog. St. Rep.* **4**: 37–48.

Wasz-Höckert, O., Lind, J., Vuorenkoski, V., Partanen, T. J., & Valanne, E. (1968). *The infant cry. A spectrographic and auditory analysis.* London: Spastics International Publications/Heinemann.

Wasz-Höckert, O., Partanen, T., Vuorenkoski, V., Valanne, E. & Michelsson, K. (1964). Effect of training on ability to identify preverbal vocalizations. *Devl Med. Child Neurol.* **6**: 393–396.

Wasz-Höckert, O., Similä, S., Rosberg, G., Vuorenkoski, V. & Lind, J. (1969). El sindrome de *Smith-Lemli-Opitz* en dos ninas con especial atencion a los patrones de sus gritos de dolor. *Revta mex. Pediat.* **38**: 63–68.

Yao, A., Lind, J. & Vuorenkoski, V. (1971). Expiratory grunting in the late clamped normal neonate. *Pediatrics* **48**: 865–870.

Symp. zool. Soc. Lond. (1975) No. 37, 339–340.

CLOSING REMARKS TO SYMPOSIUM

J. D. PYE

Department of Zoology and Comparative Physiology, Queen Mary College, London, England

Composing an introductory review for a meeting such as this may be an embarrassing task, since long-cherished notions might be contradicted by the papers that follow. One excuse for a summing up would be the opportunity it gives for acknowledging such errors. That no major retreat appears to be necessary in this case is quite a relief, but it would be a poor meeting that left one with just the same views as before. An attempt to assess the impact of each contribution would be presumptive and of personal rather than of general value; nevertheless a few comments may be of some relevance.

This has been a stimulating meeting and the aims of the organizers in achieving variety have been met in more ways than was perhaps envisaged. Not only have the approaches to the subject varied tremendously, but so have the backgrounds of the contributors and the nature of the papers themselves. Instead of just "clinical versus zoological approaches" we have also heard engineers, physicists, pathologists and psychologists present papers that ranged from scholarly reviews to new experimental work of beautiful directness.

What has been particularly impressive has been the success with which the different contributors have been able to communicate both in the formal papers and in discussion. Every discipline (however large or small that taxon of science is assumed to be) develops its own language, not only for expressing itself but also as a tool in formulating its own concepts. This was made particularly clear when Dr Salomon, in an informal discussion of the introductory review, pointed out that computers are not only useful as mechanical aids but also have a conceptual value, since their design and operation present us with new and often fruitful ways of thinking about operations within the nervous system itself.

But if the frustration of Babel is to be avoided, meetings of this kind are essential. Despite its apparently restrictive title, the meeting turned out to be far from specialized. Beforehand many contributors expressed apprehensions to the organizers about their ability to communicate with "zoologists" or "doctors" from the "other side of the fence". That such fears were allayed in the event was shown by the wide cross-referencing that was introduced to the presentations and the lively discussions that developed afterwards. This was extremely encouraging and fully justified the original concept of the meeting.

If a zoologist may express a single disappointment with the way things turned out, it is simply that animals are too often regarded almost as surrogate humans rather than as separate organisms finely adapted to their own way of life. Results can only be fully meaningful if related to the biological needs of the organism concerned and, to return to a hobbyhorse ridden in the introduction, the differences between organisms may be as revealing as their similarities. It sometimes seems that this is better appreciated in vision research than in hearing—perhaps the implications are plainer there.

It remains only to thank again all those who helped to make this meeting so enjoyable and to hope that the published proceedings will convey, to readers who were not present, an impression of how stimulating it really was.

AUTHOR INDEX

Numbers in italics refer to pages in the References at the end of each article.

I

J

K

L

TAXONOMIC INDEX

This index consists of those chordates (excluding man) which are mentioned in this Symposium. They are grouped in taxonomic order.

SUBJECT INDEX